Biodiversity of Mites

Biodiversity of Mites

Editor

Dariusz J. Gwiazdowicz

MDPI • Basel • Beijing • Wuhan • Barcelona • Belgrade • Manchester • Tokyo • Cluj • Tianjin

Editor
Dariusz J. Gwiazdowicz
Faulty of Forestry and Wood Technology
Poznan University of Life Sciences
Poznan
Poland

Editorial Office
MDPI
St. Alban-Anlage 66
4052 Basel, Switzerland

This is a reprint of articles from the Special Issue published online in the open access journal *Diversity* (ISSN 1424-2818) (available at: www.mdpi.com/journal/diversity/special_issues/biodiversity_mites).

For citation purposes, cite each article independently as indicated on the article page online and as indicated below:

LastName, A.A.; LastName, B.B.; LastName, C.C. Article Title. *Journal Name* **Year**, *Volume Number*, Page Range.

ISBN 978-3-0365-1854-1 (Hbk)
ISBN 978-3-0365-1853-4 (PDF)

Contents

diversity

MDPI

Editorial

Biodiversity of Mites

Dariusz J. Gwiazdowicz

Faculty of Forestry, Poznań University of Life Sciences, Wojska Polskiego 71c, 60-625 Poznań, Poland; dariusz.gwiazdowicz@up.poznan.pl

Nature is a highly complex, intricate system constructed of a network of interdependencies between individual species, their communities and their habitats. When climatic factors (such as temperature and humidity) and environmental factors (for example, latitude) are additionally considered, it comes as no surprise that scientists investigating this complicated and wondrous world face considerable problems. An important role in this complex ecological system is played by mites, which are actively involved in the flow of energy, matter and information. As is commonly known, these tiny, ubiquitous invertebrates are found both in terrestrial and aquatic ecosystems, while additionally being relatively abundant in the aeroplankton. The richness of the mite fauna may be analysed in terms of their zoogeography, ecology, taxonomy, or parasitology, amongst others, or even paleontology. The species composition and the character of their assemblages vary greatly, beginning from pioneering species found in postglacial or volcanic regions and ending with the mind-boggling, still only partly known, richness in tropical forests. In view of all the above-mentioned factors, researchers need to adopt a unique approach to the studied research topics and precisely identify their aims; that is, the scientific problems. A specific methodology will need to be applied to solve such problems ranging from, for example, the morphological analysis of mite specimens embedded in amber to the application of state-of-the-art molecular techniques. A platform for discussions on the biodiversity of mites and the presentation of current challenges faced in acarological studies has been created by Diversity in this Special Issue.

check for updates

Citation: Gwiazdowicz, D.J. Biodiversity of Mites. *Diversity* **2021**, *13*, 80. https://doi.org/10.3390/d13020080

Received: 9 February 2021
Accepted: 10 February 2021
Published: 15 February 2021

At present, researchers specialising in life sciences focus their studies on the increasingly evident and pervasive effects of climate change. This phenomenon has been the subject of numerous public debates and political decisions. The consequences of climate change are particularly visible in polar regions, in which melting and retreating glaciers expose ground on which primary succession may develop and pioneer species may establish. This line of research is represented in this issue by a study on assemblages of the oribatid mite fauna of the sub-polar Urals [1], especially pertinent since these invertebrates may be good bioindicators of environmental changes caused, for example, by climate change. Melekhina [1], in her paper, identified 163 species belonging to 45 families, among which the leading families in the fauna include Crotoniidae, Ceratozetidae and Oppiidae. The share of Palearctic species is low (23.4%), which separates the fauna of the tundra zone from the taiga zone.

Another study, even more strongly associated with this theme, was conducted in the Svalbard archipelago located in the High Arctic [2]. With rapidly changing climatic and habitat conditions, these islands are an excellent region to observe changes in the ranges of certain species, including mites. Seniczak et al. [2] summarised over 100 years of studies on the Svalbard archipelago by listing 178 mite species (1 Ixodida, 36 Mesostigmata, 43 Trombidiformes, and 98 Sarcoptiformes) and providing their localities on each island of this archipelago. They observed that, in contrast to Trombidiformes and Sarcoptiformes, which are dominated in Svalbard by species with wide geographical distributions, the Mesostigmata include many Arctic species (39%). Such information may prove significant for future studies considering climate change and resulting alterations in the species composition or character of mite assemblages. Moreover, these authors stated that a large number of new species (42 spp.) have been described from Svalbard, including 15 that have so far been found there exclusively. Yet, it is uncertain if any of these latter species are endemic: six are recent findings, the others are old records and, in most cases, impossible to verify.

An interesting problem of endemism was also presented in the paper by Błoszyk and Napierała [3]. They studied the geographical distribution of mites from the suborder Uropodina gathered from literature sources and their own studies conducted on approximately 40,000 samples collected from various geographical regions. On this basis, they attempted to identify a common endemism in this group of mites, or rather, if the distribution patterns observed are the result of sparce sampling and poor identification. By presenting several examples, the authors highlighted the necessity to provide more accurate, precise descriptions of new species, including individual variation, and to conduct extensive taxonomic studies on all the continents.

Acarologists have occasionally focused on species colonising trees, living, for example, in various parts of the stem, branches or leaves. However, limited research has been carried out on the evolution of these mites and their adaptation to the arboreal lifestyle, such as phenotypic adaptations. Fortunately, this research problem was investigated by Schäffer et al. [4], who suggested that an arboreal life evolved independently and that tree-living is more common in evolutionary younger taxa with strong sclerotization, sexual reproduction, and capitate sensilla. Based on the analysis of certain morphological traits of oribatid mites, these researchers stated that it is the capitate sensillus that appears to be a morphological (pre)-adaptation to life on trees (arboreal sensu lato), potentially functioning as a gravity receptor. Moreover, they concluded that only taxa that live on the bark of stems and branches without a layer of lichens or mosses, in the canopy or on twigs and leaves and undergo their whole life cycle on the tree should be regarded as arboreal sensu stricto.

The occurrence of mites in specific habitats and microhabitats depends on their locomotion, transport, or on the established manner of their dispersion. One of these is connected with phoresy, that is, a specific form of zoochory when a mite uses other animals, for example, mammals, birds or insects, as a means to spread and disperse. This phenomenon was investigated by Konwerski et al. [5]. They analysed almost a thousand beetle individuals, identifying on their bodies over 25 thousand deutonymphs of mites from the suborder Uropodina. It was observed that, depending on the beetle species, different mite species are transported. Moreover, mites show preferences for different attachment locations on the beetle's body. In the case of the *Oodinychus ovalis* (mite) and *Monochamus sartor urussovii* (beetle) association, the deutonymphs were found mostly on the pronotum and dorsal surface of the elytra, while the highest number of deutonymphs of *Trichouropoda shcherbakae* was on *Tetropium castaneum*, and *T. fuscum* were observed on the legs.

For thousands of years, humans have been converting the natural environment, for example, through the production of arable fields or pastures. Transformation of the forest environment into agricenoses may lead to a reduction in biodiversity as a source of natural richness. Such changes are monitored using various bioindicators, among which are mites. Azevedo et al. [6] analysed the species composition and population size of Gamasina mites in soils covered by natural vegetation and in agroecosystems in the Cerrado region of northern Brazil. A greater species richness of the natural vegetation was clear since, of the 45 identified species, as many as 36 were recorded in soils covered by natural vegetation. This finding persuaded these authors to conclude that arable fields are a threat to the species diversity of mites since many of them are unable to respond to the rapidly changing environmental conditions. The only method to preserve biodiversity may be to establish reserves or protected areas.

Farmers sometimes face the adverse effects of mite pests damaging crops. On the leaves of many crop species (for example, grape, avocado, cassava, cotton), we may find numerous mites which damage these leaves and which are pests reducing the yields of these crops. In turn, other mite species are beneficial as they are predators and control harmful pest species. Situngu et al. [7] focused on the diversity and composition of mites on an economically important plant host (*Coffea aribica*) compared to mites found in a neighbouring natural forest in South Africa. Their results showed that the coffee plantations were associated with only predatory mites, some of which are indigenous to South Africa. This indicates that coffee plantations may be successfully colonised by indigenous beneficial mites.

In Norway, strawberry producers use cereal straw mulching to prevent berries from contacting the soil and to control weeds. Neves Esteca et al. [8] experimentally confirmed the hypothesis that organic matter such as straw mulch also favours the maintenance of predatory mites, which visit strawberry plants at night time. It was found that the dominant species of predatory mites belonged to three families: Melicharidae, Blattisociidae and Phytoseiidae.

Water mites represent the most diverse and abundant group of Arachnida in freshwater ecosystems, with about 6000 species described. However, it is estimated that this number represents only 30% of the total expected species. Despite having strong biotic interactions with their community and having the potential to be exceptional bioindicators, they are frequently excluded from studies of water quality or ecology due to actual and perceived difficulties of the taxonomic identification of this group. These difficulties in the identification of water mite species may be overcome using molecular and genetic techniques, which opens a new chapter in the development of taxonomy. A particular region of the mitochondrial COI (cytochrome c oxidase I) gene, one of the groups known as the DNA barcode region, is the most common sequence used in water mite taxonomy research. Public databases, including the Barcode of Life Database (www.boldsystems.org) or GenBank (www.ncbi.nlm.nih.gov), as well as the use of new bioinformatic tools, represent a breakthrough in species identification. This line of research is represented by the studies of Montes-Ortiz and Elías-Gutiérrez [9], who investigated water mites in the Yucatan Peninsula of Mexico. Applying the above-mentioned method, these authors identified 77 genetic groups or putative species corresponding to 18 genera.

A slightly different subject was investigated by Pozojević et al. [10], who attempted to detect significant differences in water mite assemblages between rheocrene (river-forming springs with dominant riffle habitats) and limnocrene springs (lake-forming springs with dominant pool habitats) in Croatia. As a result, many species not previously observed in that region were reported for the first time.

The paper by Vasquez et al. [11] is a review of the problems of the relationships of aquatic mites being predators or parasites of mosquitoes. In the future, such knowledge concerning these relationships may be crucial for human health since mosquitoes are vectors of many dangerous diseases.

Demodecidae are the most specialized parasitic mites of mammals which typically inhabit the skin, but they have been found in other tissues and organs as well. They can cause demodecosis (a disease which is hazardous and difficult to cure) in humans, domestic animals and livestock. They are parasites with high host and topical specificity, they have been found for most orders of mammals, and they are common in the populations of numerous host species. Therefore, they constitute not only an important subject of veterinary and medical studies, but also an excellent model for faunistic and parasitological studies of different aspects of the functioning and evolution of the host–parasite relationship. This line of research is represented by the study of Izdebska and Rolbiecki [12], who, based on the literature from the years 1884–2020, prepared a list of 122 parasite species of demodecid mites, including their hosts and geographical distribution.

This diversity of subjects presented in the Special Issue Biodiversity of Mites reflects the complexity of research problems currently faced by researchers. Nevertheless, it may also be a source of inspiration for further challenges to gain even greater insight into biodiversity, and thus to be able to more effectively protect the natural richness of our planet.

Funding: This research received no external funding.

Acknowledgments: Thanks to Stephen J. Coulson for comments on a draft of the manuscript.

Conflicts of Interest: The author declares no conflict of interest.

References

1. Melekhina, E.N. Analysis of Oribatid Fauna of the East European Tundra with First Reported Data of Subpolar Urals. *Diversity* **2020**, *12*, 235. [CrossRef]
2. Seniczak, A.; Seniczak, S.; Schwarzfeld, M.D.; Coulson, S.J.; Gwiazdowicz, D.J. Diversity and Distribution of Mites (Acari: Ixodida, Mesostigmata, Trombidiformes, Sarcoptiformes) in the Svalbard Archipelago. *Diversity* **2020**, *12*, 323. [CrossRef]
3. Błoszyk, J.; Napierała, A. Endemism of Uropodina Mites: Spurious or Real? *Diversity* **2020**, *12*, 283. [CrossRef]
4. Schäffer, S.; Koblmüller, S.; Krisper, G. Revisiting the Evolution of Arboreal Life in Oribatid Mites. *Diversity* **2020**, *12*, 255. [CrossRef]
5. Konweski, S.; Gutowski, J.M.; Błoszyk, J. Patterns of Distribution of Phoretic Deutonymphs of Uropodina on Longhorn Bee-tles in Białowieża Primeval Forest, Central Europe. *Diversity* **2020**, *12*, 239. [CrossRef]
6. Azevedo, E.B.; Azevedo, L.H.; Moreira, G.F.; Santos, F.A.; Carvalho, M.A.F.; Sarmento, R.A.; Castilho, R.C. Diversity of soil gamasine mites (Acari: Mesostigmata: Gamasina) in an area of natural vegetation and cultivated areas of the Cerrado biome in Northern Brazil. *Diversity* **2020**, *12*, 331. [CrossRef]
7. Situngu, S.; Barker, N.P.; Vetter, S. A Snap-Shot of Domatial Mite Diversity of Coffea arabica in Comparison to the Adjacent Umtamvuna Forest in South Africa. *Diversity* **2020**, *12*, 79. [CrossRef]
8. Esteca, F.D.C.N.; Trandem, N.; Klingen, I.; Santos, J.C.; Júnior, I.D.; De Moraes, G.J. Cereal Straw Mulching in Strawberry—A Facilitator of Plant Visits by Edaphic Predatory Mites at Night? *Diversity* **2020**, *12*, 242. [CrossRef]
9. Montes-Ortiz, L.; Elías-Gutiérrez, M. Water Mite Diversity (Acariformes: Prostigmata: Parasitengonina: Hydrachnidiae) from Karst Ecosystems in Southern of Mexico: A Barcoding Approach. *Diversity* **2020**, *12*, 329. [CrossRef]
10. Pozojević, I.; Pešić, V.; Goldschmidt, T.; Gottstein, S. Crenal Habitats: Sources of Water Mite (Acari: Hydrachnidia) Diversity. *Diversity* **2020**, *12*, 316. [CrossRef]
11. Vasquez, A.A.; Kabalan, B.A.; Ram, J.L.; Miller, C.J. The Biodiversity of Water Mites That Prey on and Parasitize Mosquitoes. *Diversity* **2020**, *12*, 226. [CrossRef]
12. Izdebska, J.N.; Rolbiecki, L. The Biodiversity of Demodecid Mites (Acariformes: Prostigmata), Specific Parasites of Mammals with a Global Checklist and a New Finding for *Demodex sciurinus*. *Diversity* **2020**, *12*, 261. [CrossRef]

diversity

MDPI

Article

Diversity and Distribution of Mites (Acari: Ixodida, Mesostigmata, Trombidiformes, Sarcoptiformes) in the Svalbard Archipelago

Anna Seniczak [1,*], Stanisław Seniczak [2], Marla D. Schwarzfeld [3], Stephen J. Coulson [4,5] and Dariusz J. Gwiazdowicz [6]

1 Department of Natural History, University Museum of Bergen, University of Bergen, Postboks 7800, 5020 Bergen, Norway

2 Department Evolutionary Biology, Faculty of Biological Sciences, Kazimierz Wielki University, J.K. Chodkiewicza 30, 85-064 Bydgoszcz, Poland; stseni@ukw.edu.pl

3 Canadian National Collection of Insects, Arachnids and Nematodes, Agriculture and Agri-food Canada, 960 Carling Avenue, Ottawa, ON K1A 0C6, Canada; marla.schwarzfeld@canada.ca

4 Swedish Species Information Centre, Swedish University of Agricultural Sciences, SLU Artdatabanken, Box 7007, 75007 Uppsala, Sweden; stephen.coulson@slu.se

5 Department of Arctic Biology, University Centre in Svalbard, P.O. Box 156, 9171 Longyearbyen, Norway

6 Faculty of Forestry, Poznań University of Life Sciences, Wojska Polskiego 71c, 60-625 Poznań, Poland; dariusz.gwiazdoswicz@up.poznan.pl

* Correspondence: Anna.Seniczak@uib.no

Received: 21 July 2020; Accepted: 19 August 2020; Published: 25 August 2020

check for updates

Abstract: Svalbard is a singular region to study biodiversity. Located at a high latitude and geographically isolated, the archipelago possesses widely varying environmental conditions and unique flora and fauna communities. It is also here where particularly rapid environmental changes are occurring, having amongst the fastest increases in mean air temperature in the Arctic. One of the most common and species-rich invertebrate groups in Svalbard is the mites (Acari). We here describe the characteristics of the Svalbard acarofauna, and, as a baseline, an updated inventory of 178 species (one Ixodida, 36 Mesostigmata, 43 Trombidiformes, and 98 Sarcoptiformes) along with their occurrences. In contrast to the Trombidiformes and Sarcoptiformes, which are dominated in Svalbard by species with wide geographical distributions, the Mesostigmata include many Arctic species (39%); it would thus be an interesting future study to determine if mesostigmatid communities are more affected by global warming then other mite groups. A large number of new species (42 spp.) have been described from Svalbard, including 15 that have so far been found exclusively there. It is yet uncertain if any of these latter species are endemic: six are recent findings, the others are old records and, in most cases, impossible to verify. That the Arctic is still insufficiently sampled also limits conclusions concerning endemicity.

Keywords: checklist; Astigmata; Endeostigmata; Oribatida; Prostigmata; climate change; Arctic; Svalbard

1. Introduction

The invertebrate fauna of the Svalbard archipelago is amongst the best known for any region in the Arctic [1]. However, the inventory has been collated from studies over a period of 150 years and has never been subjected to a critical review and taxonomic revision. Without an improved understanding of the current fauna it is not possible to understand the current terrestrial ecosystem or determine its response to on-going environmental change.

The dramatic environmental changes that we see today may be most discernable in polar regions, which possess both a unique flora and fauna and are undergoing the greatest rates of climate change. A particularly interesting area to study this is the Svalbard archipelago that is located at 78° N, isolated, and with a distinctive and diverse flora and fauna that are poorly studied [2]. Svalbard is currently experiencing one of the fastest temperature increases and one of the highest rates of sea ice loss in the Arctic [3]. During the past 50 years, the mean air temperature in Svalbard has increased by 4 °C, the winter mean air temperature has increased by 7 °C, the snow season duration has decreased by 20 days, and precipitation has increased by 65% [4]. For example, at Svalbard airport in 2018, the mean annual temperature was −1.8 °C, winter temperature was −5.3 °C, and annual precipitation was 252.5 mm [5]. These factors, as well as others that are tightly connected to the climate (e.g., faster melting of glaciers, changes in permafrost, landslides, avalanches, and flooding), affect the ecosystems in different manners. Even though a few species benefit from a warming climate, for example vertebrates, most Arctic species in Svalbard are experiencing negative effects in the warming environment [6].

Just south of Svalbard, two sea currents meet, a northern branch of the warm North Atlantic Drift and the southerly flowing cold Arctic Current, with consequences for the local climates on the east and west coasts of the archipelago [7] and potentially for species immigration histories, thus affecting species communities. A good example is the distinctive acarofauna of Edgeøya in the east of the archipelago compared to that observed on the west coast [8] and which includes an oribatid species new to science and not so far recorded elsewhere [9].

As a consequence of the increasing temperatures and changes in precipitation patterns, many glaciers are receding and revealing new land surfaces for primary succession. Therefore, regions such as Svalbard offer a unique opportunity to identify the pioneer species that first colonize barren post-glacial habitats and to analyze long-term processes of change, e.g., in invertebrate community assembly [10].

The diversity and richness of the environment is much more than simply lists of species occurring in a given area; it is also the regional specificity of these species, for example, those species that are considered to be typically High Arctic. Changing climatic conditions that determine and stimulate changes in habitats also affect the ranges of some mite species. Several studies have considered changes in mite communities in Svalbard in space and time; their dispersal abilities [11–17], as well as their emergence in new areas or retreat from previously occupied locations [10,18–20]. The uniqueness of polar areas, with their characteristic climates and isolated island distribution that often restricts dispersal processes, limits the range of some species. In this context, knowledge of zoogeography (ranges of occurrence of species) is extremely valuable because this suggests not only the migration routes, but also the ability of selected species to colonize new areas. Nonetheless, the climate of Svalbard is changing fast. These changes will undoubtedly provide opportunities for new species to colonise the region with difficult to foresee outcomes. The consequences of human introductions of invasive and alien species (IAS) have been described in the Antarctic [21] and there are examples of similar introductions in Svalbard [18].

Understanding mite ecology is also important since mites are excellent bioindicators of environmental changes and their presence, or changes in their communities, can be used in biological monitoring of naturally occurring processes as well as the consequences of human impacts, for example, those related to industrial activities in Svalbard's natural environment [20,22].

We here present a review of the mite fauna of Svalbard. These results can be the basis for further analyzes of the acarofauna (e.g., zoogeographic, ecological, taxonomic, and parasitological).

2. Material and Methods

The data were extracted from 104 papers published between 1871 and 2020. The localities presented in Figure 1 are based on Table 1. Distribution of *Ixodes uriae* White, 1852, the only representative of the order Ixodida in Svalbard, follows [23]. The nomenclature and arrangement of the Mesostigmata families follows [24,25], while the species nomenclature follows [26] with a few modifications [27,28]. Distribution of the families follows [27–30].

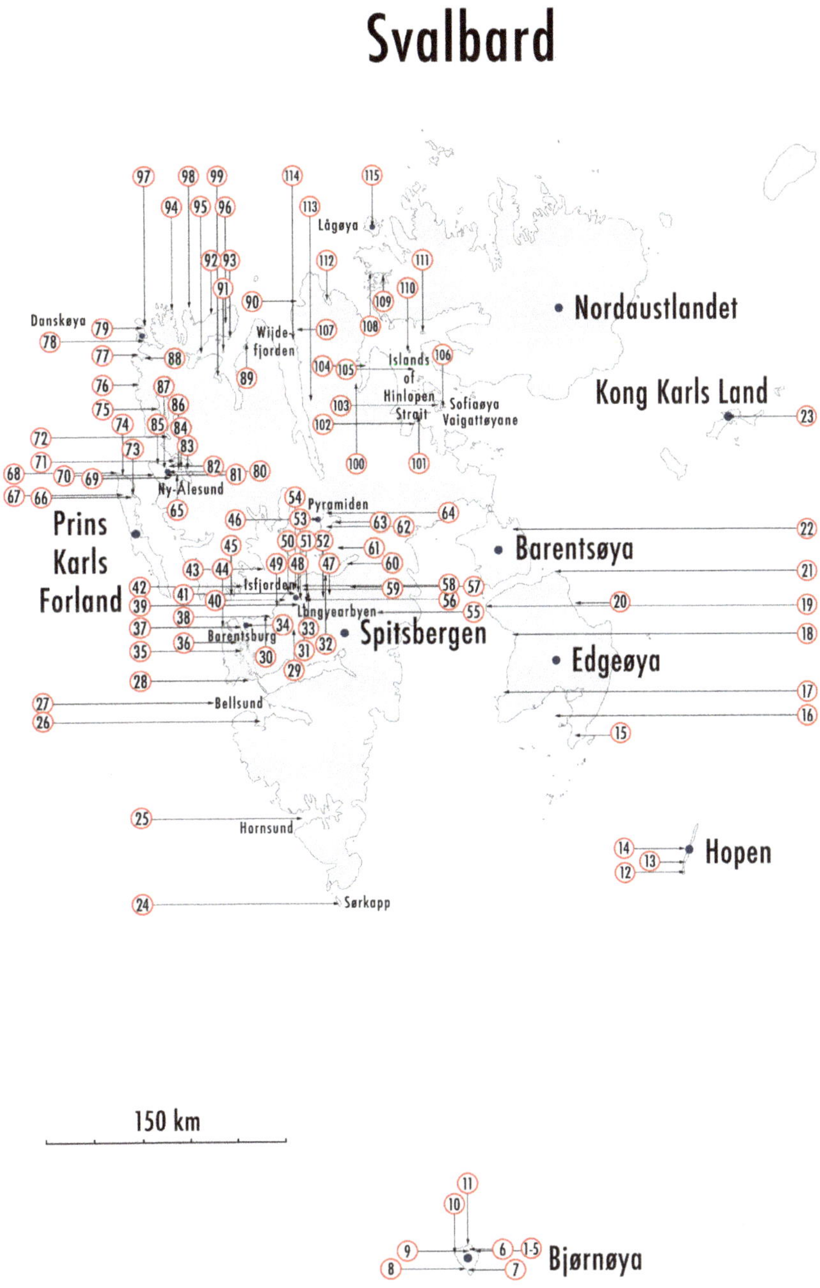

Figure 1. *Cont.*

Map Key

54 Advent City
51 Adventdalen
41 Adventfjorden
36 Aldegondabreen
7 Alfredfjellet
47 Arctowskifjellet
5 Austervåg
34 Barentsburg
22 Barentsøya
27 Bellsund
62 Billefjorden
11 Birds colony
98 Biskayerhuken
40 Bjørndalen
20 Blåbukta
71 Blomstrandhalvøya
99 Bockfjorden
43 Bohemanneset
63 Brucebyen
74 Carmichaelpynten
38 Colesbukta
114 Dirksbukta
18 Diskobukta
8 Ellasjøen
39 Endalen
1 Engelskelva
31 Engelskhytta
102 Eremitten
56 Fivelflyene
72 Fjortende Julibukta
94 Flathuken
109 Florabukta
110 Fosterøyane
68 Fuglehuken fyr
2 Fugleodden
61 Gipsdalen
35 Grønfjordbreen
37 Grønfjorden
88 Großer Trichter
49 Grumantbyen
111 Gyldénøyane
58 Hanaskogdalen
6 Hellevatnet
52 Helvetiadalen
105 Hinlopenstretet
59 Hiorthfjellet
48 Hiorthhamn
14 Hopen
13 Hopen radio
25 Hornsund
53 Isdammen
44 Isfjord radio
42 Isfjorden
21 Kapp Heuglin
30 Kapp Laila
19 Kapp Lee
70 Kiærstranda
108 Kinnvika
87 Knudsenheia
12 Koefoedodden
81 Kolhaugen
23 Kong Karls Land
82 Kongsfjorden
65 Krykkjefjellet
115 Lågøya
107 Lakssjøen
95 Liefdefjorden
104 Lomfjorden
55 Longyearbyen
67 MacKenziedale
77 Magdalenefjorden
33 Mälardalen
69 Midtre Lovénbreen
4 Mosevatnet
90 Mosselbukta
89 Mushamna
15 Negerdalen
76 Nissenfjella
29 Nordenskiöld Land
101 Nordenskiöldøya
80 Ny-Ålesund
100 Ny-Friesland
84 Observasjonsholmen
83 Ossian Sarsfjellet
64 Petuniabukta
46 Pyramiden
26 Recherchefjorden
96 Reinsdyrflya
73 Richardlaguna
113 Ringhorndalen
17 Russebukta
57 Sassendalen
32 Sassenfjorden
75 Signehamna
66 Silene Hill
97 Smeerenburg
106 Sofiaøya
78 Sørgattet
112 Sorgfjorden
24 Sørkapp
91 Station Islands at Liefdefjorden
9 Stevatnet
86 Storholmen
85 Stuphallet
92 Sven Olssonodden
10 Teltvika
60 Tempelfjorden
16 Tjuvfjorden
45 Trygghamna
3 Tunheim
28 Vårsolbukta
50 Vestpynten
79 Virgohamna
103 Von Otterøya
93 Worsleyneset

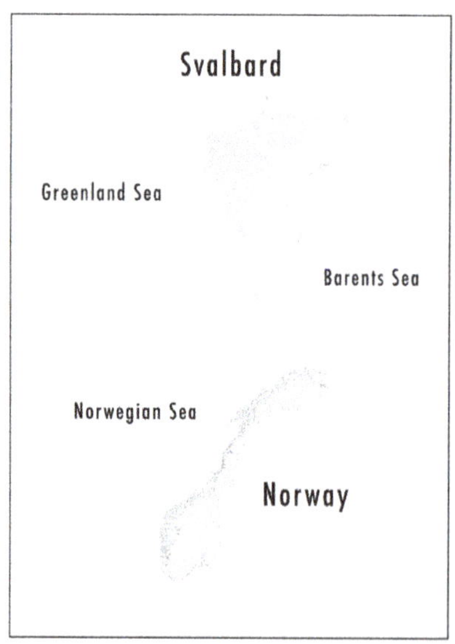

Figure 1. Localities in Svalbard where mites were found.

The arrangement of the Trombidiformes families follow [31] with the exception of the Eriophyidae that is included in the Endeostigmata, as suggested by [32,33]. Names and distributions of the Bdellidae follow [34]; Cunaxidae—[35]; Cocceupodidae and Eupodidae—[36,37] for nomenclature, and [38,39] for distributions; Penthaleidae—[40]; Penthalodidae—[41]; Rhagidiidae—[42–44]; Halacaridae—[45]; Triophtydeiidae—[46] for taxonomy, and [47,48] for distribution; Tydeidae—[49,50]; Iolinidae—[51,52]

for taxonomy, and [47,53] for distributions; Sperchontidae—[48,54,55]; Trombidiidae—[56]; Syringophilidae—[57]; Stigmaeidae—[58]; Tetranychidae—[59,60]; and Pygmephoridae—[61].

The nomenclature and arrangement of the Sarcoptiformes families follow [62,63]. For the Endeostigmata, the nomenclature and arrangement of families follow [63,64]. The distribution of Nanorchestidae follow [65]; Alicorhagiidae—[66]; and Eriophyidae—[67]. The species names of the Oribatida (without Astigmata) follow [68] with a few exceptions [69,70] and their distribution is given after [9,68,71,72]. The Astigmata species (families Acaridae, Alloptidae, and Avenzoariidae) follow [73,74] and their distribution is according to [75].

Full names of species are provided in Table 1, while in other tables and figures abbreviations are used. The species within families are organized alphabetically. The names of localities follow in alphabetical order (Table 1).

3. Results

3.1. Ixodida

Only one species, the seabird tick (*Ixodes uriae*), is known from Svalbard. This species has a wide geographical distribution (Table 1) but in Svalbard has been found only recently on Bjørnøya and Spitsbergen [76–78].

3.2. Mesostigmata

Thus far, 36 species of Mesostigmata from 13 families have been found in Svalbard (Table 1). Amongst these families, the richest in species is the Ascidae (12 spp.), followed by the Zerconidae (6 spp.); these two families contribute almost 50% of mesostigmatic alpha diversity in Svalbard (Figure 2).

The peculiarity of the mesostigmatic mite communities of Svalbard compared to other mite groups is manifested by the description of one new genus (*Arctoseius*) and seven species [*Halolaelaps coulsoni* Gwiazdowicz & Teodorowicz, 2017; *H. gerlachi* Hirschmann, 1966; *Antennoseius* (*Vitzthumia*) *oudemansi* (Thor, 1930); *Arctoseius laterincisus* Thor, 1930; *Proctolaelaps parvanalis* (Thor, 1930); *Neoseiulus grumantensis* Kolodochka & Gwiazdowicz, 2014; and *N. magnanalis* (Thor, 1930)], all first described from the largest island of the archipelago—Spitsbergen. Moreover, based on recently collected samples from this island, redescriptions of several species have been published: *Zercon solenites* Haarløv, 1942; *Antennoseius* (*Vitzthumia*) *oudemansi*; *Arctoseius haarlovi* Lindquist, 1963; *Proctolaelaps parvanalis*; *Neoseiulus ellesmerei* (Chant & Hansell, 1971); and *N. magnanalis* [27–29,79,80].

Table 1. List of mites of the Svalbard archipelago.

No.	Taxon	Distribution	Localities
	IXODIDA		
	Ixodidae		
1.	*Ixodes uriae* White, 1852	Afrotropical, Australasian, Neotropical, Holarctic	Bjørnøya [77]; Spitsbergen: Ossian Sarsfjellet [76–78]
	MESOSTIGMATA		
	Dinychidae		
2	*Uroseius acuminatus* (C.L. Koch, 1847)	Palearctic	Spitsbergen [81,82]
	Zerconidae		
3.	*Zercon andrei* Sellnick, 1958	Palearctic	Bjørnøya: Alfredfjellet, "Birds colony", Teltvika [29,30,83]; Spitsbergen: Hornsund [84]
4.	*Z. curiosus* Trägårdh, 1910	Palearctic	Spitsbergen: Longyearbyen [85]
5.	*Z. forsslundi* Sellnick, 1958	Palearctic	Edgeøya: Diskobukta, Kapp Heuglin, Kapp Lee, Negerdalen, Russebukta [8]; Hopen: Koefoedodden, Norwegian Meteorological Institute Station [86]; Nordaustlandet: Kinnvika [30,87]; Spitsbergen: Adventdalen, Billefjord, Bjørndalen, Endalen, Hornsund, Ny-Ålesund, Ossian Sarsfjellet, Petuniabukta, Storholmen, Trygghamna [10,20,30,85,88–90]
6.	*Z. solenites* Haarløw, 1942 **	Arctic	Edgeøya: Diskobukta, Kapp Lee [8]; Hopen: Koefoedodden, Norwegian Meteorological Institute Station [86]; Nordaustlandet: Kinnvika [29,87]; Spitsbergen: Adventdalen, Bjørndalen, Blomstrandhalvøya, Endalen, Fjortende Julibukta, Grønfjorden, Magdalenefjorden, Petuniabukta, Storholmen, Vårsolbukta [20,27,30,89,90]
7.	*Z. triangularis* C.L. Koch, 1836	Palearctic	Spitsbergen: Adventdalen, Barentsburg, Hanaskogdalen, Hjorthhamnn [91]
8.	*Z. zelawaiensis* Sellnick, 1944	Palearctic	Bjørnøya: Alfredfjellet [88]; Spitsbergen: Hornsund [83]
	Parasitidae		
9.	*Paragamasus* (*Aclerogamasus*) *insertus* (Micherdzinski, 1969)	Palearctic	Spitsbergen: Barentsburg [18,19]
10.	*Vulgarogamasus immanis* (Berlese, 1904)	Palearctic	Spitsbergen: Barentsburg [92]
11.	*V. remberti* (Oudemans 1912)	Palearctic	Spitsbergen: Barentsburg [18,19]
	Digamasellidae		
12.	*Dendrolaelaps foveolatus* (Leitner, 1949)	Palearctic	Spitsbergen: Pyramiden [20]
	Halolaelapidae		
13.	*Halolaelaps coulsoni* Gwiazdowicz & Teodorowicz, 2017 *,†	Arctic	Spitsbergen: Pyramiden [93]
14.	*H. gerlachi* Hirschmann, 1966 *,†	Arctic	Spitsbergen: Ny-Ålesund [94]
15.	*Saprosecans baloghi* Karg, 1964	Palearctic	Edgeøya: Diskobukta [8]

Table 1. *Cont.*

No.	Taxon	Distribution	Localities
	Ologamasidae		
16.	*Gamasellus borealis* (C.L. Koch, 1879)	Palearctic	Spitsbergen: Dirksbukta [7]
	Eviphididae		
17.	*Thinoseius spinosus* (Willmann, 1939)	Palearctic	Spitsbergen: Longyearbyen, Storholmen [90,95]
	Macrochelidae		
18.	*Macrocheles muscaedomesticae* (Scopoli, 1772)	Palearctic	Spitsbergen: Barentsburg [84]
	Ascidae		
19.	*Antennoseius* (*Vitzthumia*) *oudemansi* (Thor, 1930) *,**	Arctic	Bjørnøya: "Birds colony" [83]; Edgeøya: Diskobukta, Kapp Lee, Russebukta [8]; Hopen: Koefoedodden, Norwegian Meteorological Institute Station [86]; Nordaustlandet: Kinnvika [30,79,87]; Spitsbergen: Adventdalen, Barentsburg, Bjørndalen, Endalen, Florabukta, Hiorthhamn, Magdalenefjorden, Petuniabukta, Pyramiden, Storholmen, Vestpynten [19,20,22,30,79,85,88–91]
20.	*Arctoseius babenkoi* Makarova, 1999	Palearctic	Spitsbergen: Hornsund, Pyramiden [20,84,96]
21.	*A. cetratus* (Sellnick, 1940)	Holarctic	Spitsbergen [96]
22.	*A. haarlovi* Lindquist, 1963 **	Arctic	Bjørnøya: "Birds colony" [83]; Edgeøya: Diskobukta, Kapp Lee, Negerdalen, Russebukta [8]; Hopen: Koefoedodden, Norwegian Meteorological Institute Station [86]; Spitsbergen: Adventdalen, Barentsburg, Bjørndalen, Endalen, Hornsund, Longyearbyen, Magdalenefjorden, Mushamna, Ny-Ålesund, Petuniabukta, Pyramiden, Storholmen, Vestpynten [10,19,20,22,30,80,89,90,97]
23.	*A. laterincisus* Thor, 1930 *,†	Arctic	Spitsbergen: Adventdalen, Barentsburg, Hanaskogdalen, Hjorthhamn [91]
24.	*A. multidentatus* Evans, 1955	Arctic	Edgeøya: Blåbukta, Diskobukta, Kapp Heuglin, Kapp Lee, Negerdalen, Russebukta [8]; Hopen: Koefoedodden, Norwegian Meteorological Institute Station [86]; Nordaustlandet: Florabukta, Kinnvika [30,87]; Spitsbergen: Adventdalen, Barentsburg, Billefjorden, Endalen, Grønfjorden, Hornsund, Magdalenefjorden, Mushamna, Ny-Ålesund, Petuniabukta, Pyramiden, Storholmen, Sørkapp, Vårsolbukta [10,19,20,30,88–90,96]
25.	*A. ornatus* Lindquist, 1961	Holarctic	Spitsbergen [96]

Table 1. *Cont.*

No.	Taxon	Distribution	Localities
26.	*A. tajmyricus* Petrova & Makarova, 1991	Palearctic	Bjørnøya: Alfredfjellet [83]; Spitsbergen: Hornsund [88,96]
27.	*A. tschernovi* Makarova, 2000	Arctic	Edgeøya: Blåbukta, Kapp Heuglin, Negerdalen [8]; Nordaustlandet: Kinnvika [30,87]; Spitsbergen: Adventdalen, Endalen, Hornsund [30,89]
28.	*A. weberi* Evans, 1955	Arctic	Bjørnøya: Alfredfjellet, „Birds colony" [83]; Edgeøya: Kapp Heuglin, Kapp Lee, Negerdalen, Russebukta [8]; Spitsbergen: Adventdalen, Bjørndalen, Endalen, Ossian Sarsfjellet, Petuniabukta, Vestpynten, Vårsolbukta [20,22,30,88,89,96]
29.	*Zerconopsis labradorensis* Evans & Till, 1960	Arctic	Bjørnøya: Alfredfjellet, "Birds colony" [83]; Spitsbergen [84]
30.	*Z. muestairi* (Schweizer, 1949)	Palearctic	Bjørnøya: Teltvika [29,30]
	Melicharidae		
31.	*Proctolaelaps parvanalis* (Thor, 1930) *,**	Arctic	Bjørnøya: Alfredfjellet, "Birds colony" [83]; Edgeøya: Blåbukta, Diskobukta, Kapp Lee, Russebukta [8]; Spitsbergen: Adventdalen, Bjørndalen, Endalen, Hiorthhamn, Longyearbyen, Mälardalen, Ny-Ålesund, Petuniabukta [10,20,22,29,30,85,89,91]
	Phytoseiidae		
32.	*Neoseiulus ellesmerei* (Chant & Hansell, 1971) **	Arctic	Edgeøya: Russebukta [8,28]; Spitsbergen: Adventdalen, Endalen, Longyearbyen [28]
33.	*N. grumantensis* Kolodochka & Gwiazdowicz, 2014 *,†	Arctic	Spitsbergen: Longyearbyen, Petuniabukta [28]
34.	*N. magnanalis* (Thor, 1930) *,**,†	Arctic	Spitsbergen: Adventdalen, Hiorthhamn, Mälardalen, Ny-Ålesund, Petuniabukta [10,20,22,28,30,85,89,91]
	Dermanyssidae		
35.	*Dermnanyssus hirundinis* (Hermann, 1804)	Holarctic	Edgeøya: Blåbukta [8]; Spitsbergen: Longyearbyen, Ny-Ålesund [90,98]
	Laelapidae		
36.	*Haemogamasus ambulans* (Thorel, 1872)	Holarctic	Bjørnøya [99]; Spitsbergen: Adventdalen, Endalen, Grumantbyen, Fjortende Julibukta, Ossian Sarsfjellet, Storholmen [90,91,100]
37.	*Laelaps hilaris* C.L. Koch, 1836	Holarctic	Spitsbergen: Grumantbyen [100]
	TROMBIDIFORMES		
	Prostigmata		

Table 1. *Cont.*

No.	Taxon	Distribution	Localities
	Bdellidae		
38.	*Bdella longicornis* (Linnaeus, 1758)	Holarctic, Neotropic, Indomalayan	Bjørnøya: south [99,101]; Prins Karls Forland: MacKenziedale, Richardlaguna almost to Carmichaelpynten, Silene Hill [99,101]; Spitsbergen: Adventfjorden, Bellsund, Billefjorden, Bohemanneset, Brucebyen, Dirksbukta, Eremitten, Gipsdalen, Hiorthhamn, Sorgfjorden, Station Islands at Liefdefjorden, Salmon Lake at Dirksbukta [7,91,99,101,102]
39.	*B. semiscutata* Thor, 1930 *	Palearctic	Spitsbergen: Engelskhytta, Hiorthfjellet, Longyearbyen [91,103]
40.	*Odontoscirus lapidaria* (Kramer, 1881)	Cosmopolitan	Bjørnøya [104]; Spitsbergen: Barentsburg [91]
41.	*Cyta latirostris* (Hermann, 1804)	Cosmopolitan	Bjørnøya: Ellasjøen, south [99,101,105]; Prins Karls Forland: MacKenziedale, Silene Hill [99,101]; Spitsbergen: Adventfjorden, Barentsburg, Billefjorden, Brucebyen, Engelskhytta, "Großer Trichter", Grønfjorden, Longyearbyen, Magdalenefjorden, Salmon Lake at Dirksbukta, [7,91,99,101,103]; Svalbard [106]
42.	*Neomolgus capillatus* (Kramer, 1881)	Palearctic, Australasian	Spitsbergen: Hiorthhamn [91]
43.	*N. littoralis* (Linnaeus, 1758)	Holarctic	Bjørnøya: south [99,101,104,105]; Prins Karls Forland: Silene Hill [99,101]; Islands of Hinlopenstretet: Fosterøyane, Gyldénøyane, Vaigattøyane including Von Otterøya and Nordenskiöldøya [7]; Spitsbergen: Adventfjorden, Billefjorden, Bohemanneset, Brucebyen, Eremitten, Grønfjorden, between Hiorthhamn and Advent City, Isfjorden, Liefdefjorden, Ny-Friesland, Salmon Lake at Dirksbukta, Sassenfjorden, Sorgfjorden, south side of Reinsdyrflya [7,91,99,101,102,105]
44.	*N. pallipes* (C.L. Koch, 1879)	Holarctic	Spitsbergen: Adventfjorden, Billefjorden, Brucebyen, south side of Reinsdyrflya [7,99,101]
	Cunaxidae		
45.	*Cunaxoides croceus* (C.L. Koch, 1838)	Holarctic	Spitsbergen: "Großer Trichter" [103]
	Cocceupodidae		
46.	*Cocceupodes mollicellus* C.L. Koch 1838	Holarctic	Spitsbergen: "Großer Trichter", Hiorthhamn, Longyearbyen, Magdalenefjorden [91,103,106]
	Eupodidae		
47.	*Eupodes variegatus* C.L. Koch, 1838	Holarctic, Afrotropical	Bjørnøya: Engelskelva, Fugleodden [91]; Spitsbergen: Hanaskogdalen, Hiorthhamn, Longyearbyen? [91,103,106]
48.	*Neoprotereunetes borneri* (Thor, 1934) *,1	Holarctic	Spitsbergen: "Großer Trichter", Magdalenefjorden, Tempelfjorden [66,103]

Table 1. *Cont.*

No.	Taxon	Distribution	Localities
	Penthaleidae		
49.	*Penthaleus maior* (Dugès, 1834)	Cosmopolitan	Bjørnøya [102]; Spitsbergen: Hiorthhamn, Longyearbyen, Magdalenefjorden, Reinsdyrflya–Liefdefjorden, Station Islands at Liefdefjorden [7,91,103]
	Penthalodidae		
50.	*Penthalodes ovalis* (Dugès, 1834)	Holarctic	Bjørnøya [105]; Spitsbergen: Barentsburg, Hiorthhamn, Liefdefjorden, Longyearbyen, south side of Reinsdyrflya [7,91,103]
	Rhagidiidae		
51.	*Coccorhagidia clavifrons* (Canestrini, 1886)	Cosmopolitan	Spitsbergen: "Großer Trichter", Hiorthhamn, Tempelfjorden [91,103]
52.	*Rhagidia gelida* Thorell, 1872 *	Holarctic	Bjørnøya [101,102,104,105], south [99]; Spitsbergen: Adventfjorden, Bellsund, Billefjorden, Brucebyen, Dirksbukta, Eremitten, "Großer Trichter", Hiorthhamn, Lomfjorden, Reinsdyrflya-Liefdefjorden, Salmon Lake at Dirksbukta, south side of Reinsdyrflya [7,91,99,101–103,105]
	Halacaridae		
53.	*Bradyagaue alberti* (Trouessart, 1902) *	Arctic	Spitsbergen: north [107]; Hopen [107]
54.	*Copidognathus poucheti* (Trouessart, 1893) *	Atlantic Ocean, Arctic	Spitsbergen: western coast [81]; expedition to Spitsbergen: Bellsund, Isfjorden, and Prins Karls Forland [108]
55.	*C. reticulatus* (Trouessart, 1893) *	Atlantic Ocean, Arctic	Spitsbergen: western coast [81]; expedition to Spitsbergen: Bellsund, Isfjorden, and Prins Karls Forland [108], Hinlopenstretet [107]
56.	*C. richardi* (Trouessart, 1902) *	Atlantic Ocean, Arctic	Spitsbergen: north [107]; Hopen [107]
57.	*Halacarellus subterraneus* Schulz, 1933	Atlantic Ocean, Arctic, Palearctic	Spitsbergen [45]
58.	*H. subcrispus* Bartsch, 1978	Atlantic Ocean, Arctic	Spitsbergen [45]
59.	*Halacarus borealis* Trouessart, 1893 *	Atlantic Ocean, Arctic	Spitsbergen: western coast [81]; expedition to Spitsbergen: Bellsund, Isfjorden, and Prins Karls Forland [108]
60.	*Isobactrus levis* (Viets, 1927)	Atlantic Ocean, Arctic	Spitsbergen: Grønfjordbreen [109]
61.	*Rhombognathides spinipes* (Viets, 1933)	Atlantic Ocean, Arctic	Spitsbergen: Grønfjordbreen [109]
62.	*Rhombognathus subtilis* Bartsch, 1975	Atlantic Ocean, Arctic	Spitsbergen [45]
63.	*Thalassarachna coeca* (Trouessart, 1902) *	Arctic	Hopen [107]
64.	*T. princeps* (Trouessart, 1902) *	Arctic	Hopen [107]
	Triophtydeidae		
65.	*Triophtydeus pinicolus* (Oudemans, 1929) [2]	Holarctic, Australasian	Spitsbergen: Magdalenefjorden [103]

Table 1. *Cont.*

No.	Taxon	Distribution	Localities
	Tydeidae		
66.	*Tydeus langei* Thor, 1934 [3,*,†]	Arctic	Spitsbergen: "Großer Trichter", Magdalenefjorden, Tempelfjorden [103]
67.	*T. svalbardensis* Thor, 1932 [3,*,†]	Arctic	Bjørnøya: Engelskelva, Fugleodden [91,110]; Spitsbergen: "Großer Trichter", Hanaskogdalen, Hiorthhamn, Longyearbyen, Magdalenefjorden, Tempelfjorden [91,103]
	Iolinidae		
68.	*Tydaeolus globifer* (Thor, 1931)	Arctic	Spitsbergen: "Großer Trichter", Magdalenefjorden, Tempelfjorden [103]; Svalbard [47]
69.	*T. tenuiclaviger* (Thor, 1931)	Palearctic	Spitsbergen: Tempelfjorden [103]; Svalbard [47]
70.	*Microtydeus constans* Thor, 1931	Arctic	Spitsbergen: "Großer Trichter", Magdalenefjorden [103]; Svalbard [47]
	Sperchontidae		
71.	*Sperchon brevirostris* (Koenike, 1895)	Holarctic	Bjørnøya: Ellasjøen, Engelskelva, Hellevatnet, Mosevatnet, Stevatnet [91,99,111]
	Trombidiidae		
72.	*Podothrombium bicolor* (Hermann, 1804)	Palearctic	Spitsbergen: Hiorthhamn [91]
73.	*P. curtipalpe* (Thor, 1900) *	Holarctic	Spitsbergen: Hiorthhamn [91]
74.	*P. svalbardense* Oudemans, 1930 [*,†]	Arctic	Spitsbergen: Hiorthhamn [91]
	Syringophilidae		
75.	*Chenophila nanseni* Skoracki & Zawierucha 2016 [*,†]	Arctic	Spitsbergen: Nissenfjella [57]
	Stigmaeidae		
76.	*Eustigmaeus oudemansi* (Thor, 1930) [*,†]	Arctic	Spitsbergen: Hiorthhamn [91]
77.	*E. pulchellus* (Thor, 1930) [*,†]	Arctic	Spitsbergen: Hiorthhamn [91]
	Tetranychidae		
78.	*Bryobia borealis* Oudemans, 1930 *	Holarctic	Spitsbergen: "Großer Trichter", Tempelfjorden [103]; Svalbard [106,112]
79.	*B. praetiosa* C.L. Koch 1836	Cosmopolitan	Bjørnøya [105]; Spitsbergen: Salmon Lake at Dirksbukta, south side of Reinsdyrflya [7]
	Pygmephoridae		
80.	*Kerdabania arctica* (Thor, 1934) [4,*,†]	Arctic	Spitsbergen: Templefjorden [103]
	SARCOPTIFORMES		
	Endeostigmata		

Table 1. *Cont.*

No.	Taxon	Distribution	Localities
	Nanorchestidae		
81.	*Nanorchestes arboriger* (Berlese, 1904) [5]	Palearctic	Spitsbergen: "Großer Trichter", Hiorthhamn, Magdalenefjorden [91,103]
	Alicorhagiidae		
82.	*Alicorhagia clavipilus* (Thor, 1931) *	Arctic	Spitsbergen: "Großer Trichter", Hiorthhamn [103,113]
83.	*A. plumipilis* (Thor, 1931) *	Palearctic	Spitsbergen: "Großer Trichter" [103]
	Eriophyidae		
84.	*Aceria saxifragae* (Rostrup, 1900)	Holarctic	Spitsbergen: Hornsund [67]
85.	*Cecidophyes siedleckii* Kiedrowicz, Szydlo & Skoracka, 2016 *,†	Arctic	Spitsbergen: Hornsund [67]
	Oribatida		
	Brachychthoniidae		
86.	*Brachychthonius laetepictus* Berlese, 1910	Palaearctic	Spitsbergen: Magdalenefjorden, Pyramiden [20,103]
87.	*Eobrachychthonius borealis* Forsslund, 1942	Holarctic	Bjørnøya [83]; Spitsbergen: Barentsburg, Longyearbyen, Recherchefjorden [14,114]
88.	*E. latior* (Berlese, 1910)	Holarctic	Danskøya: Virgohamna [115]; Spitsbergen: Bockfjorden, Fivelflyene, Flathuken, "Großer Trichter", Helvetiadalen, Hiorthhamn, Isdammen, Longyearbyen, Magdalenefjorden, Reinsdyrflya, Sassendalen, Sven Olssonodden, Sørgattet [91,103,115,116]
89.	*E. oudemansi* Hammen, 1952	Holarctic, Neotropical, Indomalayan, Antarctic	Edgeøya: Kapp Lee, Russebukta [8] Spitsbergen: Longyearbyen, Pyramiden, Recherchefjorden [20,114,116]
90.	*Liochthonius alpestris* (Forsslund, 1958)	Palearctic	Spitsbergen: Hornsund [88]
91.	*L. brevis* (Michael, 1888)	Holarctic	Spitsbergen: Adventdalen, Barentsburg, Fivelflyene, Flathuken, Hiorthhamn, Isdammen, Reinsdyrflya, Worsleyneset [19,91,115,116]
92.	*L. clavatus* (Forsslund, 1942)	Palaearctic	Edgeøya: Blåbukta, Kapp Heuglin [8]
93.	*L. lapponicus* (Trägårdh, 1910)	Holarctic	Bjørnøya [83]; Spitsbergen: Barentsburg, Hornsund, Longyearbyen, Midtre Lovénbreen, Ny-Ålesund, Recherchefjorden [10,14,15,88,114,117,118]
94.	*L. muscorum* Forsslund, 1964	Palaearctic	Spitsbergen: Barentsburg, Hornsund [19,88]
95.	*L. neglectus* Moritz, 1976	Palearctic	Spitsbergen: Kongsfjorden [119]; based on Hodkinson's material stored in Liverpool [120]

Table 1. *Cont.*

No.	Taxon	Distribution	Localities
96.	*L. sellnicki* (Thor, 1930) *	Holarctic	Bjørnøya [83]; Hopen: Koefoedodden, Hopen radio [86]; Spitsbergen: Barentsburg, "Großer Trichter", Hiorthhamn, Longyearbyen, Magdalenefjorden, Petuniabukta, Pyramiden, Recherchefjorden, Vestpynten [14,19,20,22,91,103,105,114]
97.	*L. strenzkei* Forsslund, 1963	Holarctic	Edgeøya: Blåbukta, Kapp Heuglin, Kapp Lee, Negerdalen, Russebukta [8]
98.	*L. tuxeni* (Forsslund, 1957)	Palaearctic	Spitsbergen: Midtre Lovénbreen [10]
99.	*Neoliochthonius piluliferus* (Forsslund, 1942)	Holarctic	Edgeøya: Kapp Heuglin [8]
	Eniochthoniidae		
100.	*Eniochthonius minutissimus* (Berlese, 1903)	Cosmopolitan	Spitsbergen: Petuniabukta [22]
	Phthiracaridae		
101.	*Atropacarus striculus* (C.L. Koch, 1835)	Holarctic, Oriental, Neotropical, Australian	Spitsbergen: Barentsburg, Grønfjordenbreen, Isfjorden, Longyearbyen, Midtre Lovénbreen [10,13,14,16]
	Crotoniidae		
102.	*Camisia anomia* Colloff, 1993	Arctic	Bjørnøya [83]; Spitsbergen: Endalen, Kolhaugen, Kongsfjorden [121–129]
103.	*C. biverrucata* (C.L. Koch, 1839)	Holarctic	Edgeøya: Blåbukta, Kapp Heuglin [8]; Spitsbergen: Longyearbyen [103]
104.	*C. borealis* (Thorell, 1871) *	Holarctic	Danskøya: Virgohamna [115]; Spitsbergen: Barentsburg, Hornsund, Isfjorden, Longyearbyen, Recherchefjorden [14,16,102,114,115,130]
105.	*C. dictyna* Colloff, 1993	Holarctic	Svalbard [71]; Spitsbergen: Midtre Lovénbreen [10]
106.	*C. foveolata* Hammer, 1955	Holarctic	Bjørnøya [83]; Edgeøya: Blåbukta, Kapp Heuglin, Negerdalen, Russebukta [8]; Hopen: Hopen radio, Koefoedodden [86]; Spitsbergen: Aldegondabreen, Grønfjordenbreen, Hornsund, Kiærstranda, Knudsenheia, Midtre Lovénbreen [10,88,109,131]
107.	*C. horrida* (Hermann, 1804)	Holarctic, Oriental, Ethiopian, Neotropical	Bjørnøya [91]; Edgeøya: Diskobukta, Kapp Lee, Russebukta [8]; Spitsbergen: Barentsburg, "Großer Trichter", Hiorthhamn, Hiorthfjellet, Longyearbyen, Magdalenefjorden, Kongsfjorden, Reinsdyrflya, Recherchefjorden, Vestpynten, Petuniabukta, Pyramiden [20,22,91,103,105,114,115,132]
108.	*C. invenusta* (Michael, 1888)	Palaearctic	Svalbard [71]
109.	*C. lapponica* (Trägårdh, 1910)	Holarctic	Spitsbergen: Barentsburg, Hornsund, Longyearbyen [14,16,88]
110.	*C. spinifer* (C.L. Koch, 1836)	Holarctic, Oriental, Neotropical	Spitsbergen: Tempelfjorden [103]
111.	*Capillonothrus capillatus* (Berlese, 1914)	Holarctic	Spitsbergen: Arctowskifjellet [115]
112.	*Platynothrus peltifer* (C.L. Koch, 1839)	Holarctic, Oriental, Australian	Spitsbergen: Recherchefjorden, Sassendalen [114,115]
113.	*P. punctatus* (L. Koch, 1879)	Palaearctic	Spitsbergen: Barentsburg, Grønfjordenbreen, Isfjorden, Longyearbyen, Petuniabukta, Vestpynten [13,14,16,22]

Table 1. *Cont.*

No.	Taxon	Distribution	Localities
	Hermanniidae		
114.	*Hermannia reticulata* Thorell, 1871 *	Holarctic	Bjørnøya: Fugleodden, Tunheim [83,91,105,133]; Edgeøya: Blåbukta, Russebukta [134]; Prins Karls Forland: Silene Hill [99,101]; Spitsbergen: Barentsburg, Bellsund, Endalen, "Großer Trichter", Kongsfjorden, Longyearbyen, Magdalenefjorden, Midtre Lovénbreen, Ny-Ålesund, Observasjonsholmen, Recherchefjorden, Petuniabukta, Tempelfjorden, Vestpynten [10,12,15,22,91,102,103,114,121–123,125,129,132–134]
115.	*H. scabra* (L. Koch, 1879)	Holarctic	Bjørnøya [83,105]; Edgeøya: Kapp Lee [135]; Prins Karls Forland: Fuglehuken fyr [135]; Spitsbergen: Kapp Laila, Petuniabukta [135]
	Malaconothridae		
116.	*Malaconothrus monodactylus* (Michael, 1888)	Palaearctic	Spitsbergen: Barentsburg, Grønfjordenbreen, Isfjorden, Longyearbyen [13,14,16]
	Nanhermanniidae		
117.	*Nanhermannia sellnicki* Forsslund, 1958	Palearctic	Spitsbergen: Barentsburg, Longyearbyen [14]
	Nothridae		
118.	*Nothrus palustris* C.L. Koch, 1839	Holarctic	Spitsbergen: Barentsburg, Grønfjordenbreen, Isfjorden, Longyearbyen [13,16]
	Damaeidae		
119.	*Damaeus onustus* (C.L. Koch, 1844)	Palaearctic, Afrotropical	Spitsbergen: Barentsburg, Longyearbyen [16]
120.	*Kunstidamaeus arcticus* Miko & Monson, 2013 *,†	Svalbard	Bjørnøya [83]; Spitsbergen [72]
121.	*Subbelba montana* (Kulczynski, 1902)	Palearctic	Bjørnøya: Austervåg [91]
	Peloppiidae		
122.	*Ceratoppia bipilis* (Hermann, 1804)	Holarctic, Oriental, Neotropical	Prins Karls Forland: Silene Hill [99,101]; Spitsbergen: Kongsfjorden, Storholmen [7,12]
123.	*C. hoeli* Thor, 1930 *	Arctic	Bjørnøya: Tunheim [91]; Spitsbergen: Barentsburg, Bockfjorden, "Großer Trichter", Grønfjordenen, Hiorthfjellet, Hiorthhamn, Isdammen, Kolhaugen, Kongsfjorden, Longyearbyen, Magdalenefjorden, Recherchefjorden, Worsleyneset [91,103,114,115,121–125,127]
124.	*C. sphaerica* (L. Koch, 1879)	Palaearctic	Bjørnøya [83]; Edgeøya: Kapp Lee, Russebukta [8]; Hopen: Hopen radio, Koefoedodden [86]; Spitsbergen: Barentsburg, Hornsund, Longyearbyen, Petuniabukta, Pyramiden, Vestpynten [14–16,20,22,130]

Table 1. *Cont.*

No.	Taxon	Distribution	Localities
	Carabodidae		
125.	*Carabodes labyrinthicus* (Michael, 1879)	Holarctic	Spitsbergen: Barentsburg, Longyearbyen [14]
126.	*C. marginatus* (Michael, 1884)	Palearctic	Spitsbergen: Barentsburg, Longyearbyen [16]
	Autognetidae		
127.	*Autogneta kaisilai* Karppinen, 1967 *	Arctic	Spitsbergen: Biskayerhuken [115]
128	*Conchogneta dalecarlica* (Forsslund, 1947)	Palearctic	Spitsbergen: Midtre Lovénbreen [10]
	Oppiidae		
129.	*Dissorhina ornata* (Oudemans, 1900)	Holarctic	Bjørnøya: Fugleodden, Mosevatnet [91]; Spitsbergen: Adventdalen, Barentsburg, Biskayerhuken, Fivelflyene, Hiorthhamn, Isdammen, Longyearbyen, Magdalenefjorden, Petuniabukta, Pyramiden [14,20,22,91,103,115]
130.	*Lauroppia fallax* (Paoli, 1908)	Holarctic, IndoMalayan, Australian, Neotropical	Spitsbergen: Magdalenefjorden [103]
131.	*Microppia minus* (Paoli, 1908)	Cosmopolitan	Spitsbergen: Hornsund [136]
132.	*Moritziella microdentata* Gordeeva & Grishina, 1991	Palearctic	Spitsbergen: Hornsund [137]
133.	*Moritzoppia splendens* (C.L. Koch, 1841)	Holarctic	Bjørnøya [83]; Spitsbergen: Barentsburg, Grønfjordenbreen, Isfjorden, Longyearbyen [13,14,16]
134.	*M. unicarinata* (Paoli, 1908)	Holarctic	Edgeøya: Negerdalen [8]; Spitsbergen: Barentsburg, Longyearbyen [14,16]
135.	*Oppiella neerlandica* (Oudemans, 1900)	Holarctic	Edgeøya: Kapp Heuglin [8]; Spitsbergen: Barentsburg, "Großer Trichter", Hanaskogdalen, Hiorthhamn, Kongsfjord, Longyearbyen, Magdalenefjorden, Midtre Lovénbreen, Vestpynten [10,14,19,22,91,103,125]
136.	*O. nova* (Oudemans, 1902)	Cosmopolitan	Spitsbergen: Adventdalen, Barentsburg, Grønfjordenbreen, Hornsund, Isdammen, Isfjorden, Kongsfjorden, Longyearbyen, Petuniabukta, Reinsdyrflya, Worsleyneset [13,14,16,19,22,115,121,123,136]
137.	*O. translamellata* (Willmann, 1923)	Holarctic	Edgeøya: Diskobukta, Kapp Lee, Russebukta [8]; Hopen: Hopen radio, Koefoedodden [84]; Spitsbergen: Adventdalen, Arctowskifjellet, Barentsburg, Bockfjorden, Fivelflyene, Grønfjordenbreen, Hornsund, Isfjorden, Kolhaugen, Longyearbyen, Magdalenefjorden, Petuniabukta, Recherchefjorden, Reinsdyrflya, Stuphallet, Vestpynten, Worsleyneset [13,14,16,22,103,114,115,124,136,138]

Table 1. *Cont.*

No.	Taxon	Distribution	Localities
	Suctobelbidae		
138.	*Suctobelba hammerae* Krivolutsky, 1965	Holarctic	Spitsbergen: Barentsburg, Grønfjordenbreen, Isfjorden, Longyearbyen [13,14,16]
139.	*Suctobelbella sarekensis* (Forsslund, 1941)	Holarctic	Spitsbergen: Adventdalen, Fivelflyene [115]
140.	*S. subcornigera* (Forsslund, 1941)	Holarctic, Oriental, Australian	Spitsbergen: Barentsburg, Biskayerhuken, Grønfjordenbreen, Isfjorden, Longyearbyen, Reinsdyrflya, Worsleyneset [13,14,16,22,115]
	Tectocepheidae		
141.	*Tectocepheus alatus* Berlese, 1913	Palearctic	Spitsbergen: Recherchefjorden [114]
142.	*T. knuellei* Vanek, 1960	Palearctic	Spitsbergen: Barentsburg, Longyearbyen [14,16]
143.	*T. sarekensis* Trägårdh, 1910	Palearctic	Spitsbergen: Adventdalen, Blomstrandhalvøya, Bockfjorden, Isdammen, Kongsfjorden, Reinsdyrflya, Worsleyneset [115]
144.	*T. velatus* (Michael, 1880)	Cosmopolitan	Bjørnøya [83]; Edgeøya: Diskobukta, Kapp Lee, Russebukta [8]; Spitsbergen: Barentsburg, Hiorthhamn, Kolhaugen, Kongsfjorden, Longyearbyen, Midtre Lovénbreen, Petuniabukta, Pyramiden, Vestpynten [10,14,16,19,20,22,91,103,105,124,125]
	Ameronothridae		
145.	*Ameronothrus lineatus* (Thorell, 1871) *	Palaearctic	Bjørnøya: Engelskelva, Fugleodden, Tunheim [83,88,99,101,103,105,132,133,139]; Lågøya: [7]; Prins Karls Forland: MacKenziedale, Richardlaguna almost to Carmichaelpynten [99,101]; Spitsbergen: Adventdalen, Barentsburg, Bockfjorden, Colesbukta, Grønfjordenbreen, Kolhaugen, Liefdefjorden, Nordenskiöld Land, Ny-Ålesund, Observasjonsholmen, Reinsdyrflya, Smeerenburg, Worsleyneset [7,12,91,102,109,115,124,132,133,140–145]
146.	*A. nidicola* Sitnikova, 1975	Palaearctic	Spitsbergen: Hornsund [88]
147.	*A. nigrofemoratus* (L. Koch, 1879)	Palearctic	Bjørnøya [83]
	Micreremidae		
148.	*Micreremus brevipes* (Michael, 1888)	Palearctic	Spitsbergen: Recherchefjorden [114]
	Scutoverticidae		
149.	*Scutovertex minutus* (C.L. Koch, 1835)	Palearctic	Spitsbergen: Bockfjorden, Recherchefjorden [114,115]
	Achipteriidae		
150.	*Achipteria punctata* (Nicolet, 1855)	Holarctic	Spitsbergen: Barentsburg, Grønfjordenbreen, Isfjorden, Longyearbyen, Vestpynten [13,14,16,22]

Table 1. *Cont.*

No.	Taxon	Distribution	Localities
	Tegoribatidae		
151.	*Scutozetes clavatosensillus* Ermilov, Martens & Tolstikov, 2013	Palearctic	Spitsbergen: Mosselbukta [146]
	Oribatellidae		
152.	*Oribatella arctica* Thor, 1930 *	Holarctic	Edgeøya: Diskobukta, Kapp Lee [8,147]; Spitsbergen: Barentsburg, Hiorthhamn, Longyearbyen, Mälardalen, Petuniabukta, Recherchefjorden [15,22,91,103,114]
	Oribatulidae		
153.	*Oribatula exilis* (Nicolet, 1855)	Holarctic	Bjørnøya: Fugleodden, Mosevatnet, Tunheim [91]; Prins Karls Forland: Silene Hill [99,101]; Spitsbergen: Adventdalen, Barentsburg, Engelskhytta, Hanaskogdalen, Hiorthhamn, Isfjord radio, Longyearbyen [14,91]
154.	*O. tibialis* (Nicolet, 1855)	Holarctic	Bjørnøya [83]; Edgeøya: Diskobukta, Kapp Lee, Russebukta [8]; Spitsbergen: Barentsburg, Longyearbyen, Observasjonsholmen, Petuniabukta, Storholmen, Vestpynten [12,14,15,19,22]
155.	*O. venusta* Berlese, 1908	Holarctic	Bjørnøya [83]; Spitsbergen: Adventdalen, Fivelflyene, "Großer Trichter", Longyearbyen, Magdalenefjorden, Recherchefjorden, Reinsdyrflya, Worsleyneset [103,114,115]
156.	*Phauloppia lucorum* (C.L. Koch, 1841)	Palearctic	Spitsbergen: Bellsund, Isfjorden [81,133]
	Scheloribatidae		
157.	*Liebstadia similis* (Michael, 1888)	Holarctic, Indomalayan, Australian	Spitsbergen: Barentsburg, Grønfjordenbreen, Isfjorden, Longyearbyen [13,14,16]
	Ceratozetidae		
158.	*Ceratozetes spitsbergensis* Thor, 1934 *	Palearctic	Edgeøya: Blåbukta, Kapp Heuglin, Negerdalen [8]; Kong Karls Land [148]; Spitsbergen: Arctowskifjellet, Barentsburg, "Großer Trichter", Kolhaugen, Longyearbyen, Magdalenefjorden, Pyramiden, Reinsdyrflya, Worsleyneset [14,20,103,115,124]

Table 1. *Cont.*

No.	Taxon	Distribution	Localities
159.	*Diapterobates notatus* (Thorell, 1871) *	Holarctic	Bjørnøya: Fugleodden, Tunheim [14–16,83,91,99,101,105,133]; Barentsøya [130]; Edgeøya: Blåbukta, Diskobukta, Kapp Heuglin, Kapp Lee, Negerdalen, Russebukta, Tjuvfjorden [8,146]; Hopen: Hopen radio, Koefoedodden [86]; Prins Karls Forland: Fuglehuken, MacKenziedale, Silene Hill [99,101,146]; Sofiaøya [146]; Spitsbergen: Adventfjorden, Barentsburg, Bellsund, Billefjorden, Blomstrandhalvøya, Bohemanneset, Dirksbukta, Endalen, Engelskhytta, "Großer Trichter", Grønfjordenen, Hiorthfjellet, Hiorthhamn, Hornsund, Isfjorden, Kolhaugen, Krykkjefjellet, Liefdefjorden, Longyearbyen, Magdalenefjorden, Midtre Lovénbreen, Ossian Sarsfjellet, Petuniabukta, Recherchefjorden, Reinsdyrflya, Sven Olssonodden, Storholmen, Vestpynten, Worsleyneset [7,10,12,14–17,19,22,91,99,101–103,114,115,121–125,129,130,132,133,146,148]
160.	*Edwardzetes edwardsi* (Nicolet, 1855)	Holarctic	Spitsbergen: Tempelfjorden; possibly from the North Cape [103]
161.	*Fuscozetes coulsoni* A. & S. Seniczak, 2020 *,†	Arctic	Edgeøya: Negerdalen [9]
162.	*Iugoribates gracilis* Sellnick, 1944	Holarctic	Spitsbergen: Kongsfjorden [119]; based on Hodkinson's material stored in Liverpool [120]
163.	*Oromurcia lucens* (L. Koch, 1879)	Holarctic	Barentsøya [130]; Spitsbergen: Vestpynten [22]
164.	*Svalbardia paludicola* Thor, 1930 *	Holarctic	Spitsbergen: Hanaskogdalen, Hiorthhamn, Hornsund, Reinsdyrflya, Ringhorndalen, Sørkapp [91,139,146]
165.	*Trichoribates berlesei* Jacot, 1929	Holarctic	Spitsbergen: Hornsund, Storholmen [12,88]
166.	*T. novus* (Sellnick, 1928)	Holarctic	Spitsbergen: Hornsund [88]
167.	*T. setiger* (Trägårdh, 1910)	Palaearctic	Bjørnøya [83]; Spitsbergen: Recherchefjorden [114]
	Chamobatidae		
168.	*Chamobates birulai* (Kulczynski, 1902) *	Palaearctic	Spitsbergen [133]
169.	*C. borealis* (Trägårdh, 1902)	Palaearctic	Spitsbergen: Hornsund [88]
	Punctoribatidae		
170.	*Mycobates bicornis* (Strenzke, 1954)	Palaearctic	Spitsbergen: Midtre Lovénbreen [10]
171.	*M. parmeliae* (Michael, 1884)	Holarctic	Spitsbergen: Barentsburg, Longyearbyen, Midtre Lovénbreen [10,14,16]

Table 1. *Cont.*

No.	Taxon	Distribution	Localities
172.	*M. sarekensis* (Trägårdh, 1910)	Holarctic	Bjørnøya [91]; Edgeøya: Kapp Lee, Russebukta [8,149]; Spitsbergen: Adventdalen, Arctowskifjellet, Bockfjorden, Engelskhytta, Fjortende Julibukta, "Großer Trichter", Hiorthfjellet, Hiorthhamn, Hornsund, Isdammen, Kongsfjorden, Longyearbyen, Magdalenefjorden, Mosselbukta, Recherchefjorden, Sassendalen, Signehamna, Storholmen, Sørkapp, Vestpynten [22,91,103,114,115,121,123,125,149]
173.	*M. tridactylus* Willmann, 1929	Holarctic	Spitsbergen: Barentsburg, Longyearbyen [14]
174.	*Punctoribates punctum* (C.L. Koch, 1839)	Holarctic, Oriental, Australian	Spitsbergen: Barentsburg, Longyearbyen, Petuniabukta [14,16,22]
	Galumnidae		
175.	*Pergalumna nervosa* (Berlese, 1914)	Holarctic	Spitsbergen: Hornsund [136]
	Acaridae		
176.	*Sancassania mycophagus* (Mégnin, 1874)	Cosmopolitan	Spitsbergen: Hornsund [91]
	Alloptidae		
177.	*Alloptes* (*Sternalloptes*) *stercorarii* Dubinin, 1952	Arctic	Spitsbergen: Ny-Ålesund [74]
	Avenzoariidae		
178.	*Zachvatkinia isolata* Mironov, 1989	Arctic, Neotropical, Australian, Afrotropical	Spitsbergen: Ny-Ålesund [74]

Note: *—new to science; **—redescription on the base of material from Svalbard; †—so far found only on Svalbard. [1] [150] proposed the genus *Neoprotereunetes* for those species of "*Protereunetes*" that remained in the Eupodidae after the type specimen of *Protereunetes* (*P. agilis* Berlese, 1923) was transferred to *Ereynetes* [66]. It is implied that this new genus includes *Protereunetes borneri* Thor, 1934, but the combination was not published. [2] Identity and generic affiliation of this species is questionable [46]; possible record from Greenland with inconclusive identification [48]. [3] Species not included in [50], *species inquierenda*? [4] Inadequately described, not included in key to species [61]. [5] Although specimens from Svalbard were apparently examined, this species was not confirmed from Svalbard [65].

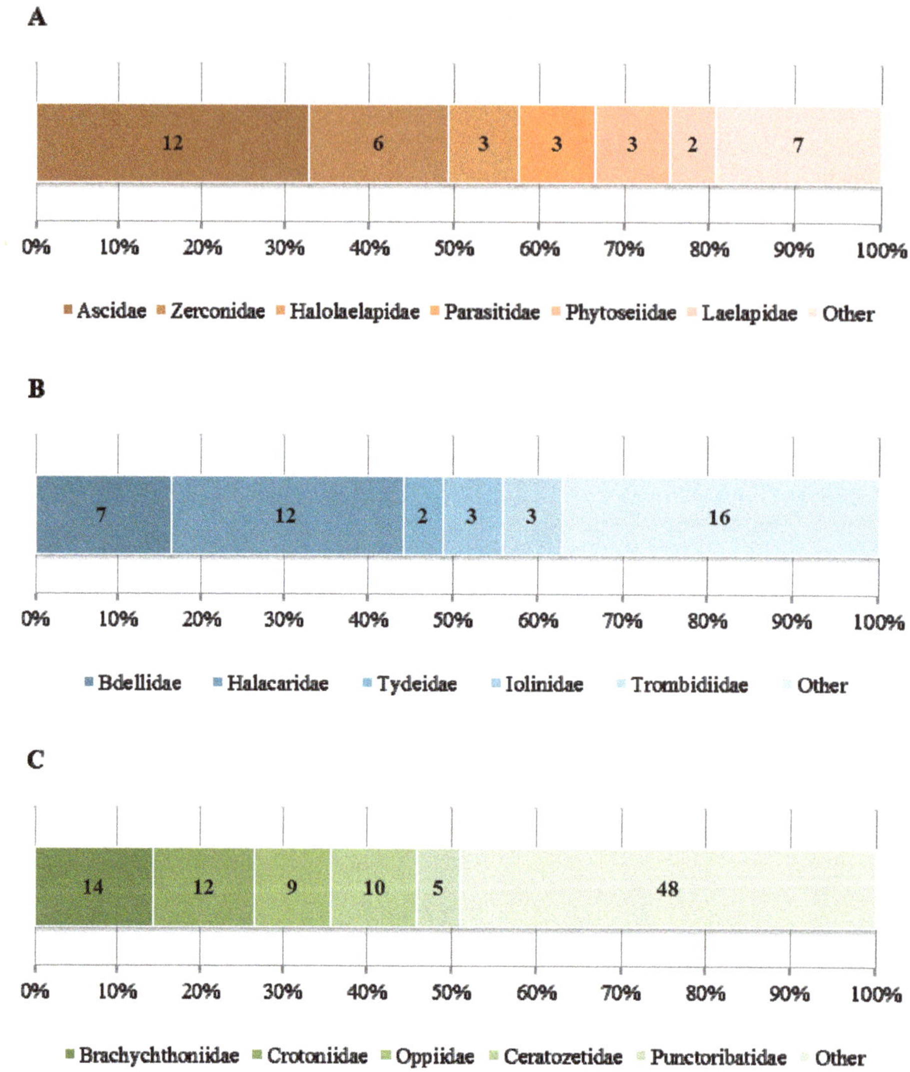

Figure 2. Diversity of the mite families in Svalbard: (**A**) Mesostigmata; (**B**) Trombidiformes; (**C**) Sarcoptiformes. The number of species occurring in Svalbard is presented on bars. No figure was made for the Ixodida, which is represented in Svalbard by one family only.

More than a half of the Mesostigmata species known from Svalbard have been recorded there only once (Table 2). The majority of these are recent findings, but four were reported only in the first half of the last century, including *Arctoseius laterincisus*, which has an Arctic distribution. Four other records come from the second half of the last century, including another species with an Arctic distribution—*Halolaelaps gerlachi*.

Table 2. Mite species found on Svalbard only once (or by the same author at a similar time).

Until 1950	1951–2000	After 2001	Reference
Mesostigmata			
U. acuminatus			[81]
		Z. curiosus	[85]
Z. triangularis			[91]
		P. (A.) insertus	[18]
		V. immanis	[92]
		V. remberti	[18]
		D. foveolatus	[20]
		H. coulsoni *	[93]
	H. gerlachi *		[94]
		S. baloghi	[8]
G. borealis			[7]
		M. muscaedomesticae	[84]
	A. cetratus		[96]
A. laterincisus *			[91]
	A. ornatus		[96]
		N. grumantensis *	[28]
	L. hilaris		[100]
Trombidiformes			
B. semiscutata *			[91,103]
N. capillatus			[91]
C. croceus			[103]
		A. saxifragae	[67]
P. borneri *			[103]
P. maior			[102]
C. clavifrons			[91,103]
B. alberti *			[107]
C. poucheti *			[81]
C. reticulatus *			[107]
C. richardi *			[107]
		H. subterraneus	[45]
		H. subcrispus	[45]
H. borealis *			[81]
		I. levis	[109]
		R. spinipes	[109]
		R. subtilis	[45]
T. coeca *			[107]
T. princeps *			[107]
T. globifer *			[103]
T. tenuiclaviger *			[103]
M. constans *			[103]
T. langei *			[103]
P. bicolor			[91]
P. curtipalpe *			[91]
P. svalbardense *			[91]
E. oudemansi *			[91]
E. pulchellus *			[91]
		C. nanseni *	[57]
K. arctica *			[103]

Table 2. *Cont.*

Until 1950	1951–2000	After 2001	Reference
Sarcoptiformes			
A. clavipilus *			[103,113]
A. plumipilis *			[103]
		C. siedleckii *	[67]
	L. alpestris		[88]
		L. clavatus	[8]
		L. neglectus	[119]
		L. tuxeni	[10]
C. spinifer			[103]
	C. capillatus		[115]
		N. sellnicki	[14]
		D. onustus	[16]
S. montana			[91]
		C. labyrinthicus	[14]
		C. marginatus	[16]
	A. kaisilai *		[115]
		C. dalecarlica	[10]
L. fallax			[103]
	M. minus		[136]
	S. sarekensis		[115]
	T. alatus		[114]
	T. sarekensis		[115]
	A. nidicola		[88]
		A. nigrofemoratus	[83]
		S. clavatosensillus	[146]
E. edwardsi			[103]
		F. coulsoni *	[9]
		I. gracilis	[119]
	T. novus		[88]
	C. birulai *		[133]
	C. borealis		[88]
		M. bicornis	[10]
	P. nervosa		[136]
S. mycophagus			[91]
		A. stercorarii	[74]
		Z. isolata	[74]

Note: *—new to science.

Most of the Mesostigmata species known from Svalbard have a Palaearctic distribution (Figure 3). Arctic species are also abundant and comprise nearly 40% of Svalbards total mesostigmatic diversity, while Holarctic species are less numerous. Five species have so far been found only in Svalbard (Table 1), including recent records (*Halolaelaps coulsoni* and *Neoseiulus grumantensis*) and historic records (*Halolaelaps gerlachi*, *Arctoseius laterincisus* and *Neoseiulus magnanalis*). Mesostigmata have been recorded from five islands in the Svalbard archipelago (Figure 1). Although most of the species are known from Spitsbergen, *Saprosecans baloghi* Karg, 1964, is known only from Edgeøya and *Zerconopsis muestairi* (Schweizer, 1949) has been found exclusively on Bjørnøya (Table 1).

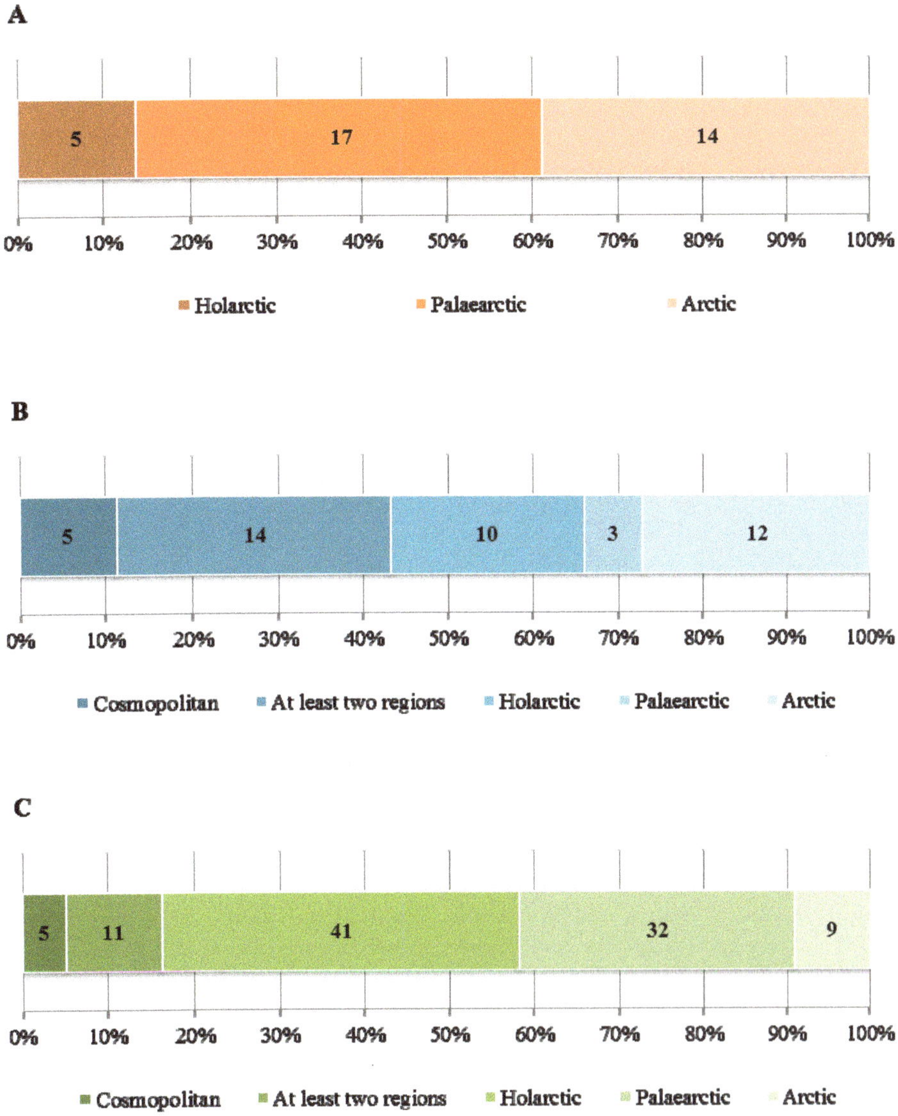

Figure 3. Distribution of mite species occurring in Svalbard: (**A**) Mesostigmata; (**B**) Trombidiformes; (**C**) Sarcoptiformes. No figure was made for the Ixodida, which are represented in Svalbard by one species with a distribution in "at least two regions" (see Table 1 for details).

3.3. *Trombidiformes*

The order Trombidiformes is represented in Svalbard only by the suborder Prostigmata with 17 families and 43 species recorded. The most diverse are the marine Halacaridae (12 spp.), followed by the terrestrial Bdellidae (7 spp.); these two families together comprise 44% of all the Trombidiformes known from Svalbard (Figure 1).

A strikingly large number of new species (19) have been described from Svalbard; i.e., nearly half of all Trombidiformes known from this archipelago: *Bdella semiscutata* Thor, 1930; *Neoprotereunetes borneri* (Thor, 1934); *Rhagidia gelida* Thorell, 1872; *Bradyagaue alberti* (Trouessart, 1902); *Copidognathus poucheti* (Trouessart, 1893); *C. reticulatus* (Trouessart, 1893); ***C.** richardi* (Trouessart, 1902); *Halacarus borealis* Trouessart, 1893; *Thalassarachna coeca* (Trouessart, 1902); *T. princeps* (Trouessart, 1902); *T. langei* Thor, 1934; *T. svalbardensis* Thor, 1932; *Podothrombium curtipalpe* (Thor, 1900); *P. svalbardense* Oudemans, 1930; *Chenophila nanseni* Skoracki & Zawierucha, 2016; *Eustigmaeus oudemansi* (Thor, 1930); *E. pulchellus* (Thor, 1930); *Bryobia borealis* Oudemans, 1930; and *Kerdabania arctica* (Thor, 1934). Moreover, 85% of the new species described from Svalbard have been recorded only once and mostly from early studies of the mite fauna (Table 2).

Species with wide geographical distributions predominate; cosmopolitan and Holarctic species, and those present in at least two zoogeographic regions (except that the Holarctic is treated as one unit) together comprise 66% of all the Trombidformes (Figure 2). Arctic species are also abundant representing 27% of the recorded species, while the fewest number of species have Palaearctic distributions. There are seven species which have only been recorded from Svalbard (Table 1), but these are mostly single old records of Sig Thor, including *Tydeus langei, T. svalbardensis, Podothrombium svalbardense, Eustigmaeus oudemansi, E. pulchellus*, and *Kerdabania arctica*. The one exception is the recently described *Chenophila nanseni*.

Trombidiformes have been found on five islands/island groups, predominantly on Spitsbergen (40 species), with others collected from Bjørnøya (11), Prins Karls Forland (6), Hopen (4), and a single species on the small islands of the Hinlopenstretet. A large number of species (27) have been found exclusively on Spitsbergen, one species [*Sperchon brevirostris* (Koenike, 1895)] only on Bjørnøya and two others (*Thalassarachna coeca* and *T. princeps*) only on Hopen (Table 1).

3.4. Sarcoptiformes

In Svalbard, 98 species of Sarcoptiformes from two suborders (Endeostigmata with 5 species and Oribatida with 93 species) have been recorded (Table 1). They represent 33 families; the richest in species is the Brachychthoniidae (14 spp.), followed by the Crotoniidae (12 spp.), Ceratozetidae (10 spp.), Oppiidae (9 spp.), and Punctoribatidae (5 spp.). Together, these five families comprise 51% of the species diversity of Sarcoptiformes in Svalbard (Figure 1). Some families are represented by only 1–4 species, yet this constitutes a significant share of their known global diversity. For example, 22% of the species belonging to the endeostigmatid family Alicorhagiidae occur in Svalbard. The oribatid families Eniochthoniidae, Peloppiidae and Tectocepheidae are also represented by large proportions of their total known species diversity, with 17%, 27%, and 24%, respectively.

Sixteen new species of Sarcoptiformes have been described from Svalbard: *Alicorhagia clavipilus* (Thor, 1931); *A. plumipilis* (Thor, 1931); *Cecidophyes siedleckii* Kiedrowicz, Szydlo & Skoracka, 2016; *Liochthonius sellnicki* (Thor, 1930); *Camisia borealis* (Thorell, 1871); *Hermannia reticulata* Thörell, 1871; *Kunstidamaeus arcticus* Miko & Monson, 2013; *Ceratoppia hoeli* Thor, 1930; *Autogneta kaisilai* Karppinen, 1967; *Ameronothrus lineatus* (Thorell, 1871); *Oribatella arctica* Thor, 1930; *Ceratozetes spitsbergensis* Thor, 1934; *Diapterobates notatus* (Thorell, 1871); *Fuscozetes coulsoni* A. & S. Seniczak, 2020; *Svalbardia paludicola* Thor, 1930; and *Chamobates birulai* (Kulczynski, 1902).

Species with wide geographic ranges together form 58% (cosmopolitan, 5%; Holarctic, 42%; and those found in at least two regions, 11%) of the sarcoptiform species (Figure 2). Species with a Palaearctic distribution also make up a substantial proportion (33%), with the remaining species having an Arctic distribution. Two of these Arctic species have only been found in the north of Spitsbergen: *Autogneta kaisilai* in Biskayerhuken and *Scutozetes clavatosensillus* Ermilov, Martens & Tolstikov, 2013, in Mosselbukta; and *Fuscozetes coulsoni* exclusively on Edgeøya (Table 1, Figure 3). Three species have been recorded only from Svalbard and these are all recently described new species (Table 1).

Sarcoptiformes have been recorded from ten islands in the Svalbard archipelago, but not on Nordaustlandet. The most commonly collected species is *Diapterobates notatus* (Table 1) found on seven islands and at nearly all collecting localities. Five species [*Camisia foveolata* Hammer, 1955; *Hermannia reticulata, H. scabra* (L. Koch, 1879); *Ceratoppia sphaerica* (L. Koch, 1879); and *Ameronothrus lineatus* (Thorell, 1871)] have been found on four islands and another seven other species [*Camisia horrida*

(Hermann, 1804); *Mycobates sarekensis* (Trägårdh, 1910); *Ceratozetes spitsbergensis*; *Oppiella translamellata* (Willmann, 1923); *Tectocepheus velatus* (Michael, 1880); *Oribatula exilis* (Nicolet, 1855); and *O. tibialis* (Nicolet, 1855)] on three islands. All of these species also have a wider distribution on Spitsbergen (Table 1, Figure 3).

Some species have been found exclusively on one island (Table 1), including 57 found only on Spitsbergen, four [*Liochthonius clavatus* (Forsslund, 1942); *L. strenzkei* Forsslund, 1963; *Neoliochthonius piluliferus* (Forsslund, 1942); and *Fuscozetes coulsoni*] on Edgeøya, and two [*Subbelba montana* (Kulczynski, 1902) and *Ameronothrus nigrofemoratus* (L. Koch, 1879)] from Bjørnøya.

Approximately one third of the sarcoptiform species have been recorded in Svalbard only once (Table 2). Seven of them were found in the first half of the last century and, 12 others, in the second half of the last century, including the descriptions of four new species.

4. Discussion

The acarofauna of the Svalbard archipelago is diverse with 178 recorded species comprising one Ixodida, 36 Mesostigmata, 43 Trombidiformes, and 98 Sarcoptiformes. However, compared with other regions of the Arctic (Alaska, Greenland, Iceland, and Taimyr), the species diversity is lower [151]. This paucity is likely to be explained by a combination of the small area of Svalbard (60% covered by permanent snow or glaciers) [4], the greater geographic isolation, and in some cases, the more northerly locality and harsher climate. As an example, from Iceland there are 428 species of mites known, with eight Ixodida, 131 Mesostigmata, 101 Trombidiformes, and 188 Sarcoptiformes [151–153]. Although Iceland has an area only 1.5 times greater than Svalbard, the island lies at a lower latitude (between 63° and 66° north latitude), glaciers cover only 11% of its area, and it lies in the path of the North Atlantic Current, which results in a more ameliorated climate. The diversity of mites in Svalbard is also much poorer in comparison to mainland Norway: Ixodida comprise 8% here with Mesostigmata—15%, Trombidiformes—13%, and Sarcoptiformes—30% [154–157]. However, these differences are not unexpected considering that the Svalbard archipelago is situated 900 km from the northern border of mainland Norway and has an area of only one fifth of that of mainland Norway [158].

The geographical isolation of islands, as a rule, results in reduced biodiversity [159], but, on the other hand, nurtures unique endemic species [160]. Svalbard has a unique flora and fauna, including endemic invertebrates: three rotifers, four tardigrades, and two aphids [86]. Fifteen mite species have been found exclusively in Svalbard, six of them being recent observations. The remaining are single historic records (i.e., from the first half of the last century) and it is consequently uncertain if any of these are endemic to Svalbard or the result of taxonomic confusion. Finding and studying the types or neotypes in Svalbard, along with more extensive sampling in the Arctic and including molecular studies, could help resolve this question.

Nevertheless, the Svalbard acarofauna is unique due to its specific Arctic elements. Amongst the Svalbard Mesostigmata in particular there are many species with an Arctic distribution (which make up nearly 40% of this group) and are similar to that observed on Greenland [151]. By contrast, the Sarcoptiformes in Svalbard are predominated by species with wide geographical ranges extending beyond the Arctic, as also seen with the Sarcoptiformes species known from Greenland [151] and northern Russia [161]. This difference between the two orders may be explained by the younger geological age of the Mesostigmata and their faster evolution rate [162]. Similarly, the Trombidiformes, which are also an ancient group [163], are dominated in Svalbard by species with wide geographic distributions. Nonetheless, Arctic species seem to be abundant among the Trombidiformes of Svalbard, comprising 27% of the species recorded, but these data should be treated with caution because they are mostly based on historic records that need to be confirmed.

Another characteristic feature of the Svalbard acarofauna is the large number of species new to science described from this region—they form approximately 25% of all mite species known from Svalbard. Seven of these species belong to the Mesostigmata, 19 to Trombidiformes, and 16 to

Sarcoptiformes. However, many of these new species were found only once and early in the study of the mite fauna of Svalbard. These species include one mesostigmatid species, 18 Trombidiformes species (i.e., 90% of all new species from this group), and two Sarcoptiformes species. Confirming these identifications is complicated by outmoded descriptions, incomplete, or too general drawings, and by the unavailability of most type material. These species require special attention in future studies. They are possibly rare or have been wrongly identified but the possibility cannot be excluded that they have disappeared from Svalbard due to environmental changes or that they were introduced and their populations have not established in Svalbard.

Out of 36 mite species found in Svalbard only once by early workers, the majority of species (24) were collected by Sig Thor, including one new species of Mesostigmata, eight new species of Trombidiformes, and two new species of Sarcoptiformes (Table 2). Despite the widespread belief that the collection of Thor along with all his valuable type material was destroyed in accordance with Thor's wishes (see, e.g., remark 2 on page 1308 in [148]), this collection exists and is kept at the Natural History Museum in Oslo, Norway [164]. However, it is in a very poor condition and it is not certain whether the Svalbard material is in a suitable condition for re-examination.

Fortunately, some of Thor's species have been found by later sampling campaigns and redescribed, for instance, *Antennoseius* (*Vitzthumia*) *oudemansi*, *Neoseiulus magnanalis*, *Proctolaelaps parvanalis*, and the specimens obtained deposited in zoological collections as neotypes [28,79]. In addition, six new species of mites have been found recently in Svalbard, including two Mesostigmata—*Neoseiulus grumantensis* and *Halolaelaps coulsoni* [28,93], one Trombidiformes—*Chenophila nanseni* [57], and three Sarcoptiformes—*Cecidophyes siedleckii*, *Kunstidamaeus arcticus*, and *Fuscozetes coulsoni* [9,67,72].

Recently, the first species of Ixodida, the seabird tick (*Ixodes uriae*), was discovered in Svalbard [76–78]. This tick is a major parasite of seabirds breeding at high latitudes and has the most extensive geographical distribution of all tick species [23], being also common throughout mainland Norway [165]. Its increasing occurence in Svalbard may be related to warming winters (the tick overwinters at the breeding sites of the seabirds) [78] as was similarly observed in Iceland with another tick species, *I. ricinus* Linnaeus, 1758, and which has become more common in recent years [153].

Studies conducted during the past 20 years have also added 16 species of Mesostigmata [8,18,20,27,28,30,85,92,93,95,98,166] and 36 species of Sarcoptiformes to the Svalbard fauna [8–10,67,71,72,74,146,167]. This indicates that despite the relatively long history of mite studies for a region in the Arctic, our knowledge remains surprisingly poor. It is striking, however, that the Trombidiformes have been much less studied in Svalbard than other mite groups (Table 1). The reference list presented here includes only 18 papers referring to the Trombidiformes (vs. 72 papers on Sarcoptiformes and 34 on Mesostigmata), most of which were published at the end of the 19th and first half of the 20th centuries. Only three come from the present century. Therefore, it is clear that in future studies this group in particular should receive more attention.

There are also some species in Svalbard that have extraordinary disjunct distribution patterns: they occupy mainly the Arctic and parts of the subarctic regions, but are also found in some distant localities, in harsh conditions. One example is an oribatid species, *Ceratozetes spitsbergensis*, which has been assumed to be an Arctic species, present in Svalbard, Alaska [148], northern Canada [168,169] and the Nordic Arctic of the Russian Far East [148,170], but was found also in the Altai Mountains in Mongolia, at an altitude of 2800–3200 m a.s.l. [171] and in the Alps in Austria, at an altitude of 3300 m a.s.l. [172]. Since it was not found at lower elevations and/or lower latitudes, its presence in the Alps was explained by one of these theories: relict distribution [a cold-adapted species that was widely distributed in Europe during the Last Glacial Maximum (LGM), but when conditions became warmer it retreated to very restricted areas at high altitudes], or the nunatak hypothesis (survival of species on ice-free refuges) [172]. Some studies suggest that parts of Svalbard, e.g., Amsterdamøya (north-west of Svalbard), remained ice-free during the LGM [173]. It is thus possible that some invertebrates survived the LGM *in situ* in these glacial refugia, although, due to the harsh conditions prevailing over an extended period of time, it seems likely that most biota could probably not survive on nunataks [86]. Another oribatid mite with an interesting distribution is *Scutozetes clavatosensillus*, which has been found only in Svalbard and in Nepal (Dhaulagiri massif, 3200–3600 m a.s.l.); its distribution may also be related to glacial history. Similarly to *S. clavatosensillus*, the mesostigmatic species, *Paragamasus* (*Aclerogamasus*) *insertus*, may also be a glacial relict since it is known only from the Gory Stolowe Mtns., Poland [174,175], where it occurred in rock crevices in very extreme conditions with long-lasting snow cover, and recently collected in Svalbard [18].

The composition of the present acarofauna of Svalbard likely results mostly from postglacial immigration [86]. The mites, being wingless, have rather low dispersal abilities. However, phoresy with migrating birds [13–17] or insects [18,95], and transport on driftwood or even direct dispersal in or on seawater with ocean currents [11], are possible dispersal pathways from the mainland to, or between, remote islands such as the Svalbard archipelago. Humans may also play a role in the dispersal of mites; as with the import of fertile agricultural soils transported from southern Russia and contemporary Ukraine to the Russian settlements on Spitsbergen (Barentsburg and Pyramiden) to enrich the soils of the greenhouses and grass lawns [18,19]. Mite communities in such transformed microhabitats differ noticeably from those of the adjacent Arctic tundra. Moreover, manure from abandoned cow sheds provided specific ameliorated environmental conditions enabling the survival of terrestrial invertebrate species not yet recorded elsewhere in Svalbard [18–20]. Introduction of new species to islands can be deleterious since the island communities may be disrupted, often resulting in the extinction of their endemic species [176–178].

Reconstructing the colonization of Svalbard after the LGM may be attempted by observing primary succession before retreating glaciers. For example, at the Midtre Lovén glacier foreland, two oribatid species, *Camisia foveolata* and *Tectocepheus velatus*, were the first colonizers. Mesostigmata appeared later, *Proctolaelaps parvanalis* being the first species recorded at this glacier foreland [10]. All these species are widely distributed in different parts of the archipelago (see Table 1). *Tectocepheus velatus* was also the earliest colonizer on a glacier foreland in southern Norway [179], while *Camisia foveolata* was one of the first colonizers on geothermally active lava fields in Iceland [180]. Oribatida are mainly saprophagous but species that are fungivores, bacteriovores, algivores, or omnivores, such as *T. velatus*, can find the appropriate food on seemingly baren ground [181]. In turn, the development of the first animal communities provides prey to the predatory Mesostigmata.

As shown in the present review, acarological studies of Svalbard are heavily geographically biased since most sampling has been carried out on the largest island—Spitsbergen; 90% of papers refer to this island and 90% of species have been found there (see Table 1). Almost 20% of papers refer to Bjørnøya and 25% of the species total for the archipelago have been found there, while 7% of papers refer to Edgeøya and 17% of the species are known from there. Nine other islands and island groups have been studied to a much lesser extent, with single records from other locations. It needs to be emphasized that some mites have been found exclusively on one island, other than Spitsbergen: five species on Edgeøya (*Saprosecans baloghi*, *Liochthonius clavatus*, *L. strenzkei*, *Neoliochthonius piluliferus*, and *Fuscozetes coulsoni*), four on Bjørnøya (*Zerconopsis muestairi*, *Sperchon brevirostris*, *Subbelba montana*, and *Ameronothrus nigrofemoratus*), and two on Hopen (*Thalassarachna coeca* and *T. princeps*), including three species new to science. It is unclear why these species have not been collected from Spitsbergen despite the more comprehensive sampling efforts on this island. The importance of further studies in different parts of the Svalbard archipelago is highlighted, in particular since the archipelago is extremely diverse geographically and climatically.

In natural conditions, the species composition and abundance of mites depends mainly on the vegetation [89], which, in turn, depends largely on the climate. For example, a relatively high mesostigmatid diversity is present along the western coast of Spitsbergen (about 30 species) [86], which experiences a comparatively mild climate for the latitude, in contrast to the polar deserts, where only five species were recorded [30]. Within the same climatic conditions, the densities of the mites also vary greatly according to the vegetation types [22,30,88,89,136].

Since mites are connected in different ways with other parts of the ecosystem, it would be particularly interesting to apply a multi-disciplinary approach to trace the effect of climate change in Svalbard. For example, one of the understudied mite groups is the family Eriophyidae. This family includes phytophagous species of great economic importance and with high invasive potential, and could be very useful for ecological studies on the effects of a changing climate at Svalbard [67]. Another poorly known group is the hyperorder Astigmata, parasitizing birds [74]. It is well-documented that climatic changes are affecting the diversity of seabirds by changing their foraging and breeding ecology, as well as increasing the abundance of temperate species [6], which could in turn affect bird-associated Astigmata.

Attention should also be paid to the least known mite order in Svalbard, the Trombidiformes, which is extremely diverse with respect to their feeding preferences (this taxon includes algivores, bacterivores, fungivores, predators, and parasites), and occupies terrestrial, freshwater, and marine habitats [182], but has not been studied at all from an ecological perspective in Svalbard.

Acarological research in the Arctic has developed rapidly in the last 20 years, markedly contributing to the knowledge of the mite communities and increasing somewhat our understanding of the factors determining these communities. Nonetheless, little is understood about the physical and morphological adaptations of mites, their adaptations to the extreme Arctic environment, or the genetic biodiversity of these isolated populations. Moreover, research has focused on the larger and more easily accessible regions of the archipelago to the detriment of the more environmentally extreme eastern and northern regions. There has also been a focus on the Sarcoptiformes and Mesostigmata while the Trombidiformes have been neglected. These areas will remain the subject of our research in the forthcoming years.

Author Contributions: Conceptualization, A.S. and D.J.G.; Investigation: Mesostigmata—D.J.G.; Trombidiformes—M.D.S.; Sarcoptiformes—A.S., S.S.; Svalbard peculiarities—S.J.C.; Methodology, A.S., S.S., M.D.S., S.J.C. and D.J.G.; Visualization, A.S.; Writing—original draft, review & editing, A.S., S.S., M.D.S., S.J.C. and D.J.G. All authors have read and agreed to the published version of the manuscript.

Funding: This research was funded by Norwegian Taxonomy Initiative: Knr 6-20.

Acknowledgments: We are grateful to Jacek Dabert (Institute of Environmental Biology, Adam Mickiewicz University, Poznań, Poland) for information about distribution of species of Astigmata, and to Reinhard Gerecke (Institute for Evolution and Ecology, University of Tübingen, Tübingen, Germany) for information about Thor's collection. This study was partly supported by the grant from The Norwegian Taxonomy Initiative – Knr 6-20, Mites of Atlantic Raised Bogs in Norway (MARB).

Conflicts of Interest: The authors declare no conflict of interest.

References

1. Gillespie, M.A.K.; Alfredsson, M.; Barrio, I.C.; Bowden, J.; Convey, P.; Coulson, S.J.; Culler, L.E.; Dahl, M.T.; Daly, K.M.; Koponen, S.; et al. Circumpolar terrestrial arthropod monitoring: A review of ongoing activities, opportunities and challenges, with a focus on spiders. *Ambio* **2020**, *49*, 704–717. [CrossRef] [PubMed]
2. Coulson, S.J. The terrestrial invertebrate fauna of the Svalbard archipelago in a changing world: History of research and challenges. *Can. Entomol.* **2013**, *145*, 131–146. [CrossRef]
3. IPCC. Summary for Policymakers. In *Global Warming of 1.5 °C. An IPCC Special Report on the Impacts of Global Warming of 1.5 °C above Pre-Industrial Levels and Related Global Greenhouse Gas Emission Pathways, in the Context of Strengthening the Global Response to the Threat of Climate Change, Sustainable Development, and Efforts to Eradicate Poverty*; Masson-Delmotte, V., Zhai, P., Pörtner, H.-O., Roberts, D., Skea, J., Shukla, P.R., Pirani, A., Moufouma-Okia, W., Péan, C., Pidcock, R., et al., Eds.; Intergovernmental Panel on Climate Change: Geneva, Switzerland, 2018; pp. 3–27.
4. Hanssen-Bauer, I.; Førland, E.J.; Hisdal, H.; Mayer, S.; Sandø, A.B.; Sorteberg, A. (Eds.) *Climate in Svalbard 2100–A Knowledge Base for Climate Adaptation*; The Norwegian Centre for Climate Services Report 1/2019; The Norwegian Centre for Climate Services: Oslo, Norway, 2019. Available online: https://www.miljodirektoratet.no/publikasjoner/2019 (accessed on 19 April 2020).

5. Norwegian Meteorological Institute. Precipitation in Svalbard, Hopen and Jan Mayen, Annual Total. Environmental Monitoring of Svalbard and Jan Mayen (MOSJ). 2020. Available online: http://www.mosj.no/en/climate/atmosphere/temperature-precipitation.html (accessed on 8 August 2020).
6. Descamps, S.; Aars, J.; Fuglei, E.; Kovacs, K.M.; Lydersen, C.; Pavlova, O.; Pedersen, Å.Ø.; Ravolainen, V.; Strøm, H. Climate change impacts on wildlife in a High Arctic. *Glob. Chang. Biol.* **2017**, *23*, 490–502. [CrossRef] [PubMed]
7. Summerhayes, V.S.; Elton, C.S. Further contribution to the ecology of Spitsbergen. *J. Ecol.* **1928**, *16*, 193–268. [CrossRef]
8. Ávila-Jiménez, M.L.; Solhøy, T.; Gwiazdowicz, D.J.; Fjellberg, A.; Dozsa-Farkas, K.; Monson, F.; De Smet, W.H.; Stur, E.; Ekrem, T.; Coulson, S.J. The terrestrial invertebrate fauna of Edgeøya, Svalbard: Arctic landscape community composition reflects biogeography patterns. *Polar Biol.* **2019**, *42*, 837–850. [CrossRef]
9. Seniczak, A.; Seniczak, S. Morphological ontogeny of *Fuscozetes coulsoni* sp. nov. (Acari: Oribatida: Ceratozetidae) from Svalbard, Norway. *Syst. Appl. Acarol.* **2020**, *25*, 680–696. [CrossRef]
10. Gwiazdowicz, D.J.; Zawieja, B.; Olejniczak, I.; Skubała, P.; Gdula, A.K.; Coulson, S.J. Changing microarthropod communities in front of a receding glacier in the High Arctic. *Insects* **2020**, *11*, 226. [CrossRef]
11. Coulson, S.J.; Hodkinson, I.D.; Block, W.; Webb, N.R.; Harrison, J.A. Survival of terrestrial soil-dwelling arthropods on and in seawater: Implications for trans-oceanic dispersal. *Funct. Ecol.* **2002**, *16*, 353–356. [CrossRef]
12. Coulson, S.J.; Moe, B.; Monson, F.; Gabrielsen, G.W. The invertebrate fauna of High Arctic seabird nests: The microarthropod community inhabiting nests on Spitsbergen, Svalbard. *Polar Biol.* **2009**, *32*, 1041–1046. [CrossRef]
13. Lebedeva, N.V.; Krivolutsky, D.A. Birds spread soil microarthropods to Arctic Islands. *Dokl. Biol. Sci.* **2003**, *391*, 329–332. [CrossRef]
14. Lebedeva, N.V.; Lebedev, V.D.; Melekhina, E.N. New data on the oribatid mite (Oribatei) fauna of Svalbard. *Dokl. Biol. Sci.* **2006**, *407*, 182–186. [CrossRef] [PubMed]
15. Lebedeva, N.V.; Melekhina, E.N.; Gwiazdowicz, D.J. New data on soil mites in the nests of *Larus hyperboreus* in the Spitsbergen archipelago. *Vestn. Juzn. Naucn. Cent. Ran* **2012**, *8*, 70–75.
16. Lebedeva, N.V.; Lebedev, V.D. Transport of oribatid mites to the polar areas by birds. In *Integrative Acarology*; Bertrand, M., Kreiter, S., McCoy, K.D., Migeon, A., Navajas, M., Tixier, M.S., Vial, L., Eds.; European Association of Acarologists: Montpellier, France, 2008; pp. 359–367.
17. Coulson, S.J. Association of the soil mite *Diapterobates notatus* (Thorell, 1871) (Acari, Oribatidae) with *Cynomya mortuorum* (Linnaeus, 1761) (Calliphoridae, Calliphorinae): Implications for the dispersal of oribatid mites. *Int. J. Acarol.* **2009**, *35*, 175–177. [CrossRef]
18. Coulson, S.J.; Fjellberg, A.; Gwiazdowicz, D.J.; Lebedeva, N.V.; Melekhina, E.N.; Solhøy, T.; Erséus, C.; Maraldo, K.; Miko, L.; Schatz, H.; et al. Introduction of invertebrates into the High Arctic via imported soils: The case of Barentsburg in the Svalbard. *Biol. Invasions* **2013**, *15*, 1–5. [CrossRef]
19. Coulson, S.J.; Fjellberg, A.; Gwiazdowicz, D.J.; Lebedeva, N.V.; Melekhina, E.N.; Solhøy, T.; Erséus, C.; Maraldo, K.; Miko, L.; Schatz, H.; et al. The invertebrate fauna of anthropogenic soils in the High-Arctic settlement of Barentsburg; Svalbard. *Polar Res.* **2013**, *32*, 19273. [CrossRef]
20. Coulson, S.J.; Fjellberg, A.; Melekhina, E.N.; Taskaeva, A.A.; Lebedeva, N.V.; Belkina, O.; Seniczak, S.; Seniczak, A.; Gwiazdowicz, D.J. Microarthropod communities of industrially disturbed or imported soils in the High Arctic; the abandoned coal mining town of Pyramiden, Svalbard. *Biodivers. Conserv.* **2015**, *24*, 1671–1690. [CrossRef]
21. Hughes, K.A.; Convey, P. The protection of Antarctic terrestrial ecosystems from inter and intra-continental transfer of non-indigenous species by human activities: A review of current systems and practices. *Glob. Environ. Chang.* **2010**, *20*, 96–112. [CrossRef]
22. Seniczak, S.; Seniczak, A.; Gwiazdowicz, D.J.; Coulson, S.J. Community structure of oribatid and gamasid mites (Acari) in moss-grass tundra in Svalbard (Spitsbergen, Norway). *Arct. Antarct. Alp. Res.* **2014**, *46*, 591–599. [CrossRef]
23. Muñoz-Leal, S.; González-Acuña, D. The tick *Ixodes uriae* (Acari: Ixodidae): Hosts, geographical distribution, and vector roles. *Ticks tick-borne Dis.* **2015**, *6*, 843–868. [CrossRef]
24. Krantz, G.W.; Walter, D.E. *A Manual of Acarology*, 3rd ed.; Texas Tech University Press: Lubbock, TX, USA, 2009; pp. 1–807.

25. Beaulieu, F.; Dowling, A.P.G.; Klompen, H.; de Moraes, G.J.; Walter, D.E. Superorder Parasitiformes Reuter, 1909. In *Animal Biodiversity: An Outline of Higher-Level Classification and Survey of Taxonomic Richness*; Zhang, Z.-Q., Ed.; Magnolia Press: Auckland, New Zealand, 2011; pp. 123–128. [CrossRef]
26. Fauna Europaea. Available online: https://fauna-eu.org (accessed on 19 April 2020).
27. Gwiazdowicz, D.J.; Teodorowicz, E.; Coulson, S.J. Redescription of *Zercon solenites* Haarløv, 1942 (Acari, Zerconidae) with a key to the Svalbard species of the genus *Zercon*. *Int. J. Acarol.* **2011**, *37* (Suppl. 1), 135–148. [CrossRef]
28. Kolodochka, L.A.; Gwiazdowicz, D.J. A new species of predaceous mite of the genus *Neoseiulus* Hughes (Acari, Phytoseiidae), with redescriptions of *N. magnanalis* (Thor) and *N. ellesmerei* (Chant & Hansell), from Svalbard, High Arctic. *Zootaxa* **2014**, *3793*, 441–452.
29. Gwiazdowicz, D.J.; Rakowski, R. Redescription of *Proctolaelaps parvanalis* (Thor, 1930) (Acari: Ascidae) from Spitsbergen. *Entomol. Fenn.* **2009**, *20*, 281–286. [CrossRef]
30. Ávila-Jiménez, M.L.; Gwiazdowicz, D.J.; Coulson, S.J. The mesostigmatid mite (Acari: Parasitiformes) fauna of Svalbard: A revised inventory of a high Arctic Archipelago. *Zootaxa* **2011**, *3091*, 33–41. [CrossRef]
31. Zhang, Z.-Q.; Fan, Q.-H.; Pesic, V.; Smit, H.; Bochkov, A.V.; Khaustov, A.A.; Baker, A.; Wohltmann, A.; Wen, T.H.; Amrine, J.W.; et al. Order Trombidiformes Reuter, 1909. In *Animal Biodiversity: An Outline of Higher-Level Classification and Survey of Taxonomic Richness*; Zhang, Z.-Q., Ed.; Magnolia Press: Auckland, New Zealand, 2011; pp. 129–138. [CrossRef]
32. Bolton, S.J.; Chetverikov, P.E.; Klompen, H. Morphological support for a clade comprising two vermiform mite lineages: Eriophyoidea (Acariformes) and Nematalycidae (Acariformes). *Syst. Appl. Acarol.* **2017**, *22*, 1096–1131. [CrossRef]
33. Klimov, P.B.; OConnor, B.M.; Chetverikov, P.E.; Bolton, S.J.; Pepato, A.R.; Mortazavi, A.L.; Tolstikov, A.V.; Bauchan, G.R.; Ochoa, R. Comprehensive phylogeny of acariform mites (Acariformes) provides insights on the origin of the four-legged mites (Eriophyoidea), a long branch. *Mol. Phylogenet. Evol.* **2018**, *119*, 105–117. [CrossRef] [PubMed]
34. Hernandes, F.A.; Skvarla, M.J.; Fisher, J.R.; Dowling, A.P.G.; Ochoa, R.; Ueckermann, E.A.; Bauchan, G.R. Catalogue of snout mites (Acariformes: Bdellidae) of the world. *Zootaxa* **2016**, *4152*, 1–83. [CrossRef] [PubMed]
35. Skvarla, M.J.; Fisher, J.R.; Dowling, A.P. A review of Cunaxidae (Acariformes, Trombidiformes): Histories and diagnoses of subfamilies and genera, keys to world species, and some new locality records. *ZooKeys* **2014**, *418*, 1–103. [CrossRef] [PubMed]
36. Khaustov, A.A. A new genus and species in the mite family Eupodidae (Acari, Eupodoidea) from Crimea. *ZooKeys* **2014**, *422*, 11–22. [CrossRef] [PubMed]
37. Szudarek-Trepto, N.; Kazmierski, A.; Dabert, M.; Dabert, J. Molecular phylogeny of Eupodidae reveals that the family Cocceupodidae (Actinotrichida; Eupodoidea) and its genus *Filieupodes* are valid taxa. *Exp. Appl. Acarol.* **2020**, *80*, 43–57. [CrossRef] [PubMed]
38. Meyer, M.K.P.; Ryke, P.A.J. Mites of the superfamily Eupodoidea (Acarina: Prostigmata) associated with South African plants. *S. Afr. J. Agric. Sci.* **1960**, *3*, 481–496.
39. Strandtmann, R.W.; Davies, L. Eupodiform mites from Possession Island, Crozet Islands, with a key to the species of *Eupodes* (Acarina: Prostigmata). *Pac. Insects* **1972**, *14*, 39–56.
40. Umina, P.A.; Hoffman, A.A.; Weeks, A.R. Biology, ecology and control of the *Penthaleus* species complex (Acari: Penthaleidae). *Exp. Appl. Acarol.* **2004**, *34*, 211–237. [CrossRef] [PubMed]
41. Jesionowska, K. A morphological study of the genus *Penthalodes* (Acari, Prostigmata, Eupodoida, Penthalodidae) with description of a new species. *Zootaxa* **2010**, *2672*, 29–49. [CrossRef]
42. Zacharda, M. Soil mites of the family Rhagidiidae (Actinedida: Eupodoidea). Morphology, sytematics, ecology. *Acta Univ. Carol. Biol.* **1980**, *516*, 489–785.
43. Zacharda, M.; Gude, M.; Kraus, S.; Hauck, C.; Molenda, R.; Růžička, V. The relict mite *Rhagidia gelida* (Acari, Rhagidiidae) as a biological cryoindicator of periglacial microclimate in European Highland Screes. *Arct. Antarct. Alp. Res.* **2005**, *37*, 402–408. [CrossRef]
44. Qin, T.K. A checklist and key to species of Eupodoidea (Acari: Prostigmata) from Australia and New Zealand and their subantarctic islands. *J. R. Soc. N. Z.* **1998**, *28*, 295–307. [CrossRef]
45. Bartsch, I. Checklist of marine and freshwater halacarid mite genera and species (Halacaridae: Acari) with notes on synonyms, habitats, distribution. *Zootaxa* **2009**, *1998*, 1–170. [CrossRef]

46. André, H.M. Redefinition of the genus *Triophtydeus* Thor, 1932 (Acari: Actinedida). *Zool. Meded.* **1985**, *59*, 189–195.
47. Thor, S. *Acarina–Tydeidae, Ereynetidae*; Das Tierreich, Walter de Gruyter: Berlin, Germany, 1933; Volume 60, pp. 1–84.
48. Makarova, O.; Behan-Pelletier, V. 18.3.6. Oribatida (=Cryptostigmata, Beetle Mites). In *The Greenland Entomofauna*; Böcher, J., Kristensen, N.P., Pape, T., Vilhelmsen, L., Eds.; An Identification Manual of Insects, Spiders and Their Allies; Koninklijke Brill NV: Leiden, The Netherlands, 2015; pp. 802–845.
49. André, H.M. In search of the true *Tydeus* (Acari, Tydeidae). *J. Nat. Hist.* **2005**, *39*, 975–1001. [CrossRef]
50. Silva, G.L.; Metzelthin, M.H.; Silva, O.S.; Ferla, N.J. Catalogue of the mite family Tydeidae (Acari: Prostigmata) with the world key to the species. *Zootaxa* **2016**, *4135*, 1–68. [CrossRef]
51. Baker, E.W. A review of the genera of the family Tydeidae (Acarina). *Adv. Acarol.* **1965**, *2*, 95–133.
52. André, H.M. A generic revision of the family Tydeidae (Acari: Actinedida). IV. Generic descriptions, keys and conclusions. *Bull. Ann. Soc. Belg. Ent.* **1980**, *116*, 103–130.
53. Makarova, O.L. Acarocenoses (Acariformes, Parasitiformes) in polar deserts: 1. Mite assemblages of the Severnaya Zemlya Archipelago: Structure of fauna and abundance. *Entomol. Rev.* **2002**, *82*, 839–856.
54. Abe, H. Annotated checklist of Japanese water mites (Acari: Prostigmata: Hydracarina). *Acta Arachnol.* **2005**, *54*, 111–145. [CrossRef]
55. Esen, Y.; Pesic, V.; Erman, O. Water mites of the genus *Sperchon* Kramer (Acari: Hydrachnidia: Sperchontidae) from Turkey, with description of two new species. *Zootaxa* **2010**, *2514*, 35–46. [CrossRef]
56. Mąkol, J.; Wohltmann, A. An annotated checklist of terrestrial Parasitengona (Actinotrichida: Prostigmata) of the world, excluding Trombiculidae and Walchiidae. *Ann. Zool.* **2012**, *62*, 359–562. [CrossRef]
57. Skoracki, M.; Zawierucha, K. *Chenophila nanseni* sp. n. (Acari: Syringophilidae) parasitising the barnacle goose in Svalbard. *Pol. Polar Res.* **2016**, *37*, 121–130. [CrossRef]
58. Beron, P. Acarorum Catalogus VII. Trombidiformes: Prostigmata: Raphignathoidea. Fam. Barbutiidae, Caligonellidae, Camerobiidae, Cryptognathidae, Dasythyreidae, Dytiscacaridae, Eupalopsellidae, Homocaligidae, Mecognathidae, Raphignathidae, Stigmaeidae, Xenocaligonellididae. *Pensoft Natl. Mus. Nat. Hist. Sofia Bulg.* **2020**. [CrossRef]
59. Pritchard, A.E.; Baker, E.W. *A Revision of the Spider Mite Family Tetranychidae*; Memoirs Series; Pacific Coast Entomological Society: San Francisco, CA, USA, 1955; Volume 2.
60. Migeon, A.; Dorkeld, F. Spider Mites Web: A Comprehensive Database for the Tetranychidae. 2020. Available online: http://www1.montpellier.inra.fr/CBGP/spmweb (accessed on 2 July 2020).
61. Khaustov, A.A. A description of new genus, *Kerdabania* gen. n. with four new species (Acari: Heterostigmata: Neopygmephoridae). *Acarina* **2009**, *17*, 171–188.
62. Schatz, H.; Behan-Pelletier, V.M.; O'Connor, B.M.; Norton, R.A. Suborder Oribatida van der Hammen, 1968. In *Animal Biodiversity: An Outline of Higher-Level Classification and Survey of Taxonomic Richness*; Zhang, Z.-Q., Ed.; Magnolia Press: Auckland, New Zealand, 2011; pp. 141–148.
63. Beaulieu, F.; Knee, W.; Nowell, V.; Schwarzfeld, M.; Lindo, Z.; Behan-Pelletier, V.M.; Lumley, L.; Young, M.R.; Smith, I.; Proctor, H.C.; et al. Acari of Canada. In *The Biota of Canada–A Biodiversity Assessment. Part 1: The Terrestrial Arthropods*; Langor, D.W., Sheffield, C.S., Eds.; ZooKeys: Sofia, Bulgaria, 2019; Volume 819, pp. 77–168. [CrossRef]
64. Walter, D.E.; Bolton, S.; Uusitalo, M.; Zhang, Z.-Q. Suborder Endeostigmata Reuter, 1909. In *Animal Biodiversity: An Outline of Higher-Level Classification and Survey of Taxonomic Richness*; Zhang, Z.-Q., Ed.; Magnolia Press: Auckland, New Zealand, 2011; pp. 139–140.
65. Uusitalo, M. Terrestrial species of the genus *Nanorchestes* (Endeostigmata: Nanorchestidae) in Europe. In *Trends in Acarology*; Sabelis, M.W., Bruin, J., Eds.; Springer: Dordrecht, The Netherlands, 2010; pp. 161–166. [CrossRef]
66. Thor, S.; Willmann, C. Acarina. 7la Eupodidae, Penthalodidae, Penthaleidae, Rhagidiidae, Pachygnathidae, Cunaxidae. *Das Tierreich* **1941**, *71*, 1–186.
67. Kiedrowicz, A.; Rector, B.G.; Zawierucha, K.; Szydło, W.; Skoracka, A. Phytophagous mites (Acari: Eriophyoidea) recorded from Svalbard, including the description of a new species. *Polar Biol.* **2016**, *39*, 1359–1368. [CrossRef]

68. Subías, L.S. Listado Sistemático, Sinonímico y Biogeográfico de los Ácaros Oribátidos (Acariformes, Oribatida) del Mundo (excepto fósiles) (15ª actualisación). *Graellsia* **2004**, *2020*, 60, (número extraordinario), 3–305 (2004), actualized pdf in 2020. Available online: http://bba.bioucm.es/cont/docs/RO_1.pdf (accessed on 14 April 2020).
69. Seniczak, S.; Seniczak, A. Differentiation of external morphology of Oribatulidae (Acari: Oribatida) in light of the ontogeny of three species. *Zootaxa* **2012**, *3184*, 1–34. [CrossRef]
70. Seniczak, A.; Seniczak, S. Morphological ontogeny of *Achipteria punctata* (Acari: Oribatida: Achipteriidae). *Zootaxa* **2018**, *4540*, 54–72. [CrossRef]
71. Bayartogtokh, B.; Schatz, H.; Ekrem, T. Distribution and diversity of the soil mites of Svalbard, with redescriptions of three known species (Acari: Oribatida). *Int. J. Acarol.* **2011**, *37*, 467–484. [CrossRef]
72. Miko, L.; Monson, F.D. Two interesting damaeid mites (Acari, Oribatida, Damaeidae Berlese, 1896) from the British Isles and Svalbard (Spitsbergen, Norway), with a description of *Kunstidamaeus arcticus* n.sp. *Acarologia* **2013**, *53*, 89–100. [CrossRef]
73. Klimov, P.B. A review of acarid mites of the tribe Caloglyphini (Acaridae, Acariformes) with description of a new genus and species from Siberia and Russian Far East. *Vestn. Zool.* **2000**, *34*, 27–35.
74. Dabert, M.; Coulson, S.J.; Gwiazdowicz, D.J.; Moe, B.; Hanssen, S.A.; Biersma, E.; Pilskog, H.E.; Dabert, J. Differences in speciation progress in feather mites (Analgoidea) inhabiting the same host: The case of *Zachvatkinia* and *Alloptes* living on arctic and long-tailed skuas. *Exp. Appl. Acarol.* **2015**, *65*, 163–179. [CrossRef]
75. Dabert, J. *Personal Communication*; Adam Mickiewicz University: Poznan, Poland, 2020.
76. McCoy, K.D. Consequences of Dispersal in Host–Parasite Systems: Population Dynamics, Genetic Structure, and Localadaptation of the Seabird tickixodes Uriae. Ph.D. Thesis, Pierreand Marie Curie University, Paris, France, 2001.
77. Coulson, S.J.; Lorentzen, E.; Strøm, H.; Gabrielsen, G.W. The parasitic tick *Ixodes uriae* (Acari: Ixodidae) on seabirds from Spitsbergen, Svalbard. *Polar Res.* **2009**, *28*, 399–402. [CrossRef]
78. Descamps, S. Winter temperature affects the prevalence of ticks in an Arctic sea-bird. *PLoS ONE* **2013**, *8*, e65374. [CrossRef]
79. Teodorowicz, E.; Gwiazdowicz, D.J.; Coulson, S.J. Redescription of *Antennoseius* (*Vitzthumia*) *oudemansi* (Acari, Mesostigmata) from Spitsbergen, Svalbard. *Entomol. Fenn.* **2014**, *25*, 27–42. [CrossRef]
80. Gwiazdowicz, D.J.; Teodorowicz, E.; Coulson, S.J. Redescription of *Arctoseius haarlovi* Lindquist, 1963 (Acari: Ascidae) from Spitsbergen. *Entomol. Fenn.* **2011**, *22*, 140–148. [CrossRef]
81. Trouessart, E. Note sur les Acariens recueillis au Spitzberg pendant le voyage de la Manche. *Nouv. Arch. Miss. Sci. Litt.* **1893**, *5*, 255–263.
82. Trouessart, E. Révision des Acariens des regions Arctiques et description d'espèces nouvelles. *M. Soc. Natl. Sci. Nat. Math. Cherb.* **1895**, *29*, 183–206.
83. Zmudczyńska-Skarbek, K.; Barcikowski, M.; Drobniak, S.M.; Gwiazdowicz, D.J.; Richard, P.; Skubała, P.; Stempniewicz, L. Transfer of ornithogenic influence through different trophic levels of the Arctic terrestrial ecosystem of Bjørnøya (Bear Island), Svalbard. *Soil Biol. Biochem.* **2017**, *115*, 475–489. [CrossRef]
84. Makarova, O.L. Gamasid mites (Parasitiformes, Mesostigmata) of the European Arctic and their distribution patterns. *Entomol. Rev.* **2013**, *93*, 113–133. [CrossRef]
85. Gwiazdowicz, D.J.; Gulvik, M.E. Mesostigmatid mites (Acari, Mesostigmata) in Svalbard. In *Materiały XXXII Międzynarodowego Sympozjum Polarnego*; Uniwersytet Wrocławski, Wydział Nauk o Ziemi i Kształtowania Środowiska: Wrocław, Poland, 2008; pp. 32–34.
86. Coulson, S.J.; Schatz, H.; Gwiazdowicz, D.J.; Solhøy, T. On the oribatid and mesostigmatid mites (Acari) of the High Arctic island of Hopen. *Pol. Polar Res.* **2014**, *35*, 133–139. [CrossRef]
87. Coulson, S.J.; Fjellberg, A.; Snazell, R.; Gwiazdowicz, D.J.; Ávila-Jiménez, M.L. On the Collembola, Araneae and Gamasida from the Kinnvika region of Nordaustlandet, Svalbard. *Geogr. Ann. A* **2011**, *93*, 253–257. [CrossRef]
88. Byzova, J.B.; Uvarov, A.V.; Petrova, A.D. Seasonal changes in communities of soil invertebrates in tundra ecosystems of Hornsund, Spitsbergen. *Pol. Polar Res.* **1995**, *16*, 245–266.
89. Gwiazdowicz, D.J.; Coulson, S.J. High-Arctic gamasid mites (Acari, Mesostigmata): Community composition on Spitsbergen, Svalbard. *Polar Res.* **2011**, *30*, 8311. [CrossRef]

90. Pilskog, H.E.; Solhøy, T.; Gwiazdowicz, D.J.; Grytnes, J.A.; Coulson, S.J. Invertebrate communities inhabiting nests of migrating passerine, wild fowl and sea birds breeding in the High Arctic, Svalbard. *Polar Biol.* **2014**, *37*, 981–998. [CrossRef]
91. Thor, S. Beiträge zur Kenntnis der Invertebratenfauna von Svalbard. *Skr. Svalbard Ishavet* **1930**, *27*, 1–156.
92. Gwiazdowicz, D.J.; Solhøy, T.; Coulson, S.J.; Lebedeva, N.V.; Melekhina, E.N. First record of *Vulgarogamasus immanis* (Acari, Mesostigmata) in Svalbard. *Pol. Polar Res.* **2012**, *33*, 35–39. [CrossRef]
93. Gwiazdowicz, D.J.; Teodorowicz, E. Description of *Halolaelaps coulsoni* n. sp. (Acari, Halolaelapidae) from the High Arctic. *Acarologia* **2017**, *57*, 393–406. [CrossRef]
94. Hirschmann, W. Die Gattung *Halolaelaps* Berlese & Trouessart 1889 nov. comb. *Acarologia* **1966**, *9*, 21–24.
95. Gwiazdowicz, D.J.; Coulson, S.J. First record of *Thinoseius spinosus* (Acari, Eviphididae) from the High Arctic Island of Spitsbergen (Svalbard) including a key to deutonymphs of genus *Thinoseius*. *Int. J. Acarol.* **2010**, *36*, 233–236. [CrossRef]
96. Makarova, O.L. To the study of mites of the genus *Arctoseius* Thor (Parasitiformes, Ascidae) from the Far North: 3. Ranges and ecological preferences of species. *Entomol. Rev.* **2000**, *80*, 143–150.
97. Lindquist, L.L.; Makarova, O.L. Two new circumpolar mite species of the genus *Arctoseius* (Parasitiformes, Mesostigmata, Ascidae). *Entomol. Rev.* **2011**, *91*, 1054–1072. [CrossRef]
98. Gwiazdowicz, D.J.; Coulson, S.J.; Grytnes, J.A.; Pilskog, H.E. The bird ectoparasite *Dermanyssus hirundinis* (Acari, Mesostigmata) in the High Arctic; a new parasitic mite to Spitsbergen, Svalbard. *Acta Parasitol.* **2012**, *57*, 378–384. [CrossRef]
99. Summerhayes, V.S.; Elton, C.S. Contributions to the ecology of Spitsbergen and Bear Island. *J. Ecol.* **1923**, *11*, 214–286. [CrossRef]
100. Krumpál, M.; Cyprich, D.; Zejda, J.; Ambros, M. The occurrence of field vole (*Microtus arvalis* Pallas 1778) and its acarofauna on Spitsbergen (Svalbard). *Biologia* **1991**, *46*, 881–885.
101. Hull, M.A. On some land mites (Acari) from Spitzbergen and Bjørnøya. Results of the Oxford University Expedition to Spitsbergen, 1921. *Ann. Mag. Nat. Hist.* **1922**, *10*, 621–623. [CrossRef]
102. Thorell, T. Om Arachnider från Spetsbergen och Beeren-Eiland. *Öfvers. Kongl. Vetensk.-Akad. Förh.* **1871**, *6*, 683–702.
103. Thor, S. Neue Beiträge zur Kenntnis der Invertebraten-Fauna vin Svalbard. *Zool. Anz.* **1934**, *107*, 114–139.
104. Bertram, G.C.L.; Lack, D. Notes on the animal ecology of Bear Island. *J. Anim. Ecol.* **1938**, *7*, 27–52. [CrossRef]
105. Trägårdh, I. Beiträge zur Fauna der Bären-Insel. *Bihang till K. Svenska Vet. Akad. Handlingar.* **1900**, *26*, 1–24.
106. Haarløv, N. A morphologic–systematic–ecological investigation of Acarina and other representatives of the microfauna of the soil around Mørkefjord, Northeast Greenland. *Medd. Grønl.* **1942**, *128*, 1–71.
107. Trouessart, E. Note préliminaire sur les acariensmarins (Halacaridae) recueillis par S.A.S. le Prince de Monaco, dans les mersarctiques. *Bull. Soc. Zool. Fr.* **1902**, *27*, 66–70.
108. Barr, W. The Helgoland Expedition to Svalbard: Die Deutsche Expedition in das Nordliche Eismeer, 1898. *Arctic* **1988**, *41*, 203–214. [CrossRef]
109. Chertoprud, E.S.; Makarova, O.L.; Novichkova, A.A. First data on aquatic mites (Acari) of inland water bodies of West Spitsbergen, Svalbard. *Acarina* **2017**, *25*, 181–189. [CrossRef]
110. Thor, S. Norwegische Tydeidae VIII-XV, bemerkungen uber die Gattung *Tydeus* und uber Augen, Tracheen usw. *Zool. Anz.* **1932**, *98*, 69–91.
111. Soar, C.D. A species of Hydracarina found at Bear Island, June 17th, 1921–The Oxford University Expedition to Spitsbergen, 1921. *Microscopy* **1922**, *14*, 301–304.
112. Oudemans, A.C. Acarologische Aanteekeningen. XCIX. *Entomol. Ber.* **1930**, *8*, 11–20.
113. Thor, S. Norwegische Alycidae I—VII. *Zool. Anz.* **1931**, *194*, 229–238.
114. Niedbała, W. Oribatei (Acari) of Spitsbergen. *B. Pol. Acad. Sci.-Biol.* **1971**, *19*, 737–742.
115. Karppinen, E. Notes on the arthropod fauna of Spitsbergen. *Acta Entomol. Fenn.* **1967**, *33*, 18–26.
116. Forsslund, K.H. Notizen über Oribatei (Acari). I. *Ark. Zool.* **1957**, *2*, 583–593.
117. Forsslund, K.H. *Liochthonius muscorum* n. sp. und *L. lapponicus* (Träg.) (Acari, Oribatei). *Entomol. Tidskr.* **1964**, *85*, 236–239.
118. Byzova, J.B.; Uvarov, A.V. Mites of the tundra biotopes in the vicinity of bird colonies on Spitsbergen. In *The 6th all-Union Meeting on the Problems of Theoretical and Applied Acarology. Aszchabad, Turkmenistan. Abstracts*; Balashov, Y.S., Ed.; Russian Academy of Science: Saint Petersburg, Russia, 1990; pp. 152–153.

119. Coulson, S.J. Terrestrial and freshwater invertebrate fauna of the high Arctic Archipelago of Svalbard. *Zootaxa* **2007**, *1448*, 41–58. [CrossRef]
120. Coulson, S. *Personal Communication*; Swedish University of Agricultural Sciences: Uppsala, Sweden, 2020.
121. Coulson, S.J.; Hodkinson, I.D.; Strathdee, A.T.; Bale, J.S.; Block, W.; Worland, M.R.; Webb, N.R. Simulated climate change: The interaction between vegetation type and microhabitat temperatures at Ny Ålesund, Svalbard. *Polar Biol.* **1993**, *13*, 67–70. [CrossRef]
122. Coulson, S.J.; Hodkinson, I.D.; Block, W.; Webb, N.R.; Worland, M.R. Low summer temperatures: A potential mortality factor for high arctic soil microarthropods? *J. Insect Physiol.* **1995**, *41*, 783–792. [CrossRef]
123. Coulson, S.J.; Hodkinson, I.D.; Webb, N.R.; Block, W.; Bale, J.S.; Strathdee, A.T.; Worland, M.R.; Wooley, C. Effects of experimental temperature elevation on high-arctic soil microarthropod populations. *Polar Biol.* **1996**, *16*, 147–153. [CrossRef]
124. Coulson, S.J.; Leinaas, H.P.; Ims, R.A.; Søvik, G. Experimental manipulation of the winter surface ice layer–The effects on a High Arctic soil microarthropod community. *Ecography* **2000**, *23*, 299–306. [CrossRef]
125. Coulson, S.J.; Hodkinson, I.D.; Webb, N.R. Microscale distribution patterns in high Arctic soil microarthropod communities: The influence of plant species within the vegetation mosaic. *Ecography* **2003**, *26*, 801–809. [CrossRef]
126. Hodkinson, I.D.; Coulson, S.J.; Webb, N.R.; Block, W.; Strathdee, A.T.; Bale, J.S. Feeding studies on *Onychiurus arcticus* (Tullberg) (Collembola: Onychiuridae) on West Spitsbergen. *Polar Biol.* **1993**, *14*, 17–19. [CrossRef]
127. Webb, N.R.; Coulson, S.J.; Hodkinson, I.D.; Block, W.; Bale, J.S.; Strathdee, A.T. The effects of experimental temperature elevation on populations of cryptostigmatic mites in high Arctic soils. *Pedobiologia* **1998**, *42*, 298–308.
128. Hodkinson, I.D.; Bird, J.M. Anoxia tolerance in high Arctic terrestrial microarthropods. *Ecol. Entomol.* **2004**, *29*, 506–509. [CrossRef]
129. Dollery, R.; Hodkinson, I.D.; Jónsdóttir, I.S. Impact of warming and timing of snow melt on soil microarthropod assemblages associated with *Dryas*-dominated plant communities in Svalbard. *Ecography* **2006**, *29*, 111–119. [CrossRef]
130. Kulczyński, V. Zoologische Ergebnisse der russischen Expeditionen nach Spitzbergen. *Annu. Mus. Zool. Acad. Imp. Sci. St. Petersb.* **1902**, *7*, 347–354.
131. Solhøy, I.W. A redescription of *Mycobates sarekensis* (Trägårdh) (Acari: Oribatei). *Acarologia* **1997**, *38*, 69–77.
132. Block, W. Some arctic Oribatei. *Acarologia* **1966**, *8*, 161–162.
133. Trägårdh, I. Monographie der arktischen Acariden. *Fauna Arct.* **1904**, *4*, 1–78.
134. Seniczak, S.; Seniczak, A.; Coulson, S.J. Morphological ontogeny, distribution and descriptive population parameters of *Hermannia reticulata* (Acari: Oribatida: Hermanniidae), with comments on Crotonioidea. *Int. J. Acarol.* **2017**, *43*, 52–72. [CrossRef]
135. Seniczak, S.; Seniczak, A.; Coulson, S.J. Morphological ontogeny and distribution of *Hermannia scabra* (Acari: Oribatida: Hermanniidae) in Svalbard and descriptive population parameters. *Acarologia* **2017**, *57*, 877–892. [CrossRef]
136. Seniczak, S.; Plichta, W. Structural dependence of moss mite populations on patchiness of vegetation in moss-lichen-tundra at the north coast of Hornsund, West Spitsbergen. *Pedobiologia* **1978**, *18*, 145–152.
137. Gordeeva, E.V. *Moritzoppiella* gen. n.—A new genus of oribatid mites of the family Oppiidae (Oribatei, Acariformes). *Vestn. Zool.* **2000**, *34*, 17–26.
138. Hayward, S.A.L.; Worland, M.R.; Bale, J.S.; Convey, P. Temperature and the hygropreference of the Arctic collembolan *Onychiurus arcticus* and mite *Lauroppia translamellata*. *Physiol. Entomol.* **2000**, *25*, 266–272. [CrossRef]
139. Thor, S. Übersicht der norwegischen Cryptostigmata mit einzelnen Neben bemerkungen. *NYT Mag. Nat.* **1937**, *77*, 275–307.
140. Schubart, H. Morphologische Grundlagen für die Klärung der Verwandtschaftsbeziehungen innerhalb der Milbenfamilie Ameronothridae (Acari, Oribatei). *Zoologica* **1975**, *123*, 23–91.
141. Søvik, G.; Leinaas, H.P. Variation in extraction efficiency between juvenile and adult oribatid mites: *Ameronothrus lineatus* (Oribatida, Acari) in a Macfadyen high-gradient canister extractor. *Pedobiologia* **2002**, *46*, 34–41. [CrossRef]
142. Søvik, G.; Leinaas, H.P. Adult survival and reproduction in an arctic mite, *Ameronothrus lineatus* (Acari, Oribatida): Effects of temperature and winter cold. *Can. J. Zool.* **2003**, *81*, 1579–1588. [CrossRef]

143. Søvik, G.; Leinaas, H.P. Long life cycle and high adult survival in an arctic population of the mite *Ameronothrus lineatus* (Acari, Oribatida) from Svalbard. *Polar Biol.* **2003**, *26*, 500–508. [CrossRef]
144. Søvik, G. Observations on ovoviviparity and mixed-parity mode in arctic populations of *Ameronothrus lineatus* (Acari, Oribatida). *Acarologia* **2003**, *43*, 393–398.
145. Søvik, G.; Leinaas, H.P.; Ims, R.A.; Solhøy, T. Population dynamics and life history of the oribatid mite *Ameronothrus lineatus* (Acari, Oribatida) on the high Arctic Archipelago of Svalbard. *Pedobiologia* **2003**, *47*, 257–271. [CrossRef]
146. Seniczak, S.; Seniczak, A.; Graczyk, R.; Tømmervik, H.; Coulson, S.J. Distribution and population characteristics of the soil mites *Diapterobates notatus* and *Svalbardia paludicola* (Acari: Oribatida: Ceratozetidae) in High Arctic Svalbard (Norway). *Polar Biol.* **2017**, *40*, 1545–1555. [CrossRef]
147. Seniczak, S.; Seniczak, A.; Coulson, S.J. Morphology, distribution and certain population parameters of the Arctic mite *Oribatella arctica* (Acari: Oribatida: Oribatellidae). *Int. J. Acarol.* **2015**, *41*, 395–414. [CrossRef]
148. Behan-Pelletier, V.M. Ceratozetidae of the western North American Arctic. *Can. Entomol.* **1985**, *117*, 1287–1366. [CrossRef]
149. Seniczak, S.; Seniczak, A.; Coulson, S.J. Morphology, distribution and biology of *Mycobates sarekensis* (Acari: Oribatida: Punctoribatidae). *Int. J. Acarol.* **2015**, *41*, 663–675. [CrossRef]
150. Fain, A.; Camerik, A.M. Notes on the mites of the genus *Ereynetes* Berlese (Acari: Ereynetinae), with description of five new species from South Africa. *Bull. Inst. R. Sci. Nat. Belg. Entomol.* **1994**, *64*, 145–164.
151. Makarova, O.L. The fauna of free-living mites (Acari) of Greenland. *Entomol. Rev.* **2015**, *95*, 108–125. [CrossRef]
152. Gjelstrup, P.; Solhøy, T. Oribatid mites (Acari). In *The Zoology of Iceland, Steenstrupia 3*; Zoological Museum University of Copenhagen: Copenhagen, Denmark, 1994; Volume 57e, pp. 1–78.
153. Alfredsson, M.; Olafsson, E.; Eydal, M.; Unnsteinsdottir, E.R.; Hansford, K.; Wint, W.; Alexander, N.; Medlock, J.M. Surveillance of *Ixodes ricinus* ticks (Acari: Ixodidae) in Iceland. *Parasit. Vectors* **2017**, *10*, 466. [CrossRef]
154. Mehl, R. Checklist of Norwegian ticks and mites (Acari). *Fauna Norv. Ser. B* **1979**, *26*, 31–45.
155. Gwiazdowicz, D.J.; Gulvik, M.E. Checklist of Norwegian mesostigmatid mites (Acari, Mesostigmata). *Nor. J. Entomol.* **2005**, *52*, 117–125.
156. Seniczak, A.; Bolger, T.; Roth, S.; Seniczak, S.; Djursvoll, P.; Jordal, B.H. Diverse mite communities (Acari: Oribatida, Mesostigmata) from a broadleaf forest in western Norway. *Ann. Zool. Fenn.* **2019**, *56*, 121–136. [CrossRef]
157. Seniczak, A.; Seniczak, S.; Iturrondobeitia, J.C.; Solhøy, T.; Flatberg, K.I. Diverse *Sphagnum* mosses support rich moss mite communities (Acari, Oribatida) in mires of western Norway. *Wetlands* **2019**. [CrossRef]
158. The Editors of Encyclopaedia Britannica, Svalbard. Encyclopædia Britannica, Inc. Available online: https://www.britannica.com/place/Svalbard (accessed on 19 April 2020).
159. Veron, S.; Haevermans, T.; Govaerts, R.; Mouchet, M.; Pellens, R. Distribution and relative age of endemism across islands worldwide. *Sci. Rep.* **2019**, *9*, 11693. [CrossRef] [PubMed]
160. Losos, J.B.; Ricklefs, R.E. Adaptation and diversification on islands. *Nature* **2009**, *457*, 830–836. [CrossRef] [PubMed]
161. Melekhina, E.N. Taxonomic diversity and areology of oribatid mites (Oribatei) of the European North of Russia. *Proc. Komi Sci. Cent. Ural BranchRas* **2011**, *2*, 30–37.
162. Makarova, O.L.; Böcher, J. Diversity and geographical ranges of Greenland mites (Acari: Oribatida and Mesostigmata). In *Species and Communities in Extreme Environments*; Pensoft Publisher and KMK Science Press: Sofia, Moscow, Russia, 2009; pp. 165–186.
163. Dabert, M.; Witaliński, W.; Kaźmierski, A.; Olszanowski, Z.; Dabert, J. Molecular phylogeny of acariform mites (Acari, Arachnida): Strong conflict between phylogenetic signal and long-branch attraction artifacts. *Mol. Phylogenet. Evol.* **2010**, *56*, 222–241. [CrossRef]
164. Gerecke, R. *Personal Communication*; University of Tübingen: Tübingen, Germany, 2019.
165. Mehl, R.; Traavik, T. The tick *Ixodes uriae* (Acari, Ixodides) in seabird colonies in Norway. *Fauna Norv. Ser. B* **1983**, *30*, 94–107.
166. Gwiazdowicz, D.J.; Coulson, S.J.; Ávila-Jiménez, M.L. First records of *Zercon andrei* Sellnick, 1958 and *Zerconopsis muestairi* Schweizer, 1949) (Acari, Mesostigmata) from Bjørnøya, Svalbard. *Nor. J. Entomol.* **2009**, *56*, 117–119.

167. Seniczak, A.; Seniczak, S. Diversity of oribatid mites (Sarcoptiformes) in the Svalbard Archipelago: A historical overview. *Zootaxa* **2020**, *4834*, 41–65. [CrossRef]
168. Behan-Pelletier, V.M. Oribatid mites (Acari: Oribatida) of the Yukon. In *Insects of the Yukon. Biological survey of Canada (Terrestrial Arthropods)*; Danks, H.V., Downes, J.A., Eds.; Canadian Museum of Nature: Ottawa, Canada, 1997; pp. 115–149.
169. Behan-Pelletier, V.; Schatz, H. Patterns of diversity in the Ceratozetoidea (Acari: Oribatida): A North American assessment. In *Trends in Acarology*; Sabelis, M.W., Bruin, J., Eds.; Springer: Dordrecht, The Netherlands, 2010; pp. 97–104. [CrossRef]
170. Ryabinin, N.A. Oribatid mites (Acari, Oribatida) in soils of the Russian Far East. *Zootaxa* **2015**, *3914*, 201–244. [CrossRef] [PubMed]
171. Bayartogtokh, B. Fauna and communities of oribatid mites of Mongolia (Acari: Oribatida). Ph.D. Thesis, Russian Academy of Sciences, Moscow, Russia, 2007.
172. Fischer, B.M.; Schatz, H.; Querner, P.; Pauli, H. *Ceratozetes spitsbergensis* Thor, 1934: An Arctic mite new to continental Europe (Acari: Oribatida). *Int. J. Acarol.* **2016**, *42*, 135–139. [CrossRef]
173. Landvik, J.Y.; Brook, E.J.; Gualtieri, L.; Raisbeck, G.; Salvigsen, O.; Yiou, F. Northwest Svalbard during the last glaciation: Ice-free areas existed. *Geology* **2003**, *31*, 905–908. [CrossRef]
174. Micherdziński, W. *Die Familie Parasitidae Oudemans 1901 (Acarina, Mesostigmata)*; Państwowe Wydawnictwo Naukowe: Kraków, Poland, 1969; pp. 1–690.
175. Kamczyc, J.; Gwiazdowicz, D.J. Soil mites (Acari, Mesostigmata) from Szczeliniec Wielki in the Stołowe Mountains National Park (SW Poland). *Biol. Lett.* **2009**, *46*, 21–27. [CrossRef]
176. Frenot, Y.; Chown, S.L.; Whinam, J.; Selkirk, P.; Convey, P.; Skotnicki, M.; Bergstrom, D. Biological invasions in the Antarctic: Extent, impacts and implications. *Biol. Rev.* **2005**, *80*, 45–72. [CrossRef]
177. Bergstrom, D.M.; Lucieer, A.; Kiefer, K.; Wasley, J.; Belbin, L.; Pedersen, T.K.; Chown, S.L. Indirect effects of invasive species removal devastate World Heritage Island. *J. Appl. Ecol.* **2009**, *46*, 73–81. [CrossRef]
178. Convey, P. Terrestrial biodiversity in Antarctica–Recent advances and future challenges. *Polar Sci.* **2010**, *4*, 135–147. [CrossRef]
179. Hågvar, S.; Solhøy, T.; Mong, C.E. Primary succession of soil mites (Acari) in a Norwegian glacier foreland, with emphasis on oribatid species. *Arct. Antarct. Alp. Res.* **2009**, *41*, 219–227. [CrossRef]
180. Buda, J.; Olszanowski, Z.; Wierzgoń, M.; Zawierucha, K. Tardigrades and oribatid mites in bryophytes from geothermally active lava fields (Krafla, Iceland) and the description of *Pilatobius islandicus* sp. nov. (Eutardigrada). *Pol. Polar Res.* **2018**, *39*, 425–453. [CrossRef]
181. Hågvar, S.; Pedersen, A. Food choice of invertebrates during early glacier foreland succession. *Arct. Antarct. Alp. Res.* **2015**, *47*, 561–572. [CrossRef]
182. Walter, D.E.; Proctor, H.C. *Mites: Ecology, Evolution & Behaviour*; Springer: Dordrecht, The Netherlands; Heidelberg, Germany; New York, NY, USA; London, UK, 2013. [CrossRef]

diversity

Article

Analysis of Oribatid Fauna of the East European Tundra with First Reported Data of Subpolar Urals

Elena N. Melekhina

Institute of Biology, Komi Science Center, Ural Branch of Russian Academy of Sciences (IB FRC Komi SC UB RAS), Kommunisticheskaya Str. 28, 167982 Syktyvkar, Russia; melekhina@ib.komisc.ru

Received: 1 May 2020; Accepted: 6 June 2020; Published: 10 June 2020

Abstract: This study presents data on the oribatid mite fauna of the Subpolar Urals for the first time. Observations were made in the Lembekoyu River valley and 35 species of oribatid mites from 24 genera and 21 families were found. The analysis of taxonomic diversity and distribution of East European tundra oribatid mite species is presented based on available literature and the author's own research findings. The taxonomic list includes 163 species from 81 genera and 45 families. Ceratozetidae (15 species), Crotoniidae (14 species), Oppiidae (12 species), Suctobelbidae (12 species), Damaeidae (9 species), Brachychthoniidae (8 species), Phthiracaridae (5 species), Humerobatidae (5 species), Achipteriidae (5 species), Punctoribatidae (5 species), and Galumnidae (5 species) are the leading families, comprising more than 58% of all species. The zoogeographical structure of the fauna is dominated by widely distributed Holarctic, cosmopolitan, and semi-cosmopolitan species. The share of Palaearctic species is 23%. The specificity of the fauna of East European tundra manifests itself in the small group of Arctic species, both in the mainland tundra and on the Arctic islands. A complex of arctic-boreal species, widely distributed in the Eurasian sector of the Arctic, is distinguished.

Keywords: Arctic; Oribatida; faunistics; taxonomic diversity; distribution; checklist; arctic species; arctic-boreal species

---

1. Introduction

Interest in Arctic invertebrates has greatly increased in recent decades [1–14]. Coulson et al. [11] summarises data on the taxonomic composition of invertebrates from three archipelagos in the Barents Sea: Spitsbergen, Franz Josef Land, and Novaya Zemlya. One of the directions of modern research is to learn how invertebrates spread and what contributes to their faunistic diversity on the islands and archipelagos in the Arctic [15–18]. Makarova et al. [19] focused on 'eastern elements' in the invertebrate fauna of East European tundra, the so-called 'Siberian' species. An integrated assessment of biological diversity, including soil invertebrates, has been carried out in the large Arctic region, Nenets Autonomous District [20].

Attention is being paid to the study of taxonomic diversity of invertebrates, including oribatid mites, in anthropogenic altered soils in the Arctic and Sub-Arctic conditions and the application of this group of animals in bioindication [13,21–23]. Inventory of modern biodiversity and study of geographic trends of soil fauna diversity is important for biodiagnostics of natural communities under conditions of anthropogenic impact and forecasting changes in these communities in the long-term. In addition, the study of modern biodiversity is important for assessing the changes that take place in response to global warming.

In the European North-East, studies on oribatid mite fauna have been conducted both in the mainland tundra (plain and mountain tundra) and on the Arctic islands. The first data on Vaigach Island oribatid mites were presented in the form of a monograph by Koch [24], who processed the data collected by Adolf Nordenskiöld during his Arctic expedition of 1875. Koch named eight species of oribatids found on the Vaigach Island, four of which he described as new to science. Subsequently, Trägårdh [25] identified three species from Nordenskiöld's collection, which were already named by Koch. Krivolutsky et al. [26] provided a list of oribatid species in the Russian Arctic sector known at the time of publication, which included data on the Vaigach and Kolguev Islands. Melekhina et al. [27] studied oribatid fauna in the nests of Lapland Bunting *Calcarius lapponicus* and reported nine species, nine genera, and six families of oribatids new to Vaigach Island. In total, 25 species of oribatid mites from 24 genera and 21 families were found in this study. The authors compiled a list of Vaigach oribatid mites (43 species, 34 genera, 25 families) and analysed the zoogeographic structure of the fauna and species distribution [27].

Melekhina and Zinovyeva [28] collected samples in the north-western part of the Yugor Peninsula, on the Pai-Khoi Ridge. Oribatida from seven types of habitats of mainland tundra and mountain tundra, as well as one intrazonal plant community (motley grass meadow on the bank of the stream) were examined.

Data on the Bolshezemelskaya tundra have been collected mainly from around Vorkuta [29–31]. Melekhina and Krivolutsky [30] presented a list of oribatids of the Bolshezemelskaya tundra (33 species, 24 genera, 20 families). Peculiarities of the vertical distribution of oribatid mites in moss-lichen tundra with permafrost soil and in the tundra of dwarf birch with non-frost soil have also been reported [31]. Zubrii et al. [32] studied the plant communities near the thermal springs of Pym-Va-Shor located in the Bolshezemelskaya tundra in the Polar Pre-Urals. Samples from melted soil plots and watered mossy turfs were collected during the winter.

Research was also carried out in the Polar Urals in the area of Labytnangi [33], on the Rai-Iz Ridge [34]. Melekhina [35,36] obtained data in the vicinity of Lake Paga-Ty. Six types of mountain tundra habitats were surveyed: shrub-moss-lichen tundra, shrub-moss tundra, shrub-green-moss birch, grassy willow, horsetail moss willow, and motley grass in the hollow of the drain.

Sidorchuk [37] investigated the distribution of oribatid mites along with the vertical profile of the Malyy Paypudynskiy Ridge, from the floodplain to the belt of stony tundra. Samples were collected from the lower part of the Malaya Paypudyna River valley (settlement Polyarny, Labytnang district, YNAD), on the slope of the north-eastern exposure. Sidorchuk [37] found 82 species of oribatid mites in four types of plant communities, of which 46 were recorded for the first time in the region. The article provides a generalised list of oribatid mites of the Polar Urals (106 species, 61 genera, 34 families) taking into account all published data.

Melekhina conducted research at the tundra landscapes of the complex reserve 'Khrebtovy', located on the south-eastern slope of the Yenganepe Ridge (Polar Urals) [38]. Oribatida from seven types of habitats were examined: birch forest, larch sparse woodland, shrubby moss-lichen tundra, large bog, rocky outcrops, shrubby lichen tundra, and meadow complex.

Data on the mountain tundra oribatid fauna of the Northern Urals are also available. Melekhina [39] surveyed four types of plant communities (spruce, boggy grass-marsh, shrubby lichen, and stony lichen tundra) on the Yany-Pupu-Nyor mountain in the Pechora-Ilychsky Reserve. Melekhina [40] summarised her own research findings and those available in the literature on the taxonomic diversity of oribatids in the European North of Russia (in the taiga and tundra zones) and analysed the dynamics of diversity along the latitudinal gradient. In the Subpolar Urals, no research has been conducted on the oribatid mite fauna so far.

The purpose of this paper is to summarise the findings of the available literature and those of the author's own new research on the taxonomic diversity and distribution of oribatid mites in the East European tundra and to identify specific features of this fauna.

2. Materials and Methods

Observations were made in the Lembekoyu River valley (65°16′46′′ N, 60°4′51′′ E), in zonal (lichen-moss tundra, moss-marsh tundra) and intrazonal (stream floodplain) communities. Sampling was conducted in five plant communities: lichen-moss tundra 1, lichen-moss tundra 2, moss and dwarf birch tundra 1, moss and dwarf birch tundra 2 and grassy community in a creek valley. Field material was collected according to generally accepted methods [41]. Ten soil samples, each with dimensions 10 × 10 cm by 10 cm deep, were taken from all sites in June 2018. A total of 50 soil samples were collected. The soil samples were transported to the Institute of Biology (IB Komi SC UB RAS), Syktyvkar, and placed into Tullgren soil extractors. The microarthropod fauna was extracted under 40 Watt bulbs into 96% alcohol for seven-ten days until the soil was completely dry. The Oribatida were identified to species by morphological taxonomic characters [42]. A total of 2500 specimens of adult oribatid mites were identified up to the species level.

In this manuscript, the author summarises the results of her own research conducted earlier in the tundra zone of the European North-East and presented in publications [22,27–32,35,36,38,39], new data obtained in the Subpolar Urals, as well as all available literature information (Table 1, Figure A1). Taxonomies of oribatid mites and types of global distribution of the species follow Subías [43]. For the analysis of the geographical distribution of species, literary sources were used [1,7,11,30,44–50] and others. The term local fauna was used as understood by A.G. Tatarinov [51].

Table 1. Number of taxa of oribatid mites in the local fauna of East European tundra.

Local Fauna	Issue	Taxa		
		Species	Genera	Families
Kolguev Island	Krivolutsky et al., 2003	13	11	11
Vaygach Island	Koch, 1879	8	6	6
	Trägårdh, 1904	3	2	2
	Krivolutsky et al., 2003	25	23	18
	Melekhina et al., 2019	43	34	25
Yugor Peninsula	Melekhina and Zinovyeva, 2012	32	26	19
Bolshezemelskaya tundra	Melekhina, 1997	5	4	4
	Melekhina and Krivolutsky, 1999	33	24	20
	Goryachkin et al., 2011	22	22	18
	Zubrii et al., 2011	8	8	8
	Melekhina, unpublished	7	7	7
Polar Urals	Karpova and Poryadina, 1978	18	12	12
	Grishina, 1985	6	5	5
	Biodiversity, 2007	37	28	19
	Sidorchuk, 2009	106	61	34
	The biological diversity, 2010	32	27	18
Subpolar Urals	Melekhina and Selivanova, unpublished	35	24	21
Northern Urals	Melekhina, 2005	25	22	14
a complete list of species	all published and new data	163	81	45

Synonyms follow Subías [43]. Synonyms of species were given when the author of a publication mentioned the species using a different name. For example, the species *Eupelops plicatus* (Koch, 1835) (=*Pelops auritus* Koch, 1839) was noted in Bolshezemelskaya tundra [30] and was named *Eupelops auritus* Koch, 1839. In some cases, a synonym was given if the authors of several publications cited synonyms.

3. Results and Discussion

3.1. Taxonomic Diversity

3.1.1. Subpolar Urals

This is the first report of oribatid mite fauna of the Subpolar Urals. In total, 35 species, 24 genera, and 21 families of oribatid mites were found (Table 2, Appendix B). The largest number of species was recorded in the families Crotoniidae, Oppiidae and Suctobelbidae. In the studied five plant communities, 11 to 18 species were found. *Heminothrus (H.) longisetosus, Nanhermannia (N.) sellnicki, Tectocepheus velatus* and *Melanozetes sellnicki* were common in different types of communities. The species *Malaconothrus (M.) monodactylus, Heminothrus (P.) peltifer, Hypochthonius rufulus, Atropacarus (A.) striculus* and *Neoribates (N.) aurantiacus* were recorded only in the floodplain of the stream. The first two species are known to prefer high humidity habitats.

Table 2. The species composition of oribatid mites in the plant communities of the Subpolar Urals.

N	Species	L-m 1	L-m 2	M-d 1	M-d 2	Gras. Com.
1	*Liochthonius (L.) sellnicki*	+	+	+	–	–
2	*Hypochthonius rufulus*	–	–	–	–	+
3	*Atropacarus (A.) striculus*	–	–	–	–	+
4	*Malaconothrus (M.) monodactylus*	–	–	–	–	+
5	*Nothrus borussicus*	+	+	–	–	–
6	*Nothrus pratensis*	–	–	+	–	+
7	*Camisia (C.) biurus*	+	–	+	–	–
8	*Camisia (C.) segnis*	–	–	+	–	–
9	*Camisia (E.) lapponica*	–	–	–	+	–
10	*Heminothrus (H.) longisetosus*	+	+	+	+	–
11	*Heminothrus (P.) peltifer*	–	–	–	–	+
12	*Nanhermannia (N.) sellnicki*	–	+	+	+	+
13	*Tokukobelba compta*	–	+	–	–	–
14	*Damaeus (E.) bituberculatus*	–	+	–	–	–
15	*Ceratoppia bipilis*	+	–	–	+	–
16	*Ceratoppia quadridentata*	–	–	–	+	–
17	*Rhinoppia (R.) subpectinata*	+	+	–	–	–
18	*Oppiella (O.) nova*	+	–	+	–	–
19	*Oppiella (M.) neerlandica*	+	+	+	–	–
20	*Quadroppia (Q.) quadricarinata*	–	+	–	–	–
21	*Suctobelbella (S.) acutidens duplex*	–	+	–	–	–
22	*Suctobelbella (S.) latirostris*	+	–	–	–	–
23	*Suctobelbella (S.) singularis*	+	+	–	–	–
24	*Carabodes (C.) labyrinthicus*	–	+	–	–	–
25	*Carabodes (C.) marginatus*	–	+	+	+	–
26	*Carabodes (C.) subarcticus*	–	+	–	+	–
27	*Tectocepheus velatus*	+	+	+	+	+
28	*Parachipteria punctata*	–	–	+	–	+
29	*Ceratozetes (C.) gracilis*	+	–	–	+	+
30	*Melanozetes sellnicki*	+	+	+	+	–
31	*Diapterobates oblongus*	+	+	–	+	–
32	*Oribatula (O.) tibialis*	–	+	+	–	–
33	*Oribatula (Z.) exilis*	+	–	+	–	–
34	*Scheloribates (S.) laevigatus*	–	–	–	+	+
35	*Neoribates (N.) aurantiacus*	–	–	–	–	+
	total	15	18	14	12	11

Notes. L-m 1 lichen-moss tundra 1; L-m 2 lichen-moss tundra 2; M-d 1 moss and dwarf birch tundra 1; M-d 2 moss and dwarf birch tundra 2; Gras. com. grassy community in a creek valley; + and – The presence or absence of a species.

3.1.2. Total Number of Taxa at Different Levels

In total, 163 species of oribatid mites from 81 genera and 45 families were found in East European tundra (Table 1, Appendix B). Oribatid species found in the East European tundra are grouped into 23 superfamilies, with the majority of the species coming under six superfamilies. The leading superfamily is Ceratozetoidea, with 28 species from four families, followed by Crotonioidea (22 species, four families), Oppioidea (16 species, four families), Trizetoidea (12 species, one family), Oripodoidea (12 species, six families), and Damaeoidea (10 species, two families) (Figure 1). Eight species represent the superfamilies Brachychthonioidea (one family), Gustavioidea (three families), and Achipterioidea (two families). A similar distribution of species among superfamilies has been observed, for example, in Svalbard [7].

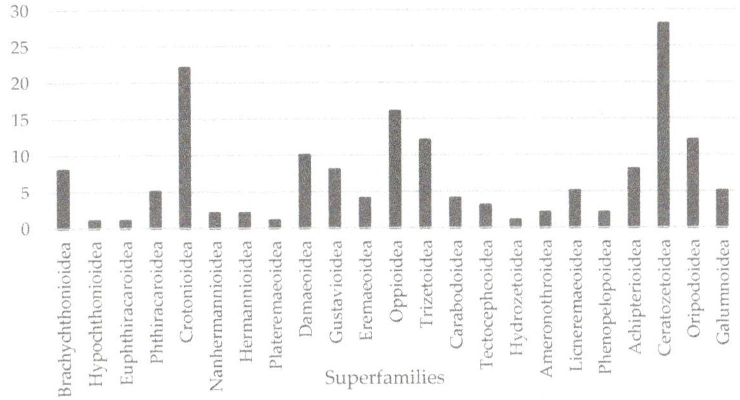

Figure 1. Distribution of species richness among different superfamilies of oribatid mites in the East European tundra (taxa are in a systematic order).

Leading families in the mite fauna are: Ceratozetidae (15 species), Crotoniidae (14 species), Oppiidae (12 species), Suctobelbidae (12 species), Damaeidae (nine species), Brachychthoniidae (eight species), Phthiracaridae (five species), Humerobatidae (five species), Achipteriidae (five species), Punctoribatidae (five species), and Galumnidae (five species). These 11 families comprise 58.6% of all species (95 species). Most families are represented by a small number of species (one to three) (Figure 2).

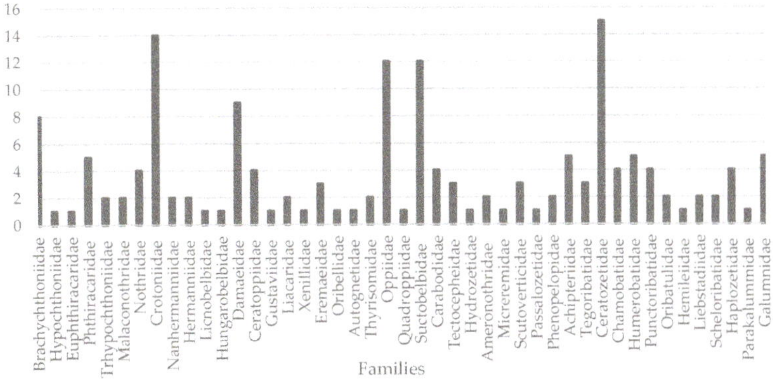

Figure 2. Distribution of species richness among different families of oribatid mites in the East European tundra (taxa are in a systematic order).

3.1.3. Characteristic Families Typical of Tundra Zone Species

It is possible to distinguish species characteristics of different families for the East European tundra. From Crotoniidae, *Heminothrus (Platynothrus) punctatus* and *Camisia (C.) horrida* are consistently found in the local tundra fauna. Both species are often found in the Eurasian sector of the Arctic [30]. *Camisia (C.) biverrucata*, *C. (C.) invenusta*, and *C. (Ensicamisia) lapponica*, are associated with high latitude. *Hermannia reticulata* and *H. scabra* Hermanniidae and *Ceratoppia bipilis* and *C. sphaerica* Ceratoppiidae are also characteristic of high latitudes. In the tundra zone, the holarctic species *Nothrus borussicus* Nothridae was often observed.

From Carabodidae, the most common species in the tundra are the circumpolar *Carabodes labyrinthicus* and the Palearctic *C. subarcticus* and *C. marginatus*. For high latitudes, there are Holarctic species *Oppiella (M.) neerlandica*, *O. (O.) splendens*, and *Moritzoppia unicarinata unicarinata* Oppiidae, and *Suctobelbella acutidens acutidens* Suctobelbidae. From Ceratozetidae, *Edwardzetes edwardsi* and *Ceratozetella sellnicki* are common at high latitudes. *Diapterobates notatus* and *Svalbardia paludicola* Humerobatidae are widely distributed in the Eurasian sector of the Arctic; they are found on the Arctic islands and archipelagos [11–13,17]. The first of these species has been noted as the most abundant in various plant communities in Svalbard [11–13].

This complex of species is characteristic of local fauna of the East European tundra, as well as of the tundra zone of Eurasia as a whole. The Holarctic species *Oribatula (O.) tibialis* and *O. (Z.) exilis*, the cosmopolitan *Tectocepheus velatus* and *Oppiella nova*, and the semi-cosmopolitan *Quadroppia quadricarinata* and *Scheloribates laevigatus laevigatus* also are widely distributed in the Arctic.

For 11 species in the list, Subías [43] indicates boreo-Alpine distribution. These are Holarctic species *C. (E.) lapponica*, *C. (C.) borealis*, *H. (P.) peltifer*, *H. (P.) humicola*, *E. edwardsi*, *Melanozetes mollicomus*, *Moritzoppia unicarinata clavigera*, *D. notatus*, *S. paludicola*, *Mycobates (Calyptozetes) sarekensis*, and *Oromurcia lucens*. For the Holarctic species *C. sphaerica*, *Melanozetes sellnicki*, *Ameronothrus lineatus*, *A. nigrofemoratus*, *Diapterobates variabilis*, and *Sphaerozetes arcticus*, boreal distribution is indicated. In conclusion, all these species are typical of the high latitudes of Eurasia.

3.2. *Types of Longitudinal Distribution*

In this section, the Holarctic, Palearctic, cosmopolitan and semi-cosmopolitan species of oribatid fauna are distinguished by their longitudinal distribution. Holarctic species predominate the fauna (Figure 3). For comparison, in Spitsbergen, Holarctic species make up 50% of all species [7,11]. It has previously been noted that the European sector of the Arctic is characterised by an increase in the proportion of Holarctic oribatid species in the latitudinal gradient, from the taiga to mainland tundra to the Arctic islands and archipelagos [40]. Thus, in the taiga zone of the European North-East, the share of Holarctic species (41.5%) is the least [40].

The share of Palearctic species (34 species) in the local fauna was much smaller (23.4%) than that of Holarctic species. For comparison, in the taiga zone of the European North-East, Palearctic species make up 37.5% of the faunal list [44]. The highest share of Palearctic species is observed in the fauna of the Polar Urals and Bolshezemelskaya tundra (Figure 3).

The composition of the Palearctic species was specific to each local fauna. In total, five Palearctic species from Damaeidae were recorded, four of which were found only in the Polar Urals. Only one species, *Damaeus (E.) bituberculatus*, was distributed in three local faunas: in the Bolshezemelskaya tundra, the Polar Urals and the Northern Urals. This species is widely distributed in the taiga zone of the European North [44].

Some Palearctic species were noted only in the Polar Urals: *Liacarus (D.) neonominatus* (Liacaridae), *Eueremaeus oblongus silvestris* (Eremaeidae), *Exochocepheus laticuspis* (Scutoverticidae), *Bipassalozetes (B.) intermedius* (Passalozetidae), and *Suctobelbella (S.) subcornigera vera* (Suctobelbidae). In turn, other palearctic species, *Scutovertex neonominatus* (Scutoverticidae), *Berniniella (B.) bicarinata* and *Lauroppia falcata* (Oppiidae), and *Suctobelbella (S.) singularis* (Suctobelbidae), were recorded only in the Bolshezemelskaya tundra.

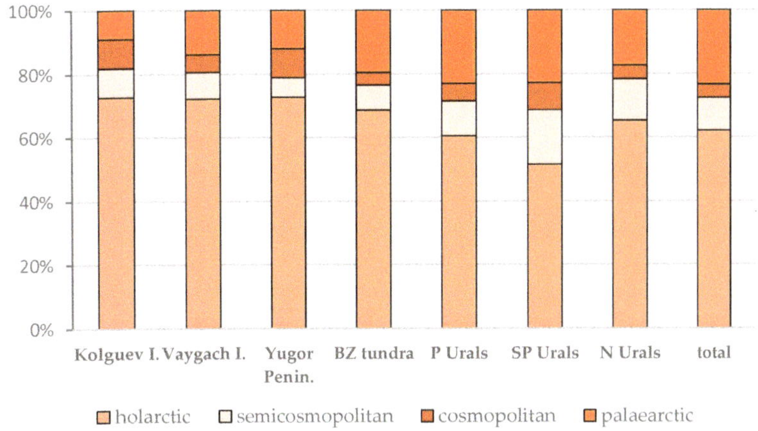

Figure 3. Relative species richness according to the longitudinal distribution of oribatid mites in the local fauna of East European tundra. Kolguev I.—Kolguev Island, Vaygach I.—Vaygach Island, Yugor Penin.—Yugor Peninsula, BZ tundra—Bolshezemelskaya tundra, P Urals—Polar Urals, SP Urals—Subpolar Urals, N Urals—North Urals.

Only a few Palearctic species were common to several local faunas. In addition to *D. (E.) bituberculatus* mentioned above, the species *Carabodes (C.) marginatus* was recorded in four mainland faunas: Polar Urals, Subpolar Urals, Yugor Peninsula, and Bolshezemelskaya tundra. The generalisate Palaearctic species for two faunas: the Yugor Peninsula and the Vaigach Island was *Minunthozetes (M.) pseudofusiger*.

The share of cosmopolitan and semi-cosmopolitan species in the total fauna structure is 14.5%. In local faunas, they account for 13.9% to 18.2%. This is much higher than their share in the taiga zone, where they account for 9.5% of the total species [40].

It can thus be concluded that the findings of previous research [40] regarding the decrease in the proportion of Palearctic species and the increase in the proportion of Holarctic, cosmopolitan, and semi-cosmopolitan species in the oribatid fauna along the latitudinal gradient (from the taiga to mainland tundra to the Arctic islands and archipelagos) holds true.

A substantial number of species (20 species, 62.5%) are distributed circumpolar; they occur in high latitudes of both the Palearctic and Nearctic regions- in Alaska, Yukon, and Greenland [10,47,49]. These include *Liochthonius lapponicus*, *N. borussicus*, *Camisia horrida*, *C. biurus*, *H. (P.) punctatus*, *H. reticulata*, *C. bipilis*, *C. sphaerica*, *Moritzoppia neerlandica*, *M. unicarinata clavigera*, *S. acutidens acutidens*, *S. hammeri*, *O. tibialis*, *M. mollicomus*, *D. notatus*, *S. paludicola*, as well as cosmopolitan species *T. velatus*, *O. nova*, *C. gracilis* and semi-cosmopolitan species *Q. quadricarinata*.

3.3. *Types of Latitudinal Distribution*

3.3.1. Arctic Species

In the composition of the oribatid fauna of the East European tundra, species with arctic, arctic-boreal, temperate, and polyzonal types of latitudinal distribution have been documented. For the East European tundra, as well as for the mainland tundra of the European part of Russia as a whole, only a single Arctic oribatid species is known until now – *S. paludicola*. In addition to the Eastern European sector, *S. paludicola* was found in the Kola Peninsula [50]. In the Arctic archipelagos of the Palearctic region, it was found in Spitsbergen [7,11] and Novaya Zemlya [1,11]; in northern Siberia, in Yamal and Taimyr [34,45]; and in the northern Far East, in Chukotka [48]. The species has a circumpolar distribution and is found in Greenland [49], Alaska, and Yukon [47]. The data obtained confirm earlier conclusion about the small number of Arctic species in the Eastern European sector of the Arctic, both in the island and mainland regions [28,40].

The oribatid species *Sphaerozetes arcticus* Hammer, 1952, noted in the Polar Urals [35], should be referred to as an arctic-boreal species. In the European part of Russia, this species is rarely found. In addition to the Polar Urals, the species was recorded in the taiga zone, in the Arkhangelsk region [44]. In northern Siberia, it is more widespread, with its distribution covering the arctic-boreal zone, but most findings are in the tundra zone [45]. The species is found in Chukotka. Thus, it can be concluded that *S. arcticus* is characteristic of high latitudes. Subías [43] indicates a boreal distribution of this species.

The number of Arctic species is increasing on remote islands and archipelagos of the European Arctic sector. For example, on the Svalbard archipelago, where 81 species of oribatid mites are known [7,11], six Arctic species have been recorded. These are: *Autogneta (A.) kaisilai* Karppinen, 1967, *Sphaerozetes setiger* (Trägårdh, 1910), *Iugoribates gracilis* Sellnick, 1944, *Svalbardia paludicola, Ceratozetes (C.) spitsbergensis* Thor, 1934, and *Oribatella (O.) arctica arctica* Thor, 1930. The first of these species is noted only on Svalbard [43] and can be called conditional endemic to Svalbard.

For comparison, two Arctic species, *S. paludicola* and *Oribatella (O.) arctica arctica,* have been registered in the fauna of Novaya Zemlya. *Oribatella (O.) arctica arctica,* was observed in northern Siberia [34] and Chukotka [45], in addition to Spitsbergen. Therefore, it cannot be ruled out that this species may be found in the East European tundra in the future.

3.3.2. Arctic-Boreal Species

The tundra oribatid fauna includes a complex of arctic-boreal species occupying the Arctic islands and archipelagos, the continental part of the tundra zone, and the taiga zone of Eurasia. The base of this complex is represented by species that constitute the majority of tundra fauna, both in the European sector of the Arctic (mainland part and island part: Vaigach Island, Novaya Zemlya, and Spitsbergen), and in Siberia and the Far East, and, accordingly, contribute to the higher percentage of tundra fauna similarity in Eurasia. These species are *H. punctatus, C. sphaerica, H. reticulata,* and *D. notatus,* with a circumpolar distribution. On Vaigach Island, this list further includes *Ameronothrus lineatus, A. nigrofemoratus,* and *Oromurcia lucens*. Species *A. lineatus* in the European sector of the Arctic is also distributed in Spitsbergen and Kola tundra. Like *A. nigrofemoratus*, it is also a boreal species, while *Oromurcia lucens* (L. Koch, 1879), also found on Spitsbergen, is a boreal-alpine species, according to Subías [43].

In the biocoenoses of the Pai-Khoi Ridge (Yugor peninsula), the complex of arctic-boreal species is complemented by *Moritzoppia unicarinata clavigera, Pyroppia lanceolata,* and *Banksinoma setosa*. For the first of these species, Subías [43] indicates a boreal-alpine distribution. The species *B. setosa,* in addition to the Yugor peninsula, is noted in the North Urals. It is mainly located in the Siberian and Far Eastern sectors. The share of arctic-boreal species in different local faunas is not high, making up 7.7% to 18.6% of the total fauna. In the overall structure of the fauna, their contribution is 7.4%.

Some arctic-boreal species are also present in the Palaearctic mountain ranges, such as *D. notatus* in the Altai [46], *C. sphaerica* in the Tien Shan [46], and *M. unicarinata clavigera* and *P. lanceolata* in the Caucasus [41]. Arctic-boreal species *H. punctatus, D. notatus, Hermannia scabra, H. reticulata, A. lineatus,* and *C. sphaerica* are also found in the Western European tundra sector (Kola Peninsula) [50]. It can be thus concluded that the arctic-boreal species which are widely distributed along the longitudinal gradient of the European sector of the Arctic are all common. No arctic-boreal species specific to the Eastern European sector have been identified. In the European North-East, the complex of arctic-boreal species also includes *M. unicarinata clavigera, P. lanceolata, B. setosa, Sphaerozetes arcticus,* and *Peloribates pilosus.*

3.3.3. Temperate and Polyzonal Species

The largest number of species recorded in the East European tundra have a temperate or polyzonal type of latitudinal distribution. Temperate species typical of East European tundra are *Liochthonius lapponicus, N. borussicus, O. (M.) neerlandica, C. labyrinthicus, C. marginatus. C. subarcticus, E. edwardsi,* and *M. mollicomus.*

Polyzonal species *C. horrida, C. biurus, N. borussicus, S. acutidens acutidens, S. acutidens duplex, C. bipilis, O. (O.) tibialis,* and *O. (Z.) exilis* were often found in local faunas of the tundra zone. Cosmopolitan species *T. velatus* and *O. nova* and semi-cosmopolitan species *Q. quadricarinata* and *S. laevigatus*, which in the latitudinal aspect are also distributed polyzonal [52], are common in the tundra. These polyzonal and temperate species that are common in the tundra zone, as well as the Arctic and arctic-boreal species, mentioned above, we call 'species of northern complex'.

3.3.4. 'Southern' Elements within the Oribatid Fauna

Attention is also drawn to species which are mainly found in the lower latitudes. These species can be called 'conditionally southern'. For example, *Hydrozetes thienemanni*, with a temperate type of distribution, was found only at Yugor Peninsula in the European sector of the tundra zone [28]. In the European part of Russia, it is distributed in taiga and coniferous/broad-leaved forests [41,44,52]. In the tundra zone of Eurasia, *H. thienemanni* was previously found only in Chukotka by Grishina [34].

Holarctic species *Malaconothrus (Trimalaconothrus) tardus*, according to Subías [43], is absent in the northern Palaearctic region. It is found in the Polar Urals [35]. In the European part of Russia, it is found in the northernmost regions. This species was observed mainly in broad-leaved forests and steppe zone [41,44].

Palaearctic species *Eueremaeus oblongus silvestris* found in the Polar Urals [37] is not characteristic of the tundra zone. In the European part of Russia, it is mainly found in the zone of broad-leaved and coniferous/broad-leaved forests [41,52]. In Siberia, the species was observed in taiga forests and the Altai [45].

4. Summary

This publication presents a generalised taxonomic list of oribatid mites of East European tundra, based on available literature and new data. The checklist of East European tundra oribatid mites includes 163 species, 81 genera, and 45 families. This study presents data on the oribatid mite fauna of the Subpolar Urals for the first time. To date, 35 species, 24 genera, and 21 families of oribatids have been registered from this region.

The leading families in the fauna structure are Crotoniidae, Ceratozetidae, Oppiidae, Suctobelbidae, Damaeidae, Brachychthoniidae, Phthiracaridae, Humerobatidae, Achipteriidae, Punctoribatidae, and Galumnidae. The greatest number of species is Holarctic. Circumpolar distributed species make 12.3% of the total. The share of Palaearctic species is low (23.4%), which distinguishes the fauna of the tundra zone from the taiga zone.

The specificity of the oribatid fauna of East European tundra manifests itself in the small group of Arctic species, both in the mainland tundra and on the Arctic islands. In the majority of local fauna, there is only one Arctic species, *Svalbardia paludicola*. *Sphaerozetes arcticus*, noted in the Polar Urals, that has been classified as an arctic-boreal species. It could also be called 'conditionally arctic', as it sometimes penetrates into the taiga zone.

The fauna of the East European tundra is characterised by a complex of arctic-boreal species, based on circumpolar species common with the Western European sector of the Arctic, as well as with the Siberian and Far Eastern sectors, such as *Heminothrus punctatus, Ceratoppia sphaerica, Hermannia reticulata,* and *Diapterobates notatus*. In different local faunas, this complex is complemented by species *Ameronothrus lineatus, A. nigrofemoratus, Banksinoma setosa, Pyroppia lanceolata, Moritzoppia unicarinata clavigera, Peloribates pilosus,* and *Oromurcia lucens.*

The largest number of species in the East European tundra is polyzonal. No species specific to the East European sector of the tundra zone was identified.

Funding: This work was supported by the RFBR No. 18-29-05028 mk and State Assignment AAAA-A17-117112850235-2.

Acknowledgments: The author thanks Leonid N. Rybin from the Institute of Biology, Komi Science Center, Ural Branch of RAS, Syktyvkar (Russia) for map.

Conflicts of Interest: The author declares no conflict of interest.

Appendix A

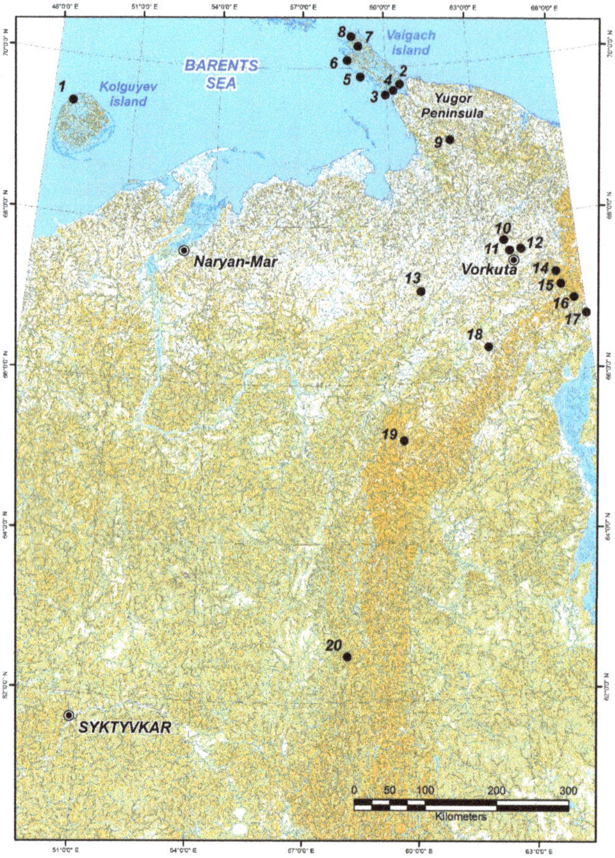

Figure A1. Localities within the region under consideration (see Table 1 for explanations). 1—Kolguev Island (Krivolutsky et al., 2003), 2—Vaigach Island, Yugorskiy Shar (Koch, 1879), 3—Vaigach Island, Cape Greben' (Koch, 1879), 4—Vaigach Island (Trägårdh, 1904), 5 and 6—Vaigach Island (Krivolutsky et al., 2003), 7—Vaigach Island, Bolvansky mountain (Melekhina et al., 2019), 8—Vaigach Island, Old Polar Station (Melekhina et al., 2019), 9—Yugor Peninsula, Pai-Khoi Ridge (Melekhina and Zinovyeva, 2012), 10—Bolshezemelskaya tundra, Vorgashor village (Melekhina, 1997), 11—Bolshezemelskaya tundra, Vorkuta (Melekhina and Krivolutsky, 1999), 12—Bolshezemelskaya tundra, Vorkuta (Goryachkin et al., 2011), 13—Bolshezemelskaya tundra, Pym-Va-Shor (Zubrii et al., 2012), 14—Polar Urals, complex reserve 'Khrebtovy', Yenganepe Ridge (The biological . . . , 2010), 15—Polar Urals, Polyarny village (Sidorchuk, 2009), 16—Polar Urals, Rai-Iz Ridge (Grishina, 1985), 17—Polar Urals, Labytnangi village (Karpova et al., 1987), 18—Polar Urals, Paga-Ty Lake (Biodiversity . . . , 2007; Melekhina, 2008), 19—Subpolar Urals, Lembekoyu River (Melekhina and Selivanova, unpublished), 20—Northern Urals, Yany-Pupu-Nyor mountain (Melekhina, 2005).

Appendix B

Checklist of oribatid mites of the East European tundra*

*Note. After the name of a species its distribution in the East European tundra with references is given; global distribution follows Subías [42].

Brachychthoniidae Thor, 1934

Brachychthonius bimaculatus Willmann, 1936 Polar Urals [35]. Holarctic

Liochthonius (Liochthonius) lapponicus (Trägårdh, 1910) Yugor Peninsula [26], Bolshezemelskaya tundra [24]. Holarctic

Liochthonius (L.) muscorum Forsslund, 1964 Polar Urals [35]. Holarctic

Liochthonius (L.) sellnicki (Thor, 1930) Kolguev Island [29], Vaygach Island [29,30], Polar Urals [33], Subpolar Urals (new data). Holarctic

Sellnickochthonius immaculatus (Forsslund, 1942) Polar Urals [35]. Holarctic

Sellnickochthonius zelawaiensis (Sellnick, 1928) Bolshezemelskaya tundra [Melekhina, unpabl.], Polar Urals [35]. Holarctic

Eobrachychthonius sp. Vaygach Island [29].

Eobrachychthonius latior (Berlese, 1910) Bolshezemelskaya tundra [24]. Holarctic

Hypochthoniidae Berlese, 1910

Hypochthonius rufulus Koch, 1835 Polar Urals [35], Subpolar Urals (new data). Semicosmopolitan

Euphthiracaridae Jacot, 1930

Acrotritia ardua (Koch, 1841) Polar Urals [35]. Cosmopolitan

Phthiracaridae Perty, 1841

Atropacarus (Atropacarus) striculus (Koch, 1835) Vaygach Island [30], Bolshezemelskaya tundra [25]; Subpolar Urals (new data). Semicosmopolitan

Phthiracarus spp. Kolguev Island [29], Vaygach Island [30].

Phthiracarus (Phthiracarus) boresetosus Jacot, 1930 Polar Urals [35]. Semicosmopolitan

Phthiracarus (P.) laevigatus (Koch, 1844) Vaygach Island (as *Phthiracarus nitens* Nicolet, 1855) [29]. Palaearctic

Phthiracarus (Archiphthiracarus) ligneus Willmann, 1931 Vaygach Island [30]. Holarctic

Trhypochthoniidae Willmann, 1931

Trhypochthonius cladonicolus (Willmann, 1919) Bolshezemelskaya tundra [23], Polar Urals [35]. Palaearctic

Trhypochthonius tectorum s. str. (Berlese, 1896) Bolshezemelskaya tundra [25]; Polar Urals [35]. Cosmopilitan

Malaconothridae Berlese, 1916

Malaconothrus (Malaconothrus) monodactylus (Michael, 1888) Bolshezemelskaya tundra [25]; Polar Urals [35], Subpolar Urals (new data). Holarctic

Malaconothrus (Trimalaconothrus) tardus (Michael, 1888) Polar Urals [33]. Holarctic

Nothridae Berlese, 1896

Nothrus sp. Vaygach Island [30].

Nothrus borussicus Sellnick, 1928 Yugor Peninsula [26], Bolshezemelskaya tundra [23,24]; Polar Urals [33,35,36]; Subpolar Urals (new data); Northern Urals [37]. Holarctic

Nothrus palustris Koch, 1839 Bolshezemelskaya tundra [23,25]. Holarctic

Nothrus pratensis Sellnick, 1928 Bolshezemelskaya tundra [23], Subpolar Urals (new data). Holarctic

Crotoniidae Thorell, 1876

Camisia (Camisia) biurus (Koch, 1839) Yugor Peninsula [26], Bolshezemelskaya tundra [24]; Polar Urals [33,35], Subpolar Urals (new data); Northern Urals [37]. Holarctic

Camisia (C.) biverrucata (Koch, 1839) Polar Urals [35]. Holarctic

Camisia (C.) borealis (Thorell, 1871) Vaygach Island [29], Polar Urals [35]. Holarctic

Camisia (C.) horrida (Hermann, 1804) Vaygach Island [27], Yugor Peninsula [26], Bolshezemelskaya tundra [23], Polar Urals [33,35], Northern Urals [37]. Holarctic

Camisia (C.) invenusta (Michael, 1888) Polar Urals [35]. Palaearctic

Camisia (C.) segnis (Hermann, 1804) Polar Urals [35], Subpolar Urals (new data). Semicosmopolitan

Camisia (C.) spinifer (Koch, 1835) Polar Urals [35]. Semicosmopolitan

Camisia (Ensicamisia) lapponica (Trägårdh, 1910) Polar Urals [35]; Subpolar Urals (new data); Northern Urals [37]. Holarctic

Heminothrus (Capillonothrus) capillatus s. str. (Berlese, 1914) Kolguev Island (as *Heminothrus septentrionalis* Sellnick, 1944) [29]. Holarctic

Heminothrus (Capillonothrus) thori (Berlese, 1904) Polar Urals [35]. Holarctic

Heminothrus (Heminothrus) longisetosus (Willmann, 1925) Polar Urals [33,35,36]; Subpolar Urals (new data); Northern Urals [37]. Holarctic

Heminothrus (Platynothrus) humicola (Forsslund, 1955) Bolshezemelskaya tundra [24]. Holarctic

Heminothrus (Platynothrus) peltifer (Koch, 1839) Bolshezemelskaya tundra [24,25]; Polar Urals [33,35,36]; Subpolar Urals (new data); Northern Urals [37]. Semicosmopolitan

Heminothrus (Platynothrus) punctatus (L. Koch, 1879) Kolguev Island [29], Vaygach Island [29,30], Yugor Peninsula [26], Bolshezemelskaya tundra [23]. Holarctic

Nanhermanniidae Sellnick, 1928

Nanhermannia (Nanhermannia) dorsalis (Banks, 1896) Polar Urals [35]. Holarctic

Nanhermannia (N.) sellnicki Forsslund, 1958 Bolshezemelskaya tundra [25]; Subpolar Urals (new data); Northern Urals [37]. Palaearctic

Hermanniidae Sellnick, 1928

Hermannia (Heterohermannia) reticulata Thörell, 1871 Vaygach Island [27–29], Yugor Peninsula [26], Holarctic

Hermannia (Heterohermannia) scabra (L. Koch, 1879) Vaygach Island [27,28]. Holarctic

Licnobelbidae Grandjean, 1965

Licnobelba latiflabellata (Paoli, 1908) Kolguev Island (as *Licnobelba alestensis* Grandjean, 1931) [29]. Palaearctic

Hungarobelbidae Miko y Travé, 1996

Tokukobelba compta s. str. (Kulczynski, 1902) Yugor Peninsula (as *Belba (Belba) compta* (Kulczynski, 1902) [26], Polar Urals (as *B. (B.) compta* (Kulczynski 1902) [35], Subpolar Urals (new data). Palaearctic

Damaeidae Berlese, 1896

Belba spp. Kolguev Island [29], Bolshezemelskaya tundra [24]; Northern Urals [37].

Belba (Belba) rossica Bulanova-Zachvatkina, 1962 Polar Urals [33]. Palaearctic

Damaeus (Damaeus) auritus Koch, 1835 Polar Urals [35]. Palaearctic

Damaeus (Epidamaeus) sp. Polar Urals [35].

Damaeus (Epidamaeus) bituberculatus (Kulczynski, 1902) Bolshezemelskaya tundra [23], Polar Urals [35], Subpolar Urals (new data); Northern Urals [37]. Palaearctic

Damaeus (Spatiodamaeus) boreus Bulanova-Zachvatkina, 1957 Polar Urals [35]. Palaearctic

Metabelba (Metabelba) pulverulenta (Koch, 1839) Polar Urals [33]. Holarctic

Porobelba spinosa (Sellnick, 1920) Polar Urals [35]. Palaearctic

Subbelba (Subbelba) montana (Kulczynski, 1902) Polar Urals [35]. Palaearctic

Ceratoppiidae Grandjean, 1954

Ceratoppia bipilis s. str. (Hermann, 1804) Yugor Peninsula [26], Bolshezemelskaya tundra [23,24]; Polar Urals [33,35], Subpolar Urals (new data); Northern Urals [37]. Holarctic

Ceratoppia sphaerica (L. Koch, 1879) Vaygach Island [29], Yugor Peninsula [26], Polar Urals [32,35]. Holarctic

Ceratoppia quadridentata (Haller, 1882) Bolshezemelskaya tundra [23], Polar Urals [32,35], Subpolar Urals (new data). Holarctic

Pyroppia lanceolata Hammer, 1955 Yugor Peninsula [26]. Holarctic

Gustaviidae Oudemans, 1900

Gustavia microcephala (Nicolet, 1855) Vaygach Island [30]. Palaearctic

Liacaridae Sellnick, 1928

Adoristes (A.) ovatus poppei (Oudemans, 1906) Yugor Peninsula [26]. Holarctic

Liacarus (Dorycranosus) neonominatus Subías, 2004 Polar Urals [32,35]. Palaearctic

Xenillidae Woolley et Higgins, 1966

Xenillus (Xenillus) clypeator Robineau-Desvoidy, 1839 Polar Urals (as *Cepheus latus* Nicolet, 1855 [35]. Holarctic

Eremaeidae Oudemans, 1900

Eueremaeus oblongus s. str. (Koch, 1835) Vaygach Island [30]. Holarctic

Eueremaeus oblongus silvestris (Forsslund, 1956) Polar Urals [35]. Palaearctic

Tricheremaeus sp. Polar Urals [35].

Oribellidae Kunst, 1971

Proteremaeus sp. Polar Urals [35].

Autognetidae Grandjean, 1960

Autogneta (Conchogneta) traegardhi Forsslund, 1947 Polar Urals [31,35]. Holarctic

Thyrisomidae Grandjean, 1954

Banksinoma lanceolata s. str. (Michael, 1885) Vaygach Island [30], Bolshezemelskaya tundra [24]; Polar Urals [35]. Holarctic

Banksinoma setosa Rjabinin, 1974 Yugor Peninsula [26], Northern Urals [37]. Holarctic

Oppiidae Sellnick, 1937

Rhinoppia (Rhinoppia) subpectinata (Oudemans, 1900) Polar Urals [33,35], Subpolar Urals (new data); Northern Urals (as *Medioppia tuberculata* (Bulanova-Zachvatkina, 1964) [37]. Holarctic

Microppia minus s. str. (Paoli, 1908) Polar Urals [35]. Cosmopolitan

Berniniella (Berniniella) bicarinata (Paoli, 1908) Bolshezemelskaya tundra [23]. Palaearctic

Dissorhina ornata s. str. (Oudemans, 1900) Bolshezemelskaya tundra [24]. Holarctic

Lauroppia falcata (Paoli, 1908) Bolshezemelskaya tundra [23]. Palaearctic

Lauroppia fallax (Paoli, 1908) Polar Urals [35]. Semicosmopolitan

Lauroppia maritima acuminata (Strenzke, 1951) Polar Urals [35]. Holarctic

Oppiella (Moritzoppiella) neerlandica (Oudemans, 1900) (=*Dameosoma translamellatum* Willmann, 1923) Vaygach Island (as *Oppia translamellata*) [29], Yugor Peninsula (as *Moritzoppia (Moritzoppiella) neerlandica* (Oudemans, 1900) [26], Bolshezemelskaya tundra (as *Lauroppia neerlandica* (Oudemens, 1900) [23] and [24]; Polar Urals (as *L. neerlandica* (Oudemens, 1900) [33] and [31,35,36]; Subpolar Urals (new data); Northern Urals (as *L. neerlandica* (Oudemens, 1900) [37]. Holarctic

Oppiella (Oppiella) nova s. str. (Oudemans, 1902) Vaygach Island [29,30], Yugor Peninsula [26], Polar Urals [31,35,36]; Subpolar Urals (new data). Cosmopolitan

Oppiella (O.) splendens (Koch, 1841) Kolguev Island [29], Vaygach Island [29], Polar Urals [35]. Holarctic

Moritzoppia unicarinata s. str. (Paoli, 1908) Vaygach Island [29], Polar Urals [31,35], Northern Urals [37]. Holarctic

Moritzoppia unicarinata clavigera (Hammer, 1952) Yugor Peninsula [26]. Holarctic

Quadroppiidae Balogh, 1983

Quadroppia (Quadroppia) quadricarinata (Michael, 1885) Vaygach Island [29,30], Yugor Peninsula [26], Bolshezemelskaya tundra [24]; Polar Urals [31,33,35,36]; Subpolar Urals (new data). Semicosmopolitan

Suctobelbidae Jacot, 1938

Suctobelbella (Suctobelbella) acutidens s. str. (Forsslund, 1941) Yugor Peninsula [26], Bolshezemelskaya tundra [23], Polar Urals [33,35]. Holarctic

Suctobelbella (S.) acutidens duplex (Strenzke, 1950) Yugor Peninsula (as *S. hammerae* Krivolutsky, 1965) [26], Bolshezemelskaya tundra (as *S. hammerae* Krivolutsky, 1965) [25]; Polar Urals [33], Subpolar Urals (new data). Holarctic

Suctobelbella (S.) acutidens sarekensis (Forsslund, 1941) Polar Urals [35] Holarctic

Suctobelbella (S.) arcana Moritz, 1970 Polar Urals [35]. Holarctic

Suctobelbella (S.) latirostris (Strenzke, 1950) Subpolar Urals (new data). Palaearctic

Suctobelbella (S.) longicuspis s. str. Jacot, 1937 Polar Urals [35]. Semicosmopolitan

Suctobelbella (S.) longirostris (Forsslund, 1941) Polar Urals [35]. Holarctic

Suctobelbella (S.) palustris (Forsslund, 1951) Bolshezemelskaya tundra [23]. Holarctic

Suctobelbella (S.) singularis (Strenzke, 1950) Bolshezemelskaya tundra [23], Subpolar Urals (new data). Palaearctic

Suctobelbella (S.) subcornigera s. str. (Forsslund, 1941) Polar Urals [35]. Semicosmopolitan

Suctobelbella (S.) subcornigera vera (Moritz, 1964) Polar Urals [35]. Palaearctic

Suctobelba spp. Vaygach Island [30], Bolshezemelskaya tundra [Melekhina, unpabl.].

Suctobelba trigona (Michael, 1888) Vaygach Island [29]. Holarctic

Carabodidae Koch, 1843

Carabodes (Carabodes) areolatus Berlese, 1916 Vaygach Island [30], Bolshezemelskaya tundra [23], Polar Urals [31–33]. Holarctic

Carabodes (C.) labyrinthicus (Michael, 1879) Kolguev Island [29], Vaygach Island [29], Polar Urals [33,35,36]; Subpolar Urals (new data). Holarctic

Carabodes (C.) marginatus (Michael, 1884) Yugor Peninsula [26]; Bolshezemelskaya tundra [24]; Polar Urals [31]; Subpolar Urals (new data). Palaearctic

Carabodes (C.) subarcticus Trägardh, 1902 Yugor Peninsula [26], Polar Urals [33,35,36]; Subpolar Urals (new data); Northern Urals [37]. Palaearctic

Tectocepheidae Grandjean, 1954

Tectocepheus minor Berlese, 1903 Kolguev Island [29]. Semicosmopolitan

Tectocepheus velatus s. str. (Michael, 1880) Kolguev Island [29], Vaygach Island [29,30], Yugor Peninsula [26], Bolshezemelskaya tundra [23], Polar Urals [31,33,35,36]; Subpolar Urals (new data); Northern Urals [37]. Cosmopolitan

Tectocepheus velatus sarekensis Trägårdh, 1910 Polar Urals [31,35]. Cosmopolitan

Hydrozetidae Grandjean, 1954

Hydrozetes thienemanni Strenzke, 1943 Yugor Peninsula [26]. Holarctic

Ameronothridae Vitzthum, 1943

Ameronothrus lineatus (Thorell, 1871) Vaygach Island [27]. Holarctic

Ameronothrus nigrofemoratus (L. Koch, 1879) Vaygach Island [27,28]. Holarctic

Micreremidae Grandjean, 1954

Micreremus brevipes (Michael, 1888) Polar Urals [35]. Palaearctic

Scutoverticidae Grandjean, 1954

Scutovertex minutus (Koch, 1835) Vaygach Island [30]. Holarctic

Scutovertex neonominatus Subías, 2004 Bolshezemelskaya tundra [Melekhina, unpabl.]. Palaearctic

Exochocepheus laticuspis (Balogh et Mahunka, 1965) Polar Urals [35]. Palaearctic

Passalozetidae Grandjean, 1954

Bipassalozetes (Bipassalozetes) intermedius (Mihelčič, 1954) Polar Urals [35]. Palaearctic

Phenopelopidae Petrunkevitch, 1955

Eupelops plicatus (Koch, 1835) (=*Pelops auritus* Koch, 1839) Bolshezemelskaya tundra (as *Eupelops auritus* Koch, 1839) [23], [24]; Polar Urals [33,35], Northern Urals [37]. Holarctic

Eupelops torulosus s. str. (Koch, 1839) Polar Urals [35]. Palaearctic

Achipteriidae Thor, 1929

Campachipteria (Triachipteria) nivalis (Hammer, 1952) (Achipteria) Polar Urals [33]. Holarctic

Parachipteria punctata (Nicolet, 1855) Vaygach Island [29], Polar Urals [33], Subpolar Urals (new data). Holarctic

Achipteria sp. Vaygach Island [30].

Achipteria (Achipteria) nitens (Nicolet, 1855) Polar Urals [35]. Holarctic

Anachipteria (Anachipteria) howardi (Berlese, 1908) Polar Urals [35]. Holarctic

Tegoribatidae Grandjean, 1954

Scutozetes lanceolatus Hammer, 1952 Bolshezemelskaya tundra [23]. Holarctic

Tegoribates latirostris (Koch, 1844) Bolshezemelskaya tundra [23]. Palaearctic

Ceratozetidae Jacot, 1925

Ceratozetella (Ceratozetella) sellnicki (Rajski, 1958) Vaygach Island [29], Polar Urals [35]. Palaearctic

Ceratozetes sp. Polar Urals [36].

Ceratozetes (Ceratozetes) gracilis s. str. (Michael, 1884) Yugor Peninsula [26], Subpolar Urals (new data). Cosmopolitan

Edwardzetes (Edwardzetes) edwardsi (Nicolet, 1855) Yugor Peninsula [26], Bolshezemelskaya tundra [23], Polar Urals [35,36]. Holarctic

Fuscozetes fuscipes (Koch, 1844) Bolshezemelskaya tundra [25]; Polar Urals [33]. Holarctic

Fuscozetes pseudosetosus Shaldybina, 1975 Bolshezemelskaya tundra [Melekhina, unpabl.]. Holarctic

Fuscozetes setosus (Koch, 1839) Polar Urals [31]. Holarctic

Melanzetes sp. Polar Urals [35].

Melanozetes mollicomus (Koch, 1839) Yugor Peninsula [26], Northern Urals [37]. Holarctic

Melanozetes sellnicki (Hammer, 1952) Kolguev Island (as *Fuscozetes sellnicki* Hammer, 1952) [29], Vaygach Island (as *Fuscozetes sellnicki* Hammer, 1952) [29], Polar Urals [35], Subpolar Urals (new data). Holarctic

Trichoribates sp. Polar Urals (as *Murcia (M.)* sp.) [35].

Trichoribates (T.) novus s. str. (Sellnick, 1928) Yugor Peninsula (as *Murcia nova* Sellnick, 1928) [26], Bolshezemelskaya tundra [24]; Polar Urals (as *Murcia* (*M.*) *nova* Sellnick 1928) [33]. Holarctic

Trichoribates (T.) berlesei (Jacot, 1929) Vaygach Island [30], Bolshezemelskaya tundra (as *Trichoribates trimaculatus* Koch, 1835) [23]. Holarctic

Oromurcia lucens (L. Koch, 1879) Vaygach Island [27]. Holarctic

Trichoribates (Latilamellobates) incisellus s. str. (Kramer, 1897) Bolshezemelskaya tundra [Melekhina, unpabl.], Northern Urals [37]. Holarctic

Sphaerozetes arcticus Hammer, 1952 Polar Urals [33]. Holarctic

Chamobatidae Thor, 1937

Chamobates (C.) cuspidatus (Michael, 1884) Bolshezemelskaya tundra [24]; Polar Urals [33]. Holarctic

Chamobates (C.) lapidarius (Lucas, 1849) Vaygach Island [29]. Palaearctic

Chamobates (C.) pusillus (Berlese, 1895) Vaygach Island (as *C. borealis*) [28], [29]. Palaearctic

Chamobates (Xiphobates) voigtsi (Oudemans, 1902) Polar Urals [35]. Palaearctic

Humerobatidae Grandjean, 1971

Diapterobates humeralis (Hermann, 1804) Polar Urals [33,35]. Holarctic

Diapterobates notatus (Thorell, 1871) Kolguev Island [29], Vaygach Island [27,29,30], Yugor Peninsula [26]. Holarctic

Diapterobates oblongus (L. Koch, 1879) Subpolar Urals (new data); Northern Urals [37]. Palaearctic

Diapterobates variabilis s. str. Hammer, 1955 Polar Urals [35]. Holarctic

Svalbardia paludicola Thor, 1930 Yugor Peninsula [26], Bolshezemelskaya tundra [Melekhina, unpabl.], Polar Urals [35]. Holarctic

Punctoribatidae Thor, 1937

Mycobates (Mycobates) monodactylus Shaldybina, 1970 Polar Urals [31,33,35]. Palaearctic

Mycobates (Calyptozetes) patrius Shaldybina, 1970 Polar Urals [33]. Palaearctic

Mycobates (Calyptozetes) sarekensis (Trägårdh, 1910) Polar Urals [35]. Holarctic

Minunthozetes (M.) pseudofusiger (Schweizer, 1922) Vaygach Island [29], Yugor Peninsula [26]. Palaearctic

Punctoribates (P.) punctum (Koch, 1839) Bolshezemelskaya tundra [23]. Semicosmopolitan

Oribatulidae Thor, 1929

Oribatula (Oribatula) tibialis (Nicolet, 1855) Vaygach Island [29], Yugor Peninsula [26], Bolshezemelskaya tundra [23], Polar Urals [31–33,35], Subpolar Urals (new data); Northern Urals [37]. Holarctic

Oribatula (Zygoribatula) exilis s. str. (Nicolet, 1855) Vaygach Island [27,30], Yugor Peninsula [26], Bolshezemelskaya tundra [23], Polar Urals [33,35], Subpolar Urals (new data); Northern Urals [37]. Holarctic

Hemileiidae J. et P. Balogh, 1984

Hemileius (Hemileius) initialis (Berlese, 1908) Bolshezemelskaya tundra (as *Scheloribates confundatus* Sellnick, 1928) [23]; Polar Urals (as *S. confundatus* Sellnick, 1928) [33] and [36]. Semicosmopolitan

Liebstadiidae J. et P. Balogh, 1984

Liebstadia (L.) pannonica (Willmann, 1951) Bolshezemelskaya tundra (as *Protoribates pannonicus* Willmann, 1951) [23]. Holarctic

Liebstadia (L.) similis (Michael, 1888) Kolguev Island [29], Vaygach Island [29,30], Polar Urals [33, 35]. Holarctic

Scheloribatidae Grandjean, 1933

Scheloribates (S.) laevigatus s. str. (Koch, 1835) Vaygach Island [30], Yugor Peninsula [26], Subpolar Urals (new data); Northern Urals [37]. Semicosmopolitan

Scheloribates (S.) pallidulus latipes (Koch, 1844) Polar Urals [33,35,36]; Northern Urals [37]. Semicosmopolitan

Haplozetidae Grandjean, 1936

Peloribates spp. Vaygach Island [29,30].

Peloribates (Peloribates) canadensis Hammer, 1952 Polar Urals [35]. Holarctic

Peloribates (P.) europaeus Willmann, 1935 Bolshezemelskaya tundra [23]. Holarctic

Peloribates (P.) pilosus Hammer, 1952 Bolshezemelskaya tundra [23], Northern Urals [37]. Holarctic

Parakalummidae Grandjean, 1936

Neoribates (Neoribates) aurantiacus (Oudemans, 1914) Polar Urals [33,35], Subpolar Urals (new data). Holarctic

Galumnidae Jacot, 1925

Galumna (Galumna) dimorpha Krivolutskaja, 1952 Bolshezemelskaya tundra [Melekhina, unpabl.]. Palaearctic

Galumna (G.) rossica Sellnick, 1926 Polar Urals [35]. Palaearctic

Pergalumna sp. Bolshezemelskaya tundra [31].

Pergalumna (P.) nervosa s. str. (Berlese, 1914) Bolshezemelskaya tundra [23]. Holarctic

Pilogalumna sp. Polar Urals [33].

References

1. Krivolutsky, D.A.; Kalyakin, V.N. *Microfauna of Soils in Ecological Control in Novaya Zemlya*; Nauka: Novaya Zemlya, Moscow, 1993; Volume 2, pp. 125–131. (In Russian)
2. Lebedeva, N.V.; Lebedev, V.D.; Melekhina, E.N. New data on the oribatid mite (Oribatei) fauna of Svalbard. *Doklady Biol. Sci.* **2006**, *407*, 182–186. [CrossRef] [PubMed]
3. Gwiazdowicz, D.J.; Gulvik, M.E. Mesostigmatid mites (Acari, Mesostigmata) in Svalbard. In Proceedings of the Materia y XXXII Sympozjum Polarnego, Toruń, Poland, 15–17 October 2020; pp. 32–34.

4. Gwiazdowicz, D.J.; Teodorowicz, E.; Coulson, S.J. Redescription of Zercon solenites Haarløv, 1942 (Acari: Mesostigmata: Zerconidae) with a key to the Svalbard species of the genus *Zercon*. *Int. J. Acarol.* **2011**, *37*, 135–148. [CrossRef]
5. Gwiazdowicz, D.J.; Zawieja, B.; Olejniczak, I.; Skubała, P.; Gdula, A.K.; Coulson, S.J. Changing Microarthropod Communities in Front of a Receding Glacier in the High Arctic. *Insects* **2020**, *11*, 226. [CrossRef] [PubMed]
6. Avila-Jimenez, M.L.; Gwiazdowicz, D.J.; Coulson, S.J. The mesostigmatid mite (Acari: Parasitiformes) fauna of Svalbard: A revised inventory of a high Arctic archipelagos. *Zootaxa* **2011**, *3091*, 33–41. [CrossRef]
7. Bayartogtokh, B.; Schatz, H.; Ekrem, T. Distribution and diversity of the soil mites of Svalbard, with redeskriptions of three known species (Acari: Oribatida). *Int. J. Acarol.* **2011**, *37*, 467–484. [CrossRef]
8. Melekhina, E.N.; Ryabinin, N.A. Analysis fauna oribatid mites (Oribatida) in tundra Russian zone. In *Biodiversity of Ecosystems of the Far North: Inventory, Monitoring, Protection*; Materials of reports of the All-Russian Conference; Institute of Biology of Komi Science Center of Ural Branch of Russian Academy of Sciences: Syktyvkar, Russia, 2013; pp. 462–463.
9. Makarova, O.L. Gamasid mites (Parasitiformes, Mesostigmata) of the European Arctic and their distribution patterns. *Entomol. Rev.* **2013**, *93*, 113–133. [CrossRef]
10. Makarova, O.L. The fauna of free-living mites (Acari) of Greenland. *Entomol. Rev.* **2015**, *95*, 108–125. [CrossRef]
11. Coulson, S.J.; Convey, P.; Aakra, K.; Aarvik, L.; Avila-Jimenez, M.L.; Babenko, A.; Biersma, E.; Bostrom, S.; Brittain, J.; Carlsson, A.M.; et al. The terrestrial and freshwater invertebrate biodiversity of the archipelagoes of the Barents Sea; Svalbard, Franz Josef Land and Novaya Zemlya. *Soil Biol. Biochem.* **2014**, *68*, 440–470. [CrossRef]
12. Coulson, S.J.; Schatz, H.; Gwiazdowicz, D.J.; Solhoy, T. The oribatid and mesostigmatid mite fauna (Acari) of the High Arctic island of Hopen. *Pol. Polar Res.* **2014**, *35*, 133–213. [CrossRef]
13. Coulson, S.J.; Fjellberg, A.; Melekhina, E.N.; Taskaeva, A.A.; Lebedeva, N.V.; Belkina, O.A.; Seniczak, S.; Seniczak, A.; Gwiazdowicz, D.J. Microarthropod communities of industrially disturbed or imported soils in the High Arctic; the abandoned coal mining town of Pyramiden, Svalbard. *Biodivers. Conserv.* **2015**, *24*, 1671–1690. [CrossRef]
14. Babenko, A.B.; Potapov, M.B.; Taskaeva, A.A. The collembolan fauna of the East- European tundra. *Russ. Entomol. J.* **2017**, *26*, 1–30. [CrossRef]
15. Lebedeva, N.V.; Krivolutsky, D.A. Bird spread soil microarthropods to Arctic Islands. *Doklady Biol. Sci.* **2003**, *391*, 329–332. [CrossRef] [PubMed]
16. Avila-Jimenez, M.L.; Coulson, S.J. A Holarctic biogeographical analysis of the Collembola (Arthropoda, Hexapoda) unravels recent Post-Glacial colonization patterns. *Insects* **2011**, *2*, 273–296. [CrossRef] [PubMed]
17. Coulson, S.J.; Fjellberg, A.; Gwiazdowicz, D.J.; Lebedeva, N.V.; Melekhina, E.N.; Solhøy, T.; Erséus, C.; Maraldo, K.; Miko, L.; Schatz, H.; et al. Introduction of invertebrates into the High Arctic via imported soils: The case of Barentsburg in the Svalbard. *Biol. Invasions* **2013**, *15*, 1–5. [CrossRef]
18. Avila-Jimenez, M.L.; Solhoy, T.; Gwiazdowicz, D.J.; Fjellberg, A.; Dózsa-Farkas, K.; Monson, F.; De Smet, W.H.; Stur, E.; Ekrem, T.; Coulson, S.J. The terrestrial invertebrate fauna of Edgeøya, Svalbard: Arctic landscape community composition reflects biogeography patterns. *Polar Biol.* **2019**, *42*, 837–850. [CrossRef]
19. Makarova, O.L.; Anufriyev, V.V.; Babenko, A.B.; Bizin, M.S.; Glazov, P.M.; Kolesnikova, A.A.; Marusik, Y.M.; Tatarinov, A.G. Fauna of the East European tundra: The input of "Siberian" species. *Bull. North-East Sci. Cent. Russia Acad. Sci. Far East Branch* **2019**, *1*, 59–71. (In Russian) [CrossRef]
20. Rozhnov, V.V.; Lavrinenko, I.A.; Razzhivin, V.Y.; Makarova, O.L.; Lavrinenko, O.V.; Anufriev, V.V.; Babenko, A.B.; Bizin, M.S.; Glazov, P.M.; Goryachkin, S.V.; et al. Biodiversity revision of a large arctic region as a basis for its monitoring and protection under conditions of active economic development (Nenetsky autonomous okrug, Russia). *Nat. Conserv. Res. Zapovednaya nauka* **2019**, *4*, 1–28. (In Russian) [CrossRef]
21. Coulson, S.J.; Fjellberg, A.; Gwiazdowicz, D.J.; Lebedeva, N.V.; Melekhina, E.N.; Solhøy, T.; Erséus, C.; Maraldo, K.; Miko, L.; Schatz, H.; et al. The invertebrate fauna of anthropogenic soils in the High Arctic settlement of Barentsburg; Svalbard. *Polar Res.* **2013**, *32*, 19273. [CrossRef]
22. Melekhina, E.N. Effect of pollucion upon the fauna of Oribatid mites in the Vorkuta tundra. In *Cryopedology'97*; Abstracts of the International Conference; Institute of Biology of Komi Science Center of Ural Branch of Russian Academy of Sciences: Syktyvkar, Russia, 1997; pp. 176–177. (In Russian)

23. Melekhina, E.N. Recovery succession soil microfauna in oil-polluted ecosystems of European Subarctic. *Biol. Bull.* **2020**, *47*, 97–105. [CrossRef]
24. Koch, L. *Arachniden aus Sibirien und Novaja Semlja eingesammelt von der schwedischen Expedition im Jahre 1875*; Kongliga Svenska Vetenskaps Academiens Handlingar: Stockholm, Sweden, 1879; pp. 1–136.
25. Trägårdh, I. Monographie der arktischen Acariden. *Fauna Arctica* **1904**, *4*, 1–78.
26. Krivolutsky, D.A.; Drozdov, N.N.; Lebedeva, N.V.; Kalyakin, V.N. Geography of soil microarthropods on the Arctic Islands. *Moscow Univ. Geogr. Bull.* **2003**, *6*, 33–40. (In Russian)
27. Melekhina, E.N.; Matyukhin, A.V.; Glazov, P.M. Oribatid mites in nests of the Lapland Bunting (Calcarius lapponicus) on the arctic island of Vaygach (with analysis of the islands fauna). *Proc. Komi Sci. Cent. Ural. Branch Russ. Acad. Sci.* **2019**, *8*, 108–122. (In Russian)
28. Melekhina, E.N.; Zinovyeva, A.N. First data on oribatid mites (Acari: Oribatida) of Pay-Khoy ridge (Yugor peninsula). *Proc. Komi Sci. Cent. Ural. Branch Russ. Acad. Sci.* **2012**, *2*, 42–50. (In Russian)
29. Melekhina, E.N. Oribatid mites. In *Tundra Animals. Vorkuta—A City on Coal, a City in the Arctic*; Getsen, M.V., Ed.; Ministry of Natural Resources and Environmental Protection of the Komi Republic: Syktyvkar, Russia, 2011; pp. 172–175. (In Russian)
30. Melekhina, E.N.; Krivolutsky, D.A. *List of the Oribatid Mites in the Republic of Komi*; Komi Science Center of Ural Branch of Russian Academy of Sciences: Syktyvkar, Russia, 1999; pp. 1–24. (In Russian)
31. Goryachkin, S.V.; Gilichinskii, G.A.; Gubin, S.V.; Lapteva, E.M.; Lupachev, A.V.; Fedorov-Davydov, D.G.; Zazovskaya, E.P.; Mazhitova, G.G.; Mergelov, N.S.; Vinogradova, Y.A.; et al. A state of the Arctic and Subarctic soils with regard to the period of International Polar Year. In *Polyarnaya Kriosfera i Vody Sushi*; Paulsen: Moscow, Russia, 2011; pp. 193–218. (In Russian)
32. Zubrii, N.A.; Kolosova, Y.C.; Taskaeva, A.A.; Melekhina, E.N. Activity of soil and herpetobiont invertebrates during winters. In *Funktsionirovanie Subarkticheskoi Gidrotermal'noi Ekosistemy v Zimnii Period*; Bogolitcin, K.G., Bolotov, I.N., Eds.; UrO RAN: Ekaterinburg, Russia, 2011; pp. 183–193. (In Russian)
33. Karpova, N.I.; Poryadina, N.M. Oribatid mites in the forest-tundra soils Tyumen region. In *Problems of Soil Zoology (Proceedings of the VI All-Union Conference)*; Science and Technology: Minsk, Belarus, 1978; pp. 108–109. (In Russian)
34. Grishina, L.G. Oribatid mites of the North of Siberia. In *Arthropods of Siberia and the Far East*; Nauka: Novosibirsk, Russia, 1985; pp. 14–24. (In Russian)
35. Getsen, M.V. *Biodiversity of the Polar Urals Ecosystems*; Institute of Biology of Komi Science Center of Ural Branch of the Russian Academy of Sciences: Syktyvkar, Russia, 2007; pp. 1–251. (In Russian)
36. Melekhina, E.N. Oribatid mites in western slopes of the Polar Urals and Northern Urals. In *Biodiversity, Environmental Problems of the Altai Mountains and Adjacent Regions: Present, Past, Future*; Materials of the International Conference; Part I; Gorno-Altaisk State University: Gorno-Altaisk, Russia, 2008; p. 324. (In Russian)
37. Sidorchuk, E.A. New data on the fauna of Oribatid mites (Acari, Oribatida) from the Polar Urals. *Entomol. Rev.* **2009**, *89*, 554–563. [CrossRef]
38. Degteva, S.V. The Biological Diversity of Komi Republic Protected Areas. Issue 7. In *Natural Complexes of "Khrebtovy" Wildlife Preserve*; Komi Science Center of Ural Branch of Russian Academy of Sciences: Syktyvkar, Russia, 2010; pp. 1–140. (In Russian)
39. Melekhina, E.N. On the fauna of the oribatid mites (Acari: Oribatida) of the Pechoro-Ilychsky Nature Reserve. In Proceedings of the Pechoro-Ilychsky Nature Reserve; Komi Science Center of Ural Branch of Russian Academy of Sciences: Syktyvkar, Russia, 2005; Volume 14, pp. 113–117. (In Russian).
40. Melekhina, E.N. Taxonomic diversity and areology of oribatid mites (Oribatei) of the European North of Russia. *Proc. Komi Sci. Cent. Ural Division Russ. Acad. Sci.* **2011**, *2*, 30–37. (In Russian)
41. Krivolutsky, D.A.; Lebren, F.; Kunst, M.; Akimov, I.A.; Bayartogtokh, B.; Vasiliu, N.; Golosova, L.D.; Grishina, L.G.; Karppinen, E.; Kramnoy, V.J.; et al. *Oribatid Mites: Morphology, Development, Phylogeny, Ecology, Research Methods, and Characteristic of the Model Species Nothrus palustris C.L. Koch, 1839*; Nauka: Moscow, Russia, 1995; pp. 1–223. (In Russian)
42. Gilyarov, M.S. The Key to Identify Soil Mites. In *Sarcoptiformes*; Nauka: Moscow, Russia, 1975; pp. 1–488. (In Russian)

43. Subías, L.S. Listado Sistemático, Sinonímico y Biogeográfico de Los Ácaros Oribátidos (Acariformes: Oribatida) del Mundo (Excepto Fósiles). Graellsia, 60 (Número Extraordinario). 2004; 305p. Available online: http://bba.bioucm.es/cont/docs/RO_1.pdf (accessed on 31 March 2019).
44. Karppinen, E.; Krivolutsky, D.A. List of oribatid mites (Acarina, Oribatei) of northern palaearctic region. 1. Europe. *Acta Entomol. Fenn.* **1982**, *41*, 1–18.
45. Golosova, L.; Karppinen, E.; Krivolutsky, D.A. List of oribatid mites (Acarina, Oribatei) of northern palaearctic region. II. Siberia and the Far East. *Acta Entomol. Fenn.* **1983**, *43*, 1–14.
46. Karppinen, E.; Krivolutsky, D.A.; Poltavskaja, M.P. List of oribatid mites (Acarina, Oribatei) of northern palaearctic region. III. Arid lands. *Ann. Entomol. Fenn.* **1986**, *52*, 81–94.
47. Behan-Pelletier, V. Oribatid mites (Acari: Oribatida) of the Yukon. In *Insects of the Yukon*; Danks, H.V., Downes, J.A., Eds.; Biological Survey of Canada: Ottawa, ON, Canada, 1997; pp. 115–149.
48. Ryabinin, N.A.; Pankov, A.N. *Catalogue of Oribatid Mites of the Far East of Russia. Part II. Continental Part of the Far East*; Far Eastern Branch of the Russian Academy of Sciences: Vladivostok-Khabarovsk, Russia, 2002; pp. 1–92.
49. Danks, H.V. *Arctic artropods. A Reviev of Systematics and Ecology with Particular Reference to the North American Fauna*; Entomological Society of Canada: Ottawa, ON, Canada, 1981; pp. 1–608.
50. Zenkova, I.V.; Zaitsev, A.S.; Zalish, L.V.; Liskovaya, A.A. Soil-dwelling oribatid mites (Acariformes: Oribatida) of the taiga and tundra zones of the Murmansk Region. *Proc. Komi Sci. Cent. Ural. Branch Russ. Acad. Sci.* **2011**, *1*, 54–67. (In Russian)
51. Tatarinov, A.G. *Geography of Lepidoptera Diurnalis in the European Northeast of Russia*; KMK Scientific Publications: Moscow, Russia, 2016; pp. 1–255.
52. Melekhina, E.N. Variety of fauna and geographical distribution of the oribatid mites (Oribatida) of the taiga zone of the European North-East. In *Regularities in the Zonal Organization of Complexes of the Animal Population in the European North-East*; Komi Science Center of Ural Branch of Russian Academy of Sciences: Syktyvkar, Russia, 2005; Volume 177, pp. 258–274. (In Russian)

diversity

MDPI

Article

Diversity of Soil Gamasine Mites (Acari: Mesostigmata: Gamasina) in an Area of Natural Vegetation and Cultivated Areas of the Cerrado Biome in Northern Brazil

Emiliano Brandão de Azevedo [1,2,*], Letícia Henrique Azevedo [1], Grazielle Furtado Moreira [3], Fábio Araújo dos Santos [2], Marcos Alberto Francisco de Carvalho [2], Renato de Almeida Sarmento [2] and Raphael de Campos Castilho [1,4]

1 Department of Agricultural Sciences, Faculdade de Ciências Agrárias e Veterinárias (FCAV), Universidade Estadual Paulista (UNESP), 14884-900 Jaboticabal, São Paulo, Brazil; leticiaazevedo@usp.br (L.H.A.); raphael.castilho@usp.br (R.d.C.C.)
2 Universidade Federal do Tocantins (UFT), P.O. Box 66, 77402-970 Gurupi, Tocantins, Brazil; fabioaraujomf@outlook.com (F.A.d.S.); marcos_rbk@hotmail.com (M.A.F.d.C.); rsarmento@uft.edu.br (R.d.A.S.)
3 Centro Universitário Adventista de São Paulo (UNASP), 13445-970 Engenheiro Coelho, São Paulo, Brazil; grabiologia@yahoo.com.br
4 Department of Entomology and Acarology, Escola Superior de Agricultura "Luiz de Queiroz" (ESALQ), Universidade de São Paulo (USP), 13418-900 Piracicaba, São Paulo, Brazil
* Correspondence: emilianogpi@yahoo.com.br

Received: 15 July 2020; Accepted: 26 August 2020; Published: 29 August 2020

check for updates

Abstract: The Brazilian Cerrado biome has undergone major changes, with the incorporation of new areas for agricultural production. While this can certainly provide for the worldwide growing need for agricultural products, especially food, care should be taken to prevent possible environmental degradation. Worldwide, mites of the cohort Gamasina constitute the most abundant and diverse group of soil predatory mites, usually considered important in maintaining the ecological balance of natural environments. Little is known about the abundance and diversity of Gamasina in the Cerrado. The objective of the present work was to evaluate the abundance and diversity of Gamasina in soils of natural vegetation and of agroecosystems in Cerrado areas of the northern Brazilian state of Tocantins. This is considered the first step in the determination of possible role of the local predators as biological control agents, and their potential for practical use locally and elsewhere. Soil samples were taken monthly between July 2015 and June 2016. In total, 1373 Gamasina representing 45 species of 24 genera and 9 families were collected. The most abundant Gamasina belonged to Rhodacaridae in areas of the natural vegetation and to Ascidae in the agroecosystems. Abundance and diversity were much higher in the rainy than in the dry season. Rhodacarids and ascids have not been used commercially for pest control, but investigations conducted so far suggest their potential as biological control agents. The confirmation of this possibility and the development of techniques that would allow their maintenance in agricultural areas require subsequent research efforts.

Keywords: mesofauna; edaphic mites; prospection

1. Introduction

Cerrado is the second largest biome in Brazil, occupying an area of 2,036,448 km^2 or about 22% of the Brazilian territory [1]. The vegetation of this biome is predominantly constituted by relatively sparse and short twisted trees and shrubs, with scant grass species [2]. This biome is undergoing major changes, for the expansion of the agricultural frontier from the early 2000s [1]. Large parts of the Cerrado are presently dedicated to the production of soybean, corn, rice, pasture, and cultivated forest trees, such as eucalyptus [3]. While this can certainly provide for the worldwide growing need for agricultural products, especially food, care should be taken to prevent possible environmental degradation.

Despite the efforts for the establishment of protected areas, the survival of endemic animal and plant species could be threatened in this biome [4]. Attention should be given to organisms living in different parts of the environment, including those in the soil, that are usually difficult to account, largely because of the difficulty in determining these species taxonomically. Soils of the Cerrado usually have low pH and low levels of plant nutrients, leading producers to use high inputs of lime and fertilizers in agricultural lands, modifying the natural edaphic environment [2].

Soils from natural environments generally show high mesofauna diversity and abundance. These organisms actively participate in the decomposition of organic matter, affecting soil aeration, chemical properties, and the ecological balance. Some members of the mesofauna act as predators of invertebrates, interfering in community structures and probably in the sustainability of agroecosystems [5,6].

The predominant members of the mesofauna may vary according to prevailing biotic and abiotic factors. Some studies of the edaphic fauna have been conducted to analyze the influence of agricultural practices on the main taxonomic groups, particularly mites and springtails, among other invertebrates [7–9]. Knowledge about the abundance and diversity of these organisms in a given habitat has been used as an indicator of soil quality, which can be related to levels of sustainability and affect crop yield [10].

Worldwide, the Gamasina (Mesostigmata) constitute the most abundant cohort of edaphic predatory mites [11]. These can feed on several organisms, such as other mites, small insects, and nematodes, but frequently also on fungi [11–13]. Therefore, it is believed that they can play an important role in reducing the population of agricultural pests, and some of them have been used for pest control [13].

The first step in estimating the natural effect of edaphic Gamasina on pest organisms in a specific location, as well as the potential for their practical use as biological control agents, is the determination of the components of the prevailing fauna [14]. Information on the abundance and diversity of Gamasina in agricultural crops in tropical regions is scarce, especially in the Cerrado. The objective of the present work was to determine the abundance and diversity of Gamasina in soils of natural vegetation and in agroecosystems in Cerrado areas of the northern Brazilian state of Tocantins.

2. Material and Methods

The study was conducted in four ecosystems in the municipality of Sucupira, Tocantins state, where the climate is of the Aw type, according to the Köppen-Geiger classification system [15]. The region is characterized as a subtropical savannah, with well-defined rainy (November–April, cooler months) and dry (May–October, hottest months) seasons, average annual rainfall of about 1500 mm and air relative humidity of 45–90%. Distances between ecosystems were 4.0–7.5 km.

2.1. Characterization of the Ecosystems

Natural vegetation (11°54′36′′ S, 48°51′10′′ W): A 208-ha patch of well-preserved vegetation of the Cerrado biome, with a litter layer of about 5 cm in thickness. Within the natural variations of this biome, the area selected for study is classified as Cerrado *sensu stricto* [2].

Soybean cultivation (11°54′80′′ S, 48°52′10′′ W): A 155-ha soybean plot, where the same crop has been yearly cultivated since 2003. Liming of the soil was done three months before the soybean planting. The land had been in fallow for about four months before soybean (variety M8644 IPRO from Monsoy) was seeded on 15 November 2015, under no tillage, spaced at 50 cm between lines and at a density of 13 plants/m in the lines. The area was sprayed with glyphosate (3.0 L ha^{-1}) before planting and glyphosate - potassium salt 620 (2.0 L ha^{-1}) with mineral oil (1.0 L ha^{-1}) 26 days after plant emergence. In the crop cycle, plants were sprayed with propiconazole (0.4 L ha^{-1}) for the control of Asian rust, and with thiamethoxam (250 mL L^{-1}) and acefato (300 mL L^{-1}) for the control of caterpillars and bugs. Harvesting was done mechanically on March 24, 2016, leaving the area in fallow again until the end of the study.

Pasture (11°56′80′′ S; 48°52′34′′ W): A 3.5-ha plot planted to the grass *Andropogon gayanus* Kunth in 2012. The area was sprayed with 2.4-D, Picloran (3.0 L ha^{-1}). Liming was last done in the area in August 2015 and NPK 20-00-20 + 0.1% B (120 kg ha^{-1}) was applied in November 2014. No chemical was applied for the control of pests and diseases.

Integration of cultivated forest–pasture (11°57′57′′ S; 48°54′34′′ W): A 25-ha plot was occupied by an association of eucalyptus (*Eucalyptus grandis* W. Hill) and pasture (*A. gayanus*) since 1998. Liming and application of NPK 20-00-20 + 0.1% B (120 kg ha^{-1}) were done respectively in August and November 2015. Fipronil (15 g/plant) for ant control and 2.4-D and picloran (3.0 L ha^{-1}) for weed control were applied in May 2016.

2.2. Edapho-Climatic Characterization

Soil samples were collected from each ecosystem for physicochemical analyses in the rainy (January 2016) and dry (June 2016) seasons. In each season and in each ecosystem, a sample was composed of 20 subsamples (at least 20 m apart from each other) and each taken at a depth of 10 cm. The soil of all areas of study was classified as Dystrophic Red Yellow Latosol (LVA) [16]. The results of the analyses are shown in Table 1.

Table 1. Physicochemical properties of soil samples (each composed of 20 subsamples) collected at four ecosystems in Sucupira, Tocantins, Brazil. Samples in January 2015 (rainy season) and July 2016 (dry season). Analyses conducted by Sellar Análises Agrícolas, Gurupi, Tocantins state. OMC: organic matter content; TOC: total organic carbon.

Parameters	Ecosystems			
	Natural Vegetation	**Soybean Cultivation**	**Pasture**	**Integration of Cultivated Forest–Pasture**
		Physical properties		
Clay (g.kg^{-1})	250	257	213	244
Sand (g.kg^{-1})	703	694	740	710
Silt (g.kg^{-1})	50	50	47	47
		Chemical Properties		
		pH $CaCl_2$		
Dry season	3.9	5.6	5.1	4.7
Rainy season	4.0	6.5	5.4	5.0
		Al ($cmol_c$·dm^{-3})		
Dry season	0.5	0	0	0
Rainy season	0.2	0	0	0
		OMC (dag·kg^{-1})		
Dry season	1.8	1.2	1.5	1.4
Rainy season	3.3	4.4	2.3	1.9
		TOC (dag·kg^{-1})		
Dry season	1.0	0.7	0.9	0.8
Rainy season	1.9	2.5	1.3	1.1

Table 1. *Cont.*

Parameters	Ecosystems			
	Natural Vegetation	Soybean Cultivation	Pasture	Integration of Cultivated Forest–Pasture
			P (mg·dm^{-3})	
Dry season	10.8	12.4	20.3	24.5
Rainy season	14.85	60.1	25.6	34
			K (mg·dm^{-3})	
Dry season	36	47	49	41
Rainy season	50	91	84	57
			Ca (cmol$_c$/dm^{-3})	
Dry season	0.2	1.2	1.5	1.9
Rainy season	0.3	6.5	5.4	4.6
			Mg (cmol$_c$/dm^{-3})	
Dry season	0.2	0.6	0.8	1.2
Rainy season	0.2	3.2	2.8	1.8
			S (mg.dm^{-3})	
Dry season	2	2	2	3
Rainy season	3	5	4	5
			B (mg.dm^{-3})	
Dry season	0.1	0.1	0.1	0.1
Rainy season	0.1	0.4	0.3	0.2
			Cu (mg.dm^{-3})	
Dry season	0.2	0.3	0.3	0.4
Rainy season	0.3	0.6	0.8	1.3
			Mn (mg.dm^{-3})	
Dry season	3.8	4.7	3.4	3.4
Rainy season	5.2	5.7	11.7	5.3
			Zn (mg.dm^{-3})	
Dry season	0.5	1.1	0.6	0.8
Rainy season	0.6	1.7	0.8	0.8

Climatic data during the experiment were registered at the closest meteorological station, located in Gurupi, 3–8 km from the study sites. In the dry and rainy seasons, average temperature, air relative humidity, and total rainfall were respectively 30 ± 26 °C, 52 ± 86 RH, and 1654 mm (119 and 1535 mm in the dry and rainy seasons, respectively).

2.3. Mite Sampling, Extraction, and Identification

Soil samples for the determination of the Gamasina fauna were collected monthly between July 2015 and June 2016. On each collection date, 32 sampling sites were determined in the area of each ecosystem (total of 384 samples in each ecosystem in all study). Each sample was taken with a metal cylinder (5 cm high × 9 cm in diameter, corresponding to a base area of about 63.6 cm^2 and a volume of about 318 cm^3/sample), completely introduced into the soil with the help of a hammer. Sampling sites were semi-randomized, each sampling point being selected so as to contain a litter layer as thick as possible (ranging between 3 and 15 cm deep), and at least 20 m apart from each other. Each sample was placed in a plastic bag, which in turn was packed in a polystyrene box for transport to a laboratory of Universidade Federal do Tocantins, Gurupi, Tocantins state. The temperature inside the box was maintained at 12–21 °C, with the use of freezing gel containers, to reduce mite activity.

About 4 h after the collection, the samples were placed in modified Berlese-Tullgren funnels for mite extraction in vials containing 70% ethanol [17]. The extracted material was sent to the "Laboratory of Taxonomy and Biological Control of Mites and Insects" of Universidade Estadual Paulista (UNESP), in Jaboticabal, São Paulo, Brazil. The content of each vial was transferred to a Petri dish for examination under a stereomicroscope, collecting the Gamasina to be later mounted on microscopic slides in Hoyer's medium.

The Gamasina were separated into species under an optical phase contrast microscope (Leica, DMLB, Wetzlar, Germany), and identified to family based on Lindquist et al. [11]. Then, adult females were identified to genera, especially based on the following publications: [18–21] and unpublished keys adopted by the Ohio State Acarology Summer Program. Identification to species was done (when possible) based on the original descriptions and redescriptions of the species, available in the authors' collections.

2.4. Data Analysis

Statistical analyses and graph construction were done using a RGui 3.4.0 program [22]. Shannon-Weaver, Simpson, and Equitability indexes were calculated using the Vegan package. The mean abundance was compared by the Kruskall-Wallis test ($\alpha = 0.05$), as variances were not homogenous and data were not normally distributed (Shapiro-Wilk normality test). As variances for mean richness were homogenous and data were normally distributed, values of this parameter were compared by Tukey test ($\alpha = 0.05$).

Euclidean distances were used to estimate similarity between ecosystems, based on abundances of species and physicochemical properties of soil in paired ecosystems. The analysis was performed based on the abundance of species in each paired environment plus the physico-chemical variables of the soil obtained. First, the data was standardized, and then Ward's minimum variance clustering algorithm and the Euclidean distance were used in analysis. Then, the similarity matrix was analyzed in a hierarchical cluster analysis.

Simple correspondence (CA) and correspondence canonical (CCA) analyses were conducted using the Vegan Package in R. CA analysis was used to explore the frequency of species found in each environment, while CCA analysis was used to relate the variable environmental (independent) and abundance of Gamasina (dependent). Initially, the data were subjected to the Chi-square method to standardize the frequencies of each species in relation to the environment. For this analysis, only species collected in at least two ecosystems were considered. In the graph, species found in a single ecosystem were shown next to the name of the corresponding ecosystem, within a box (only to CA). Interpretation was restricted to the ordination within the first two axes shown in the graph constructed in R.

3. Results

In total, 1373 Gamasina specimens were collected (Table 2), of which 919 were from natural vegetation, 64 from soybean cultivation, 160 from pasture, and 240 from the integration of cultivated forest–pasture. The Gamasina belonged to nine families, of which the most abundant was Rhodacaridae (49.5% of the Gamasina), followed by Ascidae (35.2%) and Laelapidae (10%). The remaining families corresponded to less than 3.0% each. The numbers of Gamasina collected in the natural vegetation, integration of cultivated forest–pasture, soybean cultivation, and pasture in the rainy season were 1.4, 1.8, 3.9, and 5.0 times higher than the corresponding numbers collected in the dry season.

Natural vegetation contained species of all families found in the study, with a predominance of Rhodacaridae (65.9%), Ascidae (20.5%), and Laelapidae (9.8%). In the other ecosystems, the predominant family was Ascidae, representing respectively 54.8%, 59.7%, and 74.9% of the Gamasina. In the soybean cultivation, the second predominant family was Laelapidae (20.3%), followed by Phytoseiidae (10.9%). In the pasture and in the integration of cultivated forest–pasture, the second predominant family was Rhodacaridae (respectively 21.3% and 18.3%), followed by Laelapidae (12.7% and 6.3%).

Table 2. Gamasina mites from soil samples collected at four ecosystems in Sucupira, Tocantins, Brazil, in the dry (July–October 2015 and May–June 2016) and rainy season (November 2015–April 2016).

Gamasina Species	Code	Ecosystems/Seasons								
		Natural Vegetation		Soybean Cultivation		Pasture		Integration of Cultivated Forest–Pasture		Total
		Dry	Rainy	Dry	Rainy	Dry	Rainy	Dry	Rainy	
Rhodacaridae										
Binodacarus n. sp.	Bino	15	65	0	2	0	0	0	0	82
Multidentorhodacarus tocantinensis Azevedo and Castilho	Musp	24	54	0	1	0	8	1	5	93
Multidentorhodacarus squamosus Karg	Msqu	225	201	0	3	5	30	20	18	502
Protogamasellopsis dioscorus (Manson)	Pdio	0	1	1	0	0	0	0	0	2
Protogamasellopsis zaheri Abo-Shnaf, Castilho, and Moraes	Pzah	1	0	1	0	0	0	0	0	2
Rhodacarus n. sp.	Rhod	2	9	0	0	0	0	0	0	11
Ologamasidae										
Neogamasellevans sp.	Neog	1	0	0	0	0	0	0	0	1
Ologamasus sp.	Olog	0	0	1	0	0	0	0	0	1
Macrochelidae										
Macrocheles muscaedomesticae (Scopoli)	Macr	0	6	0	0	0	0	0	0	6
Podocinidae										
Podocinium sagax (Berlese)	Podo	0	1	0	0	0	0	0	0	1
Laelapidae										
Androlaelaps sp.	Andr	6	2	0	0	0	0	0	0	8
Cosmolaelaps barbatus Moreira, Klompen and Moraes	Cbar	1	0	1	0	0	0	2	3	7
Cosmolaelaps guttulatus (Karg)	Cgut	0	0	0	0	3	1	0	0	4
Cosmolaelaps pampaensis Duarte, Moreira, Cunha & Moraes	Cpam	0	1	0	0	0	0	0	0	1

Table 2. *Cont.*

Gamasina Species	Code	Ecosystems/Seasons								
		Natural Vegetation		Soybean Cultivation		Pasture		Integration of Cultivated Forest–Pasture		Total
		Dry	Rainy	Dry	Rainy	Dry	Rainy	Dry	Rainy	
Cosmolaelaps sp. 1	Cos1	0	1	0	0	0	0	0	0	1
Cosmolaelaps sp. 2	Cosm	1	14	0	3	1	6	0	0	25
Gaeolaelaps sp. 1	Geo1	6	11	0	2	1	1	2	4	27
Gaeolaelaps sp. 2	Geo2	0	1	0	1	0	0	0	0	2
Gaeolaelaps sp. 3	Geo3	0	2	0	0	0	0	2	1	5
Gaeolaelaps sp. 4	Geo4	2	3	0	1	0	0	0	0	6
Gaeolaelaps sp. 5	Geo5	0	0	0	3	0	2	1	0	6
Gaeolaelaps sp. 6	Geo6	1	4	0	0	0	0	0	0	5
Gaeolaelaps sp. 7	Geo7	0	0	0	0	0	1	0	0	1
Gaeolaelaps sp. 8	Geo8	1	0	0	0	0	0	0	0	1
Gaeolaelaps sp. 9	Geo9	0	0	0	1	0	0	0	0	1
Gaeolaelaps sp. 10	Geo10	0	0	0	1	0	1	0	0	2
Laelaspisella cavitatis (Karg)	Pogo	7	5	0	0	0	0	0	0	12
Oloopticus reticulatus Karg	Oloo	5	27	0	0	0	1	0	0	33
Pseudoparasitus sp.	Pseu	0	1	0	0	0	0	0	0	1
Phytoseiidae										
Euseius citrifolius Denmark and Muma	Econ	0	1	0	0	0	0	0	0	1
Neoseiulus barkeri Hughes	Nann	0	1	0	3	0	0	0	0	4
Neoseiulus gracilis (Muma)	Nide	0	1	0	0	0	0	0	0	1
Proprioseiopsis mexicanus (Garman)	Pmex	0	0	0	1	0	0	0	0	1
Typhlodromus transvaalensis (Nesbitt)	Ttra	2	4	0	3	1	4	3	1	18

Table 2. *Cont.*

Gamasina Species	Code	Ecosystems/Seasons								Total
		Natural Vegetation		Soybean Cultivation		Pasture		Integration of Cultivated Forest–Pasture		
		Dry	Rainy	Dry	Rainy	Dry	Rainy	Dry	Rainy	
Ascidae										
Asca garmanni Hurlbutt	Ager	2	0	0	0	0	0	0	0	2
Asca sp.	Asca	35	9	0	0	2	1	8	3	58
New genus	Assp	4	30	0	0	0	8	1	7	50
Protogamasellus mica (Athias-Henriot)	Pmic	18	31	1	18	3	47	4	25	147
Protogamasellus sigillophorus Mineiro, Lindquist and Moraes	Psig	8	24	8	7	9	18	38	75	187
Protogamasellus pantanal Yamada and Moraes	Prot	7	18	0	1	0	1	1	5	33
Blattisocidae										
Cheiroseius pugiunculus Karg	Cher	0	1	0	0	0	0	0	0	1
Lasioseius n. sp.	Lasi	0	0	0	0	0	2	0	1	3
Melicharidae										
Proctolaelaps bickleyi (Bram)	Pbic	0	0	0	0	0	1	0	0	1
Proctolaelaps paulista Mineiro, Lindquist and Moraes	Ppau	2	5	0	0	0	2	3	6	18
Proctolaelaps sp.	Pctl	2	7	0	0	0	0	0	0	9
Total abundance		378	541	13	51	25	135	86	154	1383
Mean abundance (mites/sample) *		65.8 ± 9.8 A	87.3 ±26.7 a	2.3 ± 0.3 C	8.5 ± 3.2 c	4.3 ± 0.7 C	22.2 ± 5.9 b	13.7 ± 1.9 B	26.3 ± 12.9 b	
Species richness		24	31	6	16	8	18	13	13	
Mean richness (species/sample) *		12.2 ± 0.5 A	13.7 ± 2.1 a	1.8 ± 0.3 C	4.5 ± 0.9 b	3.0 ± 0.4 BC	7.2 ± 0.8 b	4.8 ± 0.7 B	6.0 ± 1.6 b	
Shannon-Weaver index		1.70	2.34	1.29	2.27	1.79	2.03	1.77	1.75	
Simpson index		0.63	0.82	0.59	0.83	0.79	0.82	0.73	0.72	
Equitability		0.54	0.68	0.72	0.82	0.86	0.69	0.69	0.68	

* Means followed by the same uppercase letters or the same lowercase letters are not significantly different at $p = 0.05$ by Kruskall-Wallis (mean abundance) and Tukey (mean richness) tests.

In total, 45 species of 24 genera were identified (Table 2). The highest diversity of genera was found in Laelapidae (five), Phytoseiidae, and Rhodacaridae (four genera each); other families contained one to three genera. The highest diversity of species was also observed in Laelapidae (19), followed by Rhodacaridae and Ascidae (six each); other families contained one to five species. At least four of the species collected are new to science, belonging to the families Ascidae and Melicharidae (one species each) and Rhodacaridae (two species).

As to the possible effect of rainfall, species richness and both diversity indexes were higher in the rainy season in natural vegetation, soybean cultivation, and pasture (Table 2). However, in the integration of cultivated forest–pasture, these parameters were exactly the same or about the same in both dry and rainy seasons.

The mean abundance was the highest in the natural vegetation in the dry (mean: 65.8, range 39–111; $X^2 = 20.57$; df = 3; $p > 0.001$) and in the rainy (mean: 87.3, range 22–195; $X^2 = 12.21$; df = 3; $p = 0.006$) seasons. The mean richness was also the highest in the natural vegetation, both in the dry (mean: 12.2 × mean: 1.8–4.8; $F_{3.20} = 74.87$; $p < 0.001$) and in the rainy (mean: 13.7 × mean: 4.5–7.2; $F_{3.20} = 7.72$; $p = 0.001$) seasons. Both parameters were lowest in the soybean cultivation and pasture.

In the dry season, Shannon-Weaver's and Simpson's indexes were lowest in soybean cultivation and similar to each other in other ecosystems, whereas in the rainy season, these indexes were lowest in the integration of cultivated forest–pasture. Equitability was lowest in the natural vegetation in the dry season and highest in soybean cultivation in the rainy season but similar to each other in other ecosystems at each of those seasons.

When pooling the data for both dry and rainy seasons, Shannon-Weaver's and Simpson's indexes were higher for soybean cultivation (respectively 2.35 and 0.84) than for the natural vegetation (2.16 and 0.76), pasture (2.09 and 0.81), or for the integration of cultivated forest–pasture (1.81 and 0.73). The equitability indexes of the pooled data were respectively 0.78, 0.71, 0.68, and 0.60 for soybean cultivation, pasture, integration of cultivated forest–pasture, and natural vegetation.

Considering the different ecosystems separately, the highest species diversity was found in natural vegetation (36 species), followed by soybean cultivation, pasture, and integration of cultivated forest–pasture, respectively, of 20, 19, and 14 species. Only seven species were found in all ecosystems (*Multidentorhodacarus tocantinensis*, *Multidentorhodacarus squamosus*, *Gaeolaelaps* sp.1, *Typhlodromus transvaalensis*, *Protogamasellus mica*, *Protogamasellus sigillophorus*, *Protogamasellus pantanal*). Nineteen species were found exclusively in the natural vegetation, while three species were found exclusively in the soybean cultivation and other three in the pasture; no species were found exclusively in the integration of cultivated forest–pasture.

In natural vegetation, the rhodacarids *M. squamosus*, *M. tocantinensis*, and a new species of *Binodacarus* were the predominant species (at least 78 specimens each). In addition, other 10 species were represented by more than 10 specimens each. In the other ecosystems, the predominant species were *p. mica* and *p. sigillophorus*, but in pasture and in the integration of cultivated forest–pasture, *M. squamosus* also predominated (at least 27 specimens in each). All of the predominant species in the different ecosystems were found in all ecosystems. All other species were represented by less than 10 specimens in each ecosystem, except *Asca* sp. (11 specimens in the integration of cultivated forest–pasture), and none of them were found in more than three ecosystems.

Similarity Analysis

In the similarity analysis (Figure 1), natural vegetation was positioned distant from the other ecosystems based on the abundances of species and physicochemical properties of soil. The highest similarity was observed between soybean cultivation and pasture, with the integration of cultivated forest–pasture in an intermediate position to the previous two ecosystems in one extreme and natural vegetation in the other.

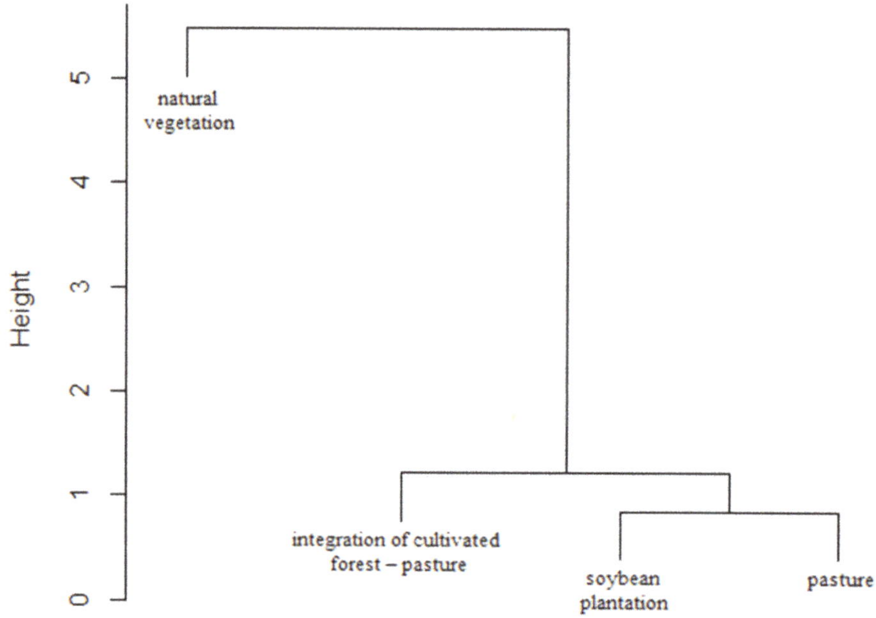

Figure 1. Similarity in mite communities between ecosystems (based on Euclidean distances) taken into account in this study: natural vegetation; soybean cultivation; pasture; integration of cultivated forest–pasture; physicochemical properties of soil, at Sucupira, Tocantins, Brazil.

In the bi-plot analysis (Figure 2), the eigenvalues of the simple correspondence analysis were significant for axes 1 ($\lambda = 0.63$) and 2 ($\lambda = 0.37$) (axes considered significant when eigenvalue is higher than 0.3, according to Dekkers et al. [23]), explaining 92.6% (axis 1: 58.5%; axis 2: 34.1%) of the variation in the data. In regard to the similarity levels between ecosystems, a greater distance was observed between soybean cultivation and pasture, as well as between any of these and natural vegetation or integration of cultivated forest–pasture. Axis 1 separated soybean cultivation and pasture from natural vegetation and integration of cultivated forest–pasture, while axis 2 separated soybean cultivation from pasture.

In the canonical correspondence analysis, both axes explained 95.32% of the variance. The first axis (CCA1) was negatively correlated with the most evaluated chemical parameters, except aluminum (Al) and manganese (Mn). It was highly correlated with total organic carbon (TOC) (−0.98), organic matter content (OMC) (−0.97), and pH (−0.71). Other parameters had lower correlations with CCA1: Boron (B, −0.45), zinc (Zn, −0.47), phosphorus (P, −0.52), and potassium (K, −0.53). TOC, MOC, and pH had a greater correlation with the soybean environment in the upper left quadrant, manganese was correlated with pasture in the lower left quadrant, copper (Cu) with the integration of cultivated forest–pasture in the upper right quadrant, and aluminum with the natural vegetation in the lower right quadrant.

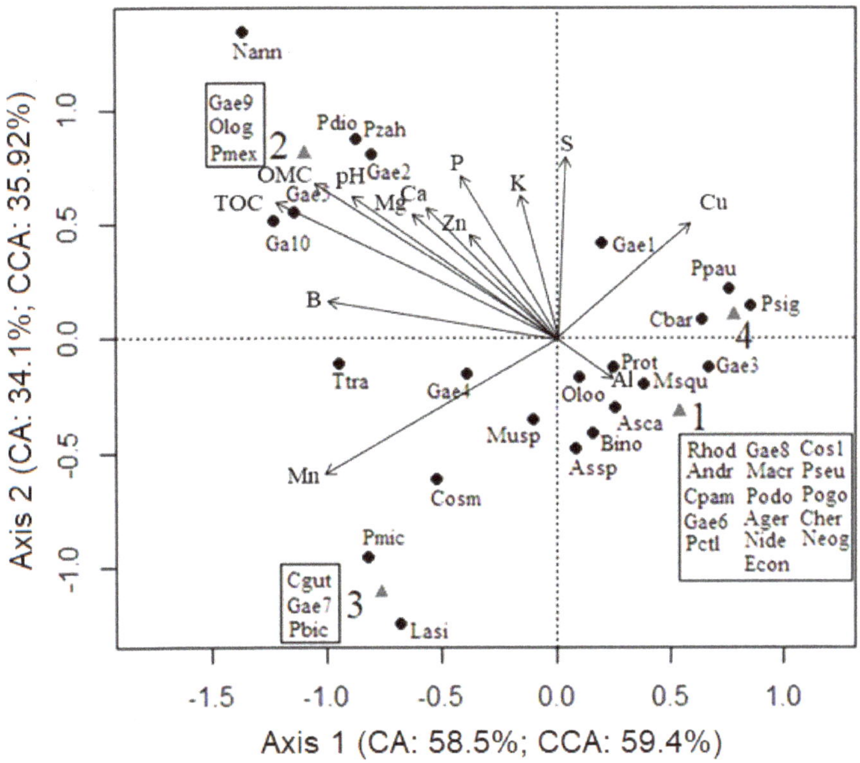

Figure 2. Biplot of simple and canonical correspondence analyses between the abundance of Gamasina and soil environmental parameters (pH; OMC, organic matter content; TOC, total organic carbon and chemical elements) in four ecosystems (1, natural vegetation; 2, soybean cultivation; 3 pasture; 4, integration of eucalyptus–pasture) at Sucupira, Tocantins, Brazil. Eigenvalues: CA: axis 1 = 0.59, axis 2 = 0.34; CCA: axis 1 = 0.59, axis 2 = 0.36. Habitats represented by triangles and species by circles. Species abbreviations shown in Table 2.

4. Discussion

The results of this work showed a major difference between the number of Gamasina in the soils of the natural vegetation and the numbers in cultivated areas, given that almost two thirds of all Gamasina specimens were collected in the former. In a distant second place were the Gamasina from the ecosystem consisting of the integration of cultivated forest–pasture, in turn followed by pasture and finally soybean cultivation. This sequence is in line with what would be expected by considering plant diversity (and diversity of other organisms) in those respective ecosystems. The strikingly low number of Gamasina in the soybean cultivation corresponded to only about 7% of the number found in the natural vegetation.

These results are compatible with findings of other authors, who also reported larger abundance and richness of mites in soils of areas of natural vegetation than in cultivated soils [24–26]. In soybean cultivation, the repeated disturbance caused by agricultural practices has been reported to result in low Gamasina abundance [26], even when the production system includes no tillage, as in the present study.

In each ecosystem, a comparison of mite abundances in rainy and dry seasons showed the smallest differences in the area of natural vegetation and of the integration of cultivated forest–pasture. In addition, the richness was higher in natural vegetation than the other ecosystems. These results suggest a relation between soil coverage (by the canopy and/or litter) with Gamasina population.

Although soil coverage was not quantified in this study, the difference between the areas was notorious. Despite the fact that soybean was cultivated under no tillage, the amount of litter in the soybean cultivation area was small, corroborating the low value of organic matter content (OMC) for this area in the dry season, when the area was bare and the lack of pronounced rain led litter to dry, exposing the soil to sunlight, and probably causing high Gamasina mortality.

However, it is curious that OMC was much higher in the rainy season in the soybean cultivation than in other areas, possibly as a result of the input of N-P-K and micronutrients in this agroecosystem in the rainy season. No tillage reduces the impact of soil management, protecting it from erosion and allowing significant increases in the levels of carbon and nitrogen, in parallel with an increase in the organic matter content [27,28], which also favors the persistence of Gamasina. In this study, difference between ecosystems was not restricted to abundance but also in terms of species richness, with the number of species in the area of natural vegetation being practically at least twice as high as in other areas.

However, quite a different picture was observed in relation to the diversity indexes, when mites of the dry and rainy seasons were pooled. The highest index for soybean cultivation, rather than reflecting species richness, reflected the highest level of equitability. The reverse can be said for the natural vegetation, where species richness was much higher than in other ecosystems but where equitability was lowest. The low equitability index in the area of natural vegetation, in turn, reflected the high predominance of a single species, *M. squamosus*, which accounted for 46% of the Gamasina in this area in comparison with only 5% in the soybean cultivation, and to 16–19% in the other areas.

Rhodacaridae has been reported as a predominant family in temperate forest soils in other parts of the world [29–31]. However, the predominance of rhodacarids in the area of natural vegetation in this study differs from the results of similar studies in other parts of Brazil, where other families have been shown as dominant [32–34]. Within the rhodacarids, *M. squamosus* had only been reported in a primary forest in Costa Rica [35] and in soybean cultivation in Mato Grosso, Brazil [36]. Another two of the dominant species were only recently described, namely *M. tocantinensis*, found in the course of this study [37], and *P. pantanal*, found in a similar study conducted in the Brazilian section of the Pantanal biome [38].

Important factors that could have contributed to determining the faunistic differences between ecosystems could obviously include the respective abiotic and biotic natural differences. However, an additional factor that might have contributed to the observed difference in Gamasina density (and possibly diversity) refers to the use of chemical products, especially in the area of soybean cultivation, but also in the area of integration of cultivated forest–pasture. Some studies have shown the negative impact of pesticides [24,39] and specifically fipronil [40,41] on edaphic organisms.

4.1. *Effect of Abiotic Soil Factors*

A high abundance of Ascidae species could be related to management practices in agroecosystems, influencing soil physical and chemical and properties and probably also affecting mite communities [24,39]. Ascid have been mentioned as common predators in soils of cultivated tropical areas [42]. In the present study, the physical characteristics of the soil of the four studied ecosystems seemed quite similar, and probably should not have affected the observed faunistic differences much. For all ecosystems, the soil consisted of about 70–74% sand, 21–26% clay, and 5% silt. However, the chemical characteristics were more variable. Thus, pH ranged between 4.7 and 6.5 in cultivated areas, where aluminum was not detected, differing from the soil of the natural vegetation area, where pH was 3.9–4.0 and where aluminum was detected. Under the prevailing circumstance of higher pH, *Protogamasellus* species were the most abundant mites. In a pasture area in Colombia where pH was 6.3, an undetermined species of *Protogamasellus* was reported as the most abundant Gamasina [43], whereas in pasture and agricultural areas of Argentina, where the pH was around 5.0–6.5, a high abundance of another *Protogamasellus* species was reported [44].

Protogamasellus mica, one of the predominant species in the present study in soybean cultivation, is apparently well adapted to agricultural areas. It has been reported in several locations in Brazil, mainly in agroecosystems of different biomes in the widespread states of São Paulo [26,32], Mato Grosso [36], and Pernambuco [45]. This species was reported as one of the predominant Mesostigmata in areas of cultivation of corn and beans [32].

The biplot analysis confirmed the highest species diversity in the natural vegetation than in other ecosystems and their positive correlation with aluminum and negative correlation with pH. The detection of that element only in the natural vegetation and not in other ecosystems is most certainly due to lime application in the latter, as usual in cultivations in Cerrado areas, where pH is usually low. Other studies have also shown a negative correlation between pH and mite abundance [46–48]. However, it is still not possible to tell the meaning of the observed correlation. In other words, does Al and/or pH have a direct effect on these mites, or is there an effect on other environmental factors that in turn affect the Gamasina? This seems an interesting subject to be evaluated in future works concerning the mites found in this study.

4.2. Interaction of Gamasina with Other Organisms

In addition to the direct effect provided by litter soil coverage (moisture retention, reduction of the incidence of light and excessive heat), an indirect effect is expected to occur, affecting the availability of other organisms, especially other mites, small insects, nematodes, and fungi, that could serve as food for predatory mites [44,49–51]. There is abundant information in the literature on the ability of Gamasina to feed on these organisms [13].

Among the organisms regularly consumed by gamasine mites, there are species potentially harmful to cultivated plants or animals. Soybean is one of the main crops in Brazil. This crop is not known to be severely damaged by edaphic mites but it is attacked by edaphic insects [52]. These insects are not known to be significantly attacked by predatory mites, as occurs for Sciaridae fly larvae in mushroom crops [53]. However, soybean can be severely damaged by nematodes [54,55], a group of organisms known to be consumed by Gamasina. Nematode species reported as important soybean parasites in Brazil are *Meloidogyne javanica* (Treub), *Meloidogyne incognita* (Kofoid and White), *Heterodera glycines* Ichinohe, and *Pratylenchus brachyurus* (Godfrey) [54,55]. Some of these could be attacked by local gamasine species, and in that case, it would be worthwhile to conserve those predatory mites as naturally occurring biological control agents.

The results of this study show the effect of anthropic activities on Gamasina diversity. A pragmatic question to be asked is whether the presently adopted cultivation practices and their effect on the local gamasine fauna is tolerable. The work carried out does not allow a conclusion in this regard, given our insufficient knowledge about the role of these mites in the Cerrado biome. This subject seems another important topic for future research.

Additionally, what would be a practical recommendation for growers in the region, in order to maintain as much as possible the natural qualities of the environment, while allowing agricultural production to take place? Most certainly, what was shown in this work is the prevailing situation in most grain production areas worldwide, and the ecological impact is expected to be related to the extent of the cultivated area. A mitigating measure would be the establishment of preservation areas. It is expected that areas of preserved natural vegetation close to the cultivation systems can serve as a reservoir of native species, from which they could migrate to neighboring areas. As verified in this study, of the total of 45 gamasine species collected, 36 occurred in the area of natural vegetation. It is possible, however, that at least part of these species cannot manage to survive in the agricultural environment, and that their presence is always restricted to soils of natural vegetation. The determination of this possibility and the development of techniques that would allow their maintenance in agricultural areas still require major research efforts.

Author Contributions: Conceptualization, E.B.d.A., L.H.A. and R.d.C.C.; Formal analysis, E.B.d.A., G.F.M. and R.D.C.C.; Methodology, E.B.d.A., G.F.M., F.A.d.S. and M.A.F.d.C.; Resources, E.B.d.A. and R.d.C.C.; Supervision, R.d.A.S. and R.d.C.C.; Visualization, L.H.A, R.d.A.S. and R.d.C.C.; Writing—original draft, E.B.d.A., R.d.A.S. and R.d.C.C.; Writing—review & editing, E.B.d.A., L.H.A. and R.d.C.C. All authors have read and agreed to the published version of the manuscript.

Funding: This research was funded by FAPESP (Fundação de Amparo à Pesquisa do Estado de São Paulo), Process 2015/14400-6, 2018/14478-3, 2019/06282-4, 2017/12004-1; and CNPq (National Council for Scientific and Technological Development), Project: 306652/2018-8.

Acknowledgments: To G.J. de Moraes, for his invaluable help in the identification of part of the species collected and in the preparation of this manuscript. To FAPESP (Fundação de Amparo à Pesquisa do Estado de São Paulo) for the scholarship to the first and second authors (Process 2015/14400-6, Process 2018/14478-3 and Process 2019/06282-4). This work is part of a BIOTA/FAPESP project (Process 2017/12004-1). We also thank the National Council for Scientific and Technological Development (CNPq)—Brazil (Project: 306652/2018-8).

Conflicts of Interest: The authors declare no conflict of interest.

References

1. MMA—Ministério do Meio Ambiente. Cerrado Biome. Available online: https://www.mma.gov.br/estruturas/chm/_arquivos/mapas_cobertura_vegetal_ingles.pdf (accessed on 10 May 2020).
2. Ribeiro, J.F.; Walter, B.M.T. As principais fitofisionomias do Bioma Cerrado. In *Cerrado Ecologia e Flora*; Sano, S.M., Almeida, S.P., Ribeiro, J.F., Eds.; Embrapa Informação Tecnológica: Brasília, Brazil, 2008; p. 406.
3. Companhia Nacional de Abastecimento (2019) Boletim da Safra de Grãos. Available online: https://www.conab.gov.br/info-agro/safras/graos/boletim-da-safra-de-graos (accessed on 18 May 2020).
4. Machado, R.B.; Ramos Neto, M.B.; Pereira, P.; Caldas, E.; Gonçalves, D.; Santos, N.; Tabor, K.; Steininger, M. *Estimativas de Perda da Área do Cerrado Brasileiro*; Conservation International do Brasil: Brasília, Brazil, 2004; relatório técnico não publicado.
5. El Titi, A. (Ed.) Implications of soil tillage for weed communities. In *Soil tillage in Agroecosystems*; CRC Press: Boca Raton, FL, USA, 2003; pp. 147–185.
6. Bedano, J.C.; Domínguez, A.; Arolfo, R.; Wall, L.G. Effect of Good Agricultural Practices under no-till on litter and soil invertebrates in areas with different soil types. *Soil Tillage Res.* **2016**, *158*, 100–109. [CrossRef]
7. Bzuneck, H.L.; Santos, H.R. Efeitos de dois sistemas de preparo do solo e sucessões de cultura, na população de ácaros Galumnidae (Cryptostigmata). *Rev. Ciência Agrária* **1991**, *11*, 1–2.
8. Lopes Assad, M.L. Fauna do solo. In *Biologia dos Solos dos Cerrados*; Vargas, M.A.T., Hungria, M., Eds.; EMBRAPA, CPAC: Planaltina, Brazil, 1997; pp. 363–444.
9. Silva, J.; Casalinho, H.; Verona, L.; Schwengber, J. Avaliação da mesofauna (colêmbolos e ácaros) do solo em agroecossistemas de base familiar no Rio Grande do Sul. *Rev. Bras. Agroecol.* **2007**, *2*, 539–542.
10. Kibblewhite, M.G.; Ritz, K.; Swift, M.J. Soil health in agricultural systems. *Phil. Trans. R Soc. B Biol. Sci.* **2008**, *363*, 685–701. [CrossRef] [PubMed]
11. Lindquist, E.E.; Krantz, G.W.; Walter, D.E. Order Mesostigmata. In *A Manual of Acarology*, 3rd ed.; Krantz, G.W., Walter, D.E., Eds.; Texas Tech University Press: Lubbock, TX, USA, 2009; p. 807.
12. Gerson, U.; Smiley, R.L.; Ochoa, R. *Mites (Acari) for Pest Control*; Blackwell Science: Oxford, MA, USA, 2003; p. 539.
13. Carrillo, D.; Moraes, G.J.; Peña, J.E. *Prospects for Biological Control of Plant Feeding Mites and Other Harmful Organisms*; Springer: London, UK, 2015; p. 344.
14. Roubinet, E.; Straub, C.; Jonsson, T.; Staudacher, K.; Traugott, M.; Ekbom, B.; Jonsson, M. Additive effects of predator diversity on pest control caused by few interactions among predator species. *Ecol. Entomol.* **2015**, *40*, 362–371. [CrossRef]
15. Peel, M.C.; Finlayson, B.L.; McMahon, T.A. Updated world map of the Köppen-Geiger climate classification. *Hydrol. Earth Syst. Sci.* **2007**, *11*, 1633–1644. [CrossRef]
16. EMBRAPA—Empresa Brasileira de Pesquisa Agropecuária. *Sistema Brasileiro de Classificação de Solos*; Embrapa Informação Tecnológica: Brasília, Brazil, 2018; p. 195.
17. Oliveira, A.R.; Moraes, G.J.; Demétrio, C.G.B.; Nardo, E.A.B. *Efeito do Vírus da Poliedrose Nuclear de Anticarsia Gemmatalis sobre Oribatida Edáficos (Arachnida: Acari) em um Campo de Soja*; Embrapa Meio Ambiente: Jaguariúna, Brazil, 2001; p. 32.

18. Chant, D.A.; McMurtry, J.A. *Illustrated Keys and Diagnoses for the Genera and Subgenera of the Phytoseiidae of the World (Acari: Mesostigmata)*; Indira Publishing House: West Bloomfield, MI, USA, 2007; p. 219.
19. Castilho, R.C.; Moraes, G.J.; Halliday, B. Catalogue of the family Rhodacaridae Oudemans, with notes on the classification of Rhodacaroidea (Acari: Mesostigmata). *Zootaxa* **2012**, *3471*, 1–69. [CrossRef]
20. Moraes, G.J.; Britto, E.P.J.; Mineiro, J.L.C.; Halliday, B. Catalogue of the mite families Ascidae Voigts & Oudemans, Blattisociidae Garman and Melicharidae Hirschmann (Acari: Mesostigmata). *Zootaxa* **2016**, *4112*, 1–299. [CrossRef]
21. Castilho, R.C.; Silva, E.S.; Moraes, G.J.; Halliday, B. Catalogue of the family Ologamasidae Ryke (Acari: Mesostigmata). *Zootaxa* **2016**, *4197*, 1–147. [CrossRef]
22. R Development Core Team. *R: A Language and Environment for Statistical Computing, v.3.0.1*; R foundation for Statistical Computing: Vienna, Austria, 2013.
23. Dekkers, T.B.; van der Werff, P.A.; van Amelsvoort, P.A.M. Soil Collembola and Acari related to farming systems and crop rotations in organic farming. *Acta Zool. Fennica* **1994**, *195*, 28–31.
24. Koehler, H. Predatory mites (Gamasina, Mesostigmata). *Agric. Ecosys. Environ.* **1999**, *74*, 395–410. [CrossRef]
25. Bedano, J.C.; Ruf, A. Soil predatory mite communities (Acari: Gamasina) in agroecosystems of Central Argentina. *Appl. Soil Ecol.* **2007**, *36*, 22–31. [CrossRef]
26. Junqueira, B.R. Diversidade de Ácaros Edáficos em um Fragmento de Mata Atlântica e Três Cultivos Agrícolas, em Jaboticabal/SP, com Ênfase nos Gamasina (Mesostigmata). Master's Thesis, Universidade Estadual Paulista (UNESP), Jaboticabal, Brazil, 2017.
27. Nelson, D.W.; Sommers, L.E. Total carbon, organic carbon and organic matter. In *Methods of Soil Analysis: Chemical and Microbiological Properties. Part 2*; Page, A.L., Miller, R.H., Keeney, D.R., Eds.; Soil Science Society of America: Madison, WI, USA, 1982; pp. 539–579.
28. Sisti, C.P.J.; Santos, H.P.; Kohhan, R.; Alves, B.J.R.; Urquiaga, S.; Boddey, R.M. Change in carbon and nitrogen stocks in soil under 13 years of conventional or zero tillage in southern Brazil. *Soil Tillage Res.* **2003**, *76*, 39–58. [CrossRef]
29. Price, D.W. Abundance and vertical distribution of microarthropods in the surface layers of a California pine forest soil. *Hilgardia* **1973**, *42*, 121–147. [CrossRef]
30. Evans, G.O.; Till, W.M. Mesostigmatic mites of Britain and Ireland (Chelicerata: Acari-Parasitiformes): An introduction to their external morphology and classification. *Trans. Zoo. Soc. Lond.* **1979**, *35*, 139–262. [CrossRef]
31. Coleman, D.C.; Crossley, D., Jr. *Fundamentals of Soil Ecology*; Academic Press, Inc.: San Diego, CA, USA, 1996; p. 205.
32. Mineiro, J.L.C.; Moraes, G.J. Gamasida (Arachnida: Acari) edáficos de Piracicaba, Estado de São Paulo. *Neotrop. Entomol.* **2001**, *30*, 379–385. [CrossRef]
33. Silva, E.S.; Moraes, G.J.; Krantz, G.W. Diversity of edaphic rhodacaroid mites (Acari: Mesostigmata: Rhodacaroidea) in natural ecosystems in the State of São Paulo, Brazil. *Neotrop. Entomol.* **2004**, *33*, 547–555. [CrossRef]
34. Santos, J.C. Ácaros (Arthropoda: Acari) Edáficos do Estado de Alagoas, com Enfase nos Mesostigmata. Master's Thesis, Universidade Estadual Paulista (UNESP), Jaboticabal, Brazil, 2013.
35. Karg, W. Zur Systematik der Raubmilbenfamilien Hypoaspididae v. Vitzthum, 1941 und Rhodacaridae Oudemans, 1902 (Acarina, Parasitiformes) mit neuen Arten aus Süd- und Mittelamerika. *Mitt. Mus. Naturkunde Berl.* **2000**, *76*, 243–262. [CrossRef]
36. Castro, M.C. Diversidade de Ácaros Gamasina (Mesostigmata) Edáficos em Sistema de Plantio Direto e Sistema de Integração Lavoura, Pecuária e Floresta (ILPF) no Estado do Mato Grosso. Master's Thesis, Universidade Estadual Paulista (UNESP), Jaboticabal, Brazil, 2019.
37. Azevedo, E.B.; Sarmento, R.A.; Castilho, R.C. A new species of *Multidentorhodacarus* (Mesostigmata: Rhodacaridae) from Brazil, complementary description of *Multidentorhodacarus squamosus* Karg and a key to the world species of the genus. *Syst. Appl. Acarol.* **2019**, *24*, 324–336. [CrossRef]
38. Yamada, M.; Moraes, G.J. A key to the species of *Protogamasellus* (Acari: Ascidae), with a new species from the Brazilian Pantanal. *Zootaxa* **2020**, *4801*, 343–354. [CrossRef]
39. Ruf, A.; Beck, L. The use of predatory soil mites in ecological soil classification and assessment concepts, with perspectives for oribatid mites. *Ecotox. Environ. Safe* **2005**, *62*, 290–299. [CrossRef] [PubMed]

40. Mesléard, F.; Garnero, S.; Beck, N.; Rosecchi, E. Uselessness and indirect negative effects of an insecticide on rice field invertebrates. *C. R. Biol.* **2005**, *328*, 955–962. [CrossRef] [PubMed]
41. San Miguel, A.; Raveton, M.; Lempérière, G.; Ravanel, P. Phenylpyrazoles impact on *Folsomia candida* (Collembola). *Soil Biol. Biochem.* **2008**, *40*, 2351–2357. [CrossRef]
42. Moraes, G.J.; Venancio, R.; Santos, V.L.V.; Paschoal, A.D. Potential of Ascidae, Blattisociidae and Melicharidae (Acari: Mesostigmata) as biological control agents of pest organisms. In *Prospects for Biological Control of Plant Feeding Mites and Other Harmful Organisms*; Carrillo, D., Moraes, G.J., Peña, J.E., Eds.; Springer: London, UK, 2015; pp. 33–75.
43. Rueda-Ramírez, D.M. Ácaros Edáficos Mesostigmata de Grandes Altitudes na Colômbia e os Possíveis Efeitos de Mudanças Edafo-Climáticas sobre as Populações destes Ácaros. Master's Thesis, Escola Superior de Agricultura "Luiz de Queiroz", Universidade de São Paulo, São Paulo, Brazil, 2012.
44. Bedano, J.C.; Ruf, A. Sensitivity of different taxonomic levels of soil Gamasina to land use and anthropogenic disturbances. *Agric. For. Entomol.* **2010**, *12*, 203–212. [CrossRef]
45. Torris, A.F. Diversidade e Flutuação Populacional de Ácaros Edáficos em um Fragmento de Caatinga e Três Cultivos Agrícolas, no Vale do São Francisco (Pernambuco), com Ênfase nos Gamasina (Mesostigmata). Master's Thesis, Universidade Estadual Paulista (UNESP), Jaboticabal, Brazil, 2019.
46. Birkhofer, K.; Schoning, I.; Alt, F.; Herold, N.; Klarner, B.; Maraun, M.; Marhan, S.; Oelmann, Y.; Wubet, T.; Yurkov, A.; et al. General relationships between abiotic soil properties and soil biota across spatial scales and different land-use types. *PLoS ONE* **2012**, *7*, e43292. [CrossRef]
47. Miller, J.; Battigelli, J.P.; Willms, W.D. Grazing protection influences soil mesofauna in ungrazed and grazed riparian and upland pastures. *Rang. Ecol. Manag.* **2014**, *67*, 429–434. [CrossRef]
48. Bedano, J.C.; Cantú, M.P.; Doucet, M.E. Influence of three different land management practices on soil mite (Arachnida: Acari) densities in relation to a natural soil. *Appl. Soil Ecol.* **2006**, *32*, 293–304. [CrossRef]
49. Edwards, C.A.; Lofty, J.R. The influence of cultivations on soil animal populations. In *Progress in Soil Zoology*; Vanek, J., Ed.; Academia Publishing House: Prague, Czech Republic, 1975; pp. 399–406.
50. Fox, C.A.; Fonseca, E.J.A.; Miller, J.J.; Tomlin, A.D. The influence of row position and selected soil attributes on Acarina and Collembola in no-till and conventional continuous corn on a clay loam soil. *Appl. Soil Ecol.* **1999**, *13*, 1–8. [CrossRef]
51. Koukoura, Z.; Mamolos, A.P.; Kalburtji, K.L. Decomposition of dominant plant species litter in a semi-arid grassland. *Appl. Soil Ecol.* **2003**, *23*, 13–23. [CrossRef]
52. Oliveira, L.J.; Roggia, S.; Salvadori, J.R.; Ávila, C.J.; Fernandes, P.M.; Oliveira, C.M. Insetos que atacam raízes e nódulos da soja. In *Soja—Manejo Integrado de Insetos e outros Artrópodes-Praga*; Hoffmann-Campo, C.B., Corrêa-Ferreira, B.S., Moscardi, F., Eds.; Embrapa: Brasília, Brazil, 2012; pp. 75–144. Available online: http://www.cnpso.embrapa.br/artropodes/Capitulo2.pdf (accessed on 10 June 2020).
53. Castilho, R.C.; Moraes, G.J.; Silva, E.S.; Freire, R.A.P.; Eira, F.C. The predatory mite *Stratiolaelaps scimitus* as a control agent of the fungus gnat *Bradysia matogrossensis* in commercial production of the mushroom *Agaricus bisporus*. *Int. J. Pest Manag.* **2009**, *53*, 181–185. [CrossRef]
54. Dias, W.P.; Garcia, A.; Silva, J.F.V.; Carneiro, G.E.S. *Nematóides em Soja: Identificação e Controle*; Embrapa Soja: Londrina, Brazil, 2010; p. 8.
55. Mattos, V.S.; Furlanetto, C.; Silva, J.G.P.; Santos, D.F.; Almeida, M.R.A.; Correa, V.R.; Moita, A.W.; Castagnone-Sereno, P.; Carneio, R.M.D.G. *Meloidogyne* spp. populations from native Cerrado and soybean cultivated areas: Genetic variability and aggressiveness. *Nematology* **2016**, *18*, 505–515. [CrossRef]

diversity

MDPI

Article

A Snap-Shot of Domatial Mite Diversity of *Coffea arabica* in Comparison to the Adjacent Umtamvuna Forest in South Africa

Sivuyisiwe Situngu [1,*,†], Nigel P. Barker [2] and Susanne Vetter [1]

1 Botany Department, Rhodes University, P.O. Box 94, Makhanda 6139, South Africa; S.Vetter@ru.ac.za
2 Department of Plant and Soil Sciences, University of Pretoria, P. Bag X20, Hatfield 0028, South Africa; nigel.barker@up.ac.za
* Correspondence: sivu.situngu@wits.ac.za; Tel.: +27-(0)11-767-6340
† Present address: School of Animal, Plant and Environmental Sciences, University of Witwatersrand, Private Bag 3, Johannesburg 2050, South Africa.

Received: 21 January 2020; Accepted: 14 February 2020; Published: 18 February 2020

check for updates

Abstract: Some plant species possess structures known as leaf domatia, which house mites. The association between domatia-bearing plants and mites has been proposed to be mutualistic, and has been found to be important in species of economic value, such as grapes, cotton, avocado and coffee. This is because leaf domatia affect the distribution, diversity and abundance of predatory and mycophagous mites found on the leaf surface. As a result, plants are thought to benefit from increased defence against pathogens and small arthropod herbivores. This study assesses the relative diversity and composition of mites on an economically important plant host (*Coffea aribica*) in comparison to mites found in a neighbouring indigenous forest in South Africa. Our results showed that the coffee plantations were associated with only predatory mites, some of which are indigenous to South Africa. This indicates that coffee plantations are able to be successfully colonised by indigenous beneficial mites. We also found an "edge effect", in that coffee trees at the edge of the plantation hosted fewer mite species. These results are a snap-shot from a single sampling period. Nonetheless, they highlight the potential importance of this mutualism in commercial crop species and the possible role of faunal exchanges between indigenous and exotic crop species. This study expands our understanding of the mite–plant mutualism in Southern Africa, a region where acarological studies are sparse.

Keywords: leaf domatia–mite mutualism; *Coffea arabica*; mite diversity; edge effect; South Africa

---

1. Introduction

The commercial coffee plant, *Coffea arabica* L. (Rubiaceae), originates from Ethiopia and is cultivated worldwide in more than 50 countries [1,2]. Leaves of this species bear pit-type domatia, and coffee is one of many economically important species that have been shown to benefit from the association with mites [3–8]. Other economically important plants known to have an association with mites include grapes, avocado, sweet pepper plants and cassava [4,6,9–11]. The mutualism with mites is important in coffee because it may provide some relief from one of the most important diseases of coffee plants, the coffee leaf rust fungus (*Hemilleia vastatrix* Berk and Boome), and also from phytophagous mites [8]. In Brazil, the top coffee-producing country in the world, these benefits are appreciated, and more studies are being undertaken in an attempt to understand the possible interactions among the different mite species found in coffee and the favourable environmental conditions for the predaceous mites on coffee plants, in order to improve their integrated pest management programs for coffee agroforest systems [12]. Furthermore, coffee plants may also benefit from an increased overall fitness due to a decrease in the damage caused by plant enemies. Agrawal et al. [4] showed that cotton plants with artificial domatia added to them had increased populations of predatory mites and that their cotton yield was enhanced compared to plants without domatia.

In many countries that produce coffee, the coffee red mite, *Oligonychus ilicis* McGregor (Tetranychidae), is a common pest [3,8,13,14]. Even so, coffee leaves with healthy domatia have been shown to harbour more beneficial mites than pest mites [11,12,15]. In North Queensland, the majority of the mites found inside the domatia of coffee were primarily predatory from the families Stigmaeidae, Phytoseiidae and Bdellidae [3]. In Brazil, predatory mites such as *Euseius citrifolius* Denmark and Muma and *Euseius concordis* (Chant) are associated with coffee, and these mites control the pest mites *Brevipalpus phoenicis* (Geijskes) and *Oligonychus ilicis* [12]. Even though these mites are currently not listed as pest mites in South Africa, their related species such as *Oligonychus coffeae* (Nietner) and *Oligonychus mangiferus* (Rahman and Sapra) are known pests.

In South Africa, coffee is not widely cultivated, and according to an official government website, only approximately 200 ha are cultivated, with a national yield of 3500 bags (60 kg per bag; however, note that this information dates back to 2014). In comparison, Brazil, which is the top coffee-producing country, supplies nearly 48 million bags [16]. Perhaps for this reason, little is known about the mites that occupy the domatia of coffee grown in South Africa, and studies on the domatia-facilitated plant–mite mutualism in Africa as a whole are rare. Despite the importance of the mites in coffee, there is little information about the fauna of predatory mites found in these plants and about the influence that neighbouring vegetation exerts as a reservoir of predatory mites, even in Brazil.

The aim of this study was to examine the mite communities found inside the domatia of coffee plants in South Africa and to compare the mite fauna with that found in leaf domatia-bearing plant species in the adjacent indigenous forest, thereby contributing to the knowledge of mites associated with domatia, from both an exotic crop species and indigenous trees from an understudied region.

Based on the literature, we hypothesised that the coffee plantations would harbour beneficial mites inside their domatia and primarily predatory species which are similar to those found in the adjacent forest. We further hypothesised that the diversity of possible host plants in the forest would result in greater mite diversity in the forest, but that some of the mite taxa may have dispersed and established themselves in the coffee plantation.

2. Materials and Methods

Beaver Creek Coffee monoculture plantation in Port Edward, KwaZulu-Natal Province, South Africa, was selected as the study site (31°02′40.86″ S; 30°10′48.18″ E). The study was conducted in February of 2015, during the summer season in South Africa. We chose to sample during this season because our results from a previous study of the effects of seasonality on the distribution of mites suggested that mites inhabit domatia during all seasons and mite abundance was highest in summer and autumn [17]. The Beaver Creek Coffee plantation was selected due to its close proximity (about 200 metres) to the indigenous Umtamvuna Forest (Figure 1). From our observation, this forest was intact, with a well-developed canopy and understorey tree layers.

Four transects were randomly laid out across the plantation: two laid in the middle of the plantation and two at the edge of the plantation. These two transect lines for both the edge and the middle were laid next to each other, parallel to the plant rows. Both the edge of the plantation closest to the forest and in the middle of the plantation were sampled to test whether there would be higher abundance and a different suite of mites at the edge of the plantation compared to the middle. Twenty leaves were sampled from each of 10 individual coffee trees that were randomly selected along the transect line. When sampling, mature leaves with fully developed domatia were chosen from all around the tree. In addition to this, forest walks in the neighbouring Umtamvuna Forest were undertaken, and twenty leaves were sampled from each plant species found to possess leaf domatia.

The leaves were placed in a reseal-able plastic bag and stored in a cooler box with ice, to keep the mites sessile. The mites were then viewed under a dissecting microscope, on the same day the sampling took place. Mites found inside domatia were counted before they were collected for identification. Mites were removed and collected from leaves using a pipette and a drop of alcohol, mounted on a slide using PVA mounting medium, and viewed under an Olympus light microscope. Some individuals were mounted on a stub, using a double-sided tape, coated with carbon, and viewed under a scanning electron microscope. Mite species abundance (average number of mites per leaf), composition and the Shannon diversity index for both sites were calculated.

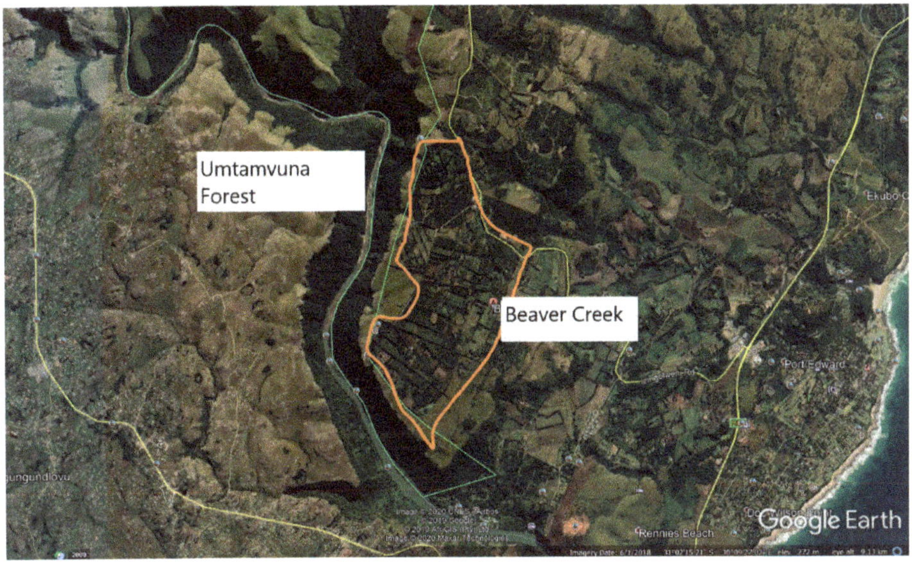

Figure 1. A Google Earth image showing the close proximity of Beaver Creek Coffee plantation to Umtamvuna forest. (Image © 2020 Maxar technologies).

A list of all mite species collected from each of the plant species from the different sampling sites was collated, and a presence–absence data matrix was produced, where "1" represented the presence of a species of mite in a sampled tree, and "0" its absence. Prior to analysis, the data were subjected to the "Absence and Presence" transformation option in PRIMER 6 software package, where each sampling unit was one plant (based on 20 leaves per plant). The presence-and-absence transformation allows for the less abundant mite species to have the same weighting as abundant ones.

A resemblance analysis (Bray–Curtis resemblance) using the software PRIMER 6 [18] was then performed. From this analysis, a dendrogram plot was produced, using the "group average" linkage to identify relationships and similarities in mite biota between the host species sampled. A similarity percentage analysis (SIMPER) was also conducted to determine which mite species are characteristic of the different vegetation types and to identify which mites contribute most to the similarity between vegetation types and within sample sites.

3. Results

Twenty-seven different plant specimens from 18 species were sampled from Umtamvuna forest, and 40 plants (10 plants per transect; four transects) were sampled from a coffee plantation that is a monoculture. Table 1 lists all the mite species collected from the coffee plantation and the neighbouring Umtamvuna forest and also gives the total abundance and frequency for each of the mite species collected. In total, nine different species of mites were collected from leaves of coffee, and 20 mite species were found on the forest tree species. Furthermore, these results show that three mite species were exclusively found in coffee and 14 mite species were found exclusively in the forest (Table 1). The most abundant species in coffee were *Euseus addoensis* and *Typhlodromus crassus* (both predaceous). In the adjacent forest, the predacious mite *Amblyseius anomalus* and an herbivorous mite *Eriophyid* species were the most abundant.

The samples from the coffee plantation formed a distinct cluster in the multivariate analysis with the exception of three samples from the edge of the plantation that were grouped with *Canthiun cilliatum* (Figure 2), which is a shrub. However, there is no clear separation between the samples from the edge and the centre of the plantation. This shows that the coffee plantation had a different suite of mite species when compared to those found in the adjacent forest patch.

Similarly, where multiple trees of the same species were sampled from the indigenous forest (five species were sampled more than once), the mite fauna did not group the specimens of these species together, suggesting that the host selection of the mites in the forest is rather eclectic.

Tables 2 and 3 give the results of the similarity percentage analysis, within group similarity and between group dissimilarity, respectively, and the top three mite species that contributed to the similarity at the different sites. This analysis allows us to account for which species explain similarities and dissimilarities between the groups. The natural forest had a low percentage similarity value (31.2%), and the coffee plantation had a slightly higher within group similarity value (middle 68.2%; edge 46.2%). The dissimilarity values when comparing different vegetation sites were high (Table 3).

Table 1. A list of the mite species collected from the coffee plantation and the neighbouring Umtamvuna forest. Here, we have indicated the total abundance and frequency (i.e., the percentage number of samples in which each mite was found), for each of the sampled locations.

Mite Species	Feeding Guild	Coffee Edge: Abundance (% Frequency)	Coffee Middle: Abundance (% Frequency)	Natural Forest: Abundance (% Frequency)	Total: Abundance (% Frequency)
Cheyletidae					
Prosocheyla hepburni **(Lawrence)**	Predacious			14 (7%)	14 (3%)
Cunaxidae					
Bunaxella quini **Den Heyer**	Predacious	15 (25%)	28 (30%)	37 (15%)	80 (24%)
Rubroscirus **sp.**	Predacious			5 (4%)	5 (1%)
Eriophyidae					
Eriophyid **sp.**	Herbivorous			104 (19%)	104 (7%)
Eupodidae					
Eupodes **sp.**	Algiphagus			17 (15%)	17 (6%)
Oribatida					
Oribatid mite	Mycophagous/Saprophytic		2 (10%)	30 (30%)	32 (12%)
Phytoseiidae					227 (22%)
Amblyseius anomalus **van der Merwe**	Predacious			227 (56%)	357 (38 %)
Euseius addoensis **(van der Merwe and Ryke)**	Predacious	212 (95%)	136 (30%)	9 (4%)	59 (7%)
Ueckermannseius sp1	Predacious	2 (10%)	59 (25%)	18 (26%)	20 (13%)
Ueckermannseius sp2	Predacious		35 (35%)	73 (41%)	73 (16%)
Ueckermannseius munsteriensis **van der Merwe**	Predacious	77 (50%)	78 (85%)	7 (7%)	7 (3%)
Typhlodromus microbullatus **van der Merwe**	Predacious			4 (4%)	35 (10%)
Typhlodromus crassus **van der Merwe**	Predacious				159 (41%)
Stigmaeidae					
Agistemus tranatalensis **Meyer**	Predacious				5 (1%)
Agistemus **sp. (probably new)**	Predacious	2 (5%)	15 (20%)	5 (4%)	15 (6%)
Mullederia centrata **(Meyer)**	Predacious				2 (1%)
Tenuipalpidae					
Brevipalpus **sp. (probably new)**	Herbivorous			8 (11%)	8 (4%)
Tetranychidae					
Oligonychus **sp. (probably new)**	Herbivorous			6 (4%)	6 (1%)
Tetranychus **sp.**	Herbivorous			33 (25%)	33 (10%)
Tetranychus nymph	Herbivorous			83 (25%)	83 (10%)
Triophtydeidae					
Tetratriophtydeus myacanthus **Ueckermann**	Predacious			5 (4%)	45 (1%)

Table 1. *Cont.*

Tydeidae ***Tydeus munsteri*** **Meyer and Ryke**	Predacious	46 (48%)	46 (19%)
Winterschmidtiidae ***Saproglyphus*** **sp.**	Mycophagous	11 (15%)	11 (6%)

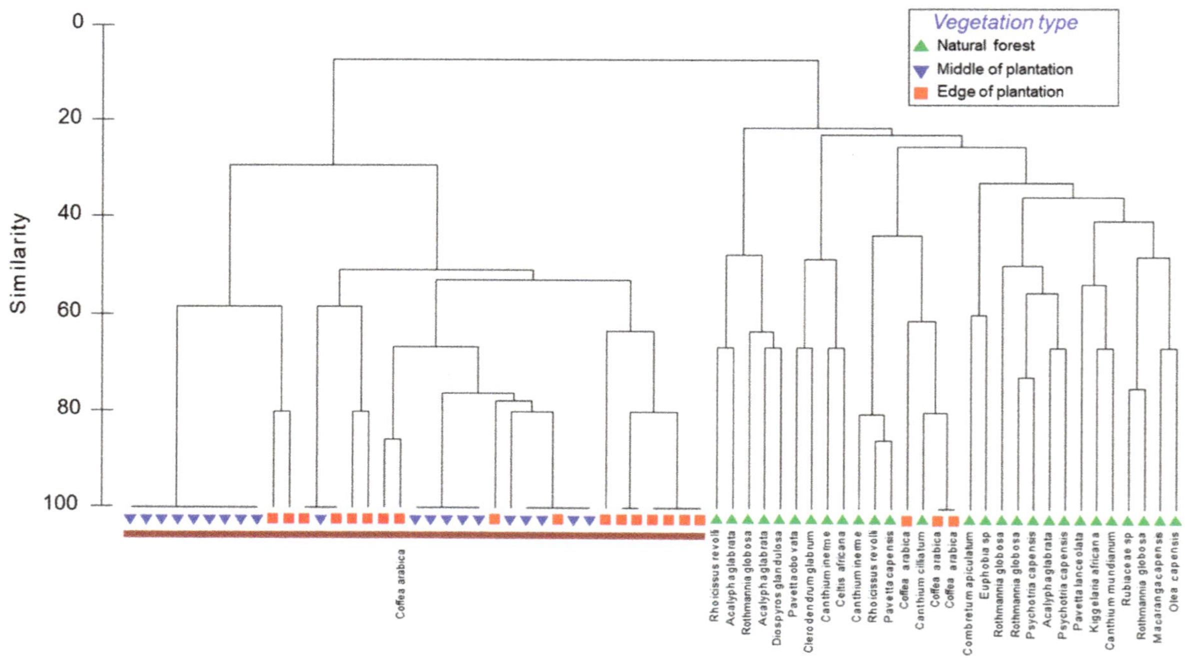

Figure 2. A group average dendrogram produced by using the Bray–Curtis similarity measure, showing similarities between coffee sample sites and the different indigenous plant species sampled from the adjacent forest.

Table 2. The contribution of the mite species sampled to the average within group similarity amongst the sites sampled.

Site	Average within Group Similarity (%)	Contribution of Species (%)
Natural forest	31.21	Phytoseiidae *Amblyseius anomalus* (23.6) *Ueckermannseius munsteriensis* (21.5) Oribatida Oribatid mite (16.8)
Middle of coffee plantation	68.27	Phytoseiidae *Euseius addoensis* (81.1) *Typhlodromus crassus* (16.0)
Edge of coffee plantation	46.17	Phytoseiidae *Typhlodromus crassus* (70.2) Cunaxidae *Bunaxella quini* (8.7) Phytoseiidae *Euseius addoensis* (8.2)

Table 3. The contribution of the mite species sampled to the average dissimilarity between the sampled sites.

Site	Average between Group Dissimilarity (%)	Contribution of Species (%)
Natural forest vs. middle of plantation	97.97	Phytoseiidae *Euseius addoensis* (17.5) *Amblyseius anomalus* (10.1) *Ueckermannseius Munsteriensis* (9.7)
Middle of coffee plantation vs. edge of plantation	62.48	Phytoseiidae *Euseius addoensis* (29.0) *Typhlodromus crassus* (23.3) Cunaxidae *Bunaxella quini* (16.1)
Edge of coffee plantation vs. natural forest	89.64	Phytoseiidae *Typhlodromus crassus* (15.0) *Amblyseius anomalus* (9.9) *Phytoseius* sp. (9.4)

The adjacent forest had a higher Shannon diversity index than the coffee plantation (Figure 3B), and when comparing the two sample sites within the coffee plantation, leaves collected in the middle of the plantation had more mites and a higher diversity index compared to samples collected at the edge (Figure 3A,B).

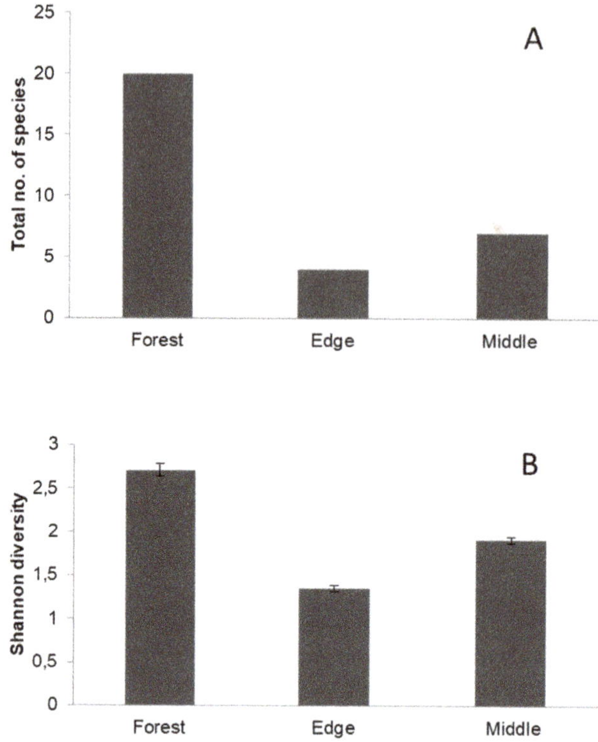

Figure 3. Shows (**A**) total number of mite species and (**B**) Shannon diversity index of mites collected from the Umtamvuna forest and two sites (middle and edge) in the *Coffea arabica* plantation.

4. Discussion

This study provides the first assessment of the mite biota found in coffee plants of South Africa. An improved understanding of mites found in coffee and how they assist in keeping the crop plant healthy is important. Our results from a single sampling period show that *Coffee arabica* is associated with more than one mite species occupying its domatia (Table 1). The coffee plantation had low mite species diversity when compared to the neighbouring indigenous forest (Figure 3B). The mites collected from the coffee plants belonged to the families Phytoseiidae, Stigmaeidae and Cunaxidae, and these are commonly found in leaf domatia all over the world. These are all predatory mites and mites from these families have previously been shown to be associated with coffee plantations [3,12]. Matos et al. [19] found that coffee domatia had a positive effect on the abundance of predatory mites, and plants with higher densities of domatia harbour more predatory mites and fewer prey mites. Mineiro at al. [20] and others [21] also found a higher number of predatory mites and phytophagous mites on coffee plants across different cropping systems. These results further support the hypothesis that domatia attract beneficial mites that act as the plant's "bodyguards" and highlight the importance of this mutualism in economically important species. In a study similar to ours, O'Connell et al. [22] found that native forests supported higher numbers of mite species than either plantation forest or pastoral grasses.

This beneficial mutualism has been shown in other commercial crop plants, including grapes, cassava and even cotton [6,13,23,24]. Avocado (*Persea americana)* plant leaves with domatia were found to be associated with more predatory mites and fewer herbivorous mites than those without domatia. Onzo et al. [6] showed that predatory mites help protect cassava plants by reducing herbivorous mites on young leaves, which are the most photosynthetically active and force herbivorous mite to move down the plant to less profitable older leaves. Norton et al. [5] found high densities of the beneficial mites *Orthotydeus lambi* on grape plants with intact domatia than on plants with blocked domatia. Both English-Loeb et al. [9] and Melidossian et al. [24] have shown that tydeid mite suppresses powdery mildew on the fruit and foliage of *Vitis vinifera* and that the mite could be an important bio-control agent for the grape powdery mildew, which is a pathogen of cultivated and wild grapes. Furthermore, grape vines with bigger domatia are resilient to pathogenic fungi because they support larger communities of fungivorous mites [23].

In our study, the mites found inside the domatia of coffee were all predatory. Other studies examining the occupants of coffee domatia also observed phytophagous mites such as *Oligonychus ilicis* and *Brevipalpus phoenicis*, which cause damage to the leaves [8,19,25]. However, we did not find any of these mites in our study site. This could be due to the fact that we only sampled the site once. Some potentially harmful mites were found in the Umtamvuna forest, but these appear not to have migrated to the plantation. Alternatively, these harmful mites could have been preyed upon by the predatory mites associated with the coffee plants. From this, we speculate that predatory mites which were found at the plantation were successful at keeping the plants healthy by preying on any harmful mite that landed on the coffee leaves. This is a widely accepted benefit that plants get as a result of producing domatia, and many studies [5,12,15,23] support the hypothesis that mites act as plant "bodyguards" and protect plant leaves from their natural enemies.

When compared to the adjacent forest patch, coffee plants were found to possess a different suite of mites (Figure 2). These results were further supported by the SIMPER analysis, which gave higher dissimilarly values when comparing the two sampled sites (Tables 2 and 3). Three mite species (*Mullederia centrata*, *Agistemus* sp. (*probably new*) and *Ueckermannseius sp1*) were uniquely found in the coffee plantation and five species (*Agistemus tranatalensis*, *Bunaxella quini*, *Euseius addoensis*, *Typhlodromus crassus* and *Ueckermannseius munsteriensis*) were found in both the adjacent forest and in the coffee plantations (Table 1). These results suggest that, to some extent, certain mites had a preference for coffee plants over the species found in the adjacent forest patch. Some of these mites (*Agistemus* sp. (probably new). *Ueckermannseius specis*, *Ueckermannseius munsteriensis*, *Euseius addoensis* and *Typhlodromus crassus*, found on coffee plants, are indigenous to South Africa, suggesting that this exotic plant is not associated with its own mite that could have been imported from another country, and that indigenous mites are able to migrate and establish on exotic species. The origin of the three mite species found exclusively on the coffee plants (two of which are indigenous) is uncertain. Either they are present in the indigenous forest but were not sampled, or they were introduced along with the coffee plants, at the time the plantation was established. In a similar study assessing plant-inhabiting mite fauna associated with sugarcane, the important role that natural vegetation plays as a source of natural mites, which are enemies of pests, was demonstrated [26,27].

Interestingly, plants at the edge of the coffee plantation had fewer mites and a lower Shannon diversity compared to plants in the middle of the plantation (Figure 3B). This suggests that mites found on plants at the edge may be subjected to negative edge effects, as they were closest to the fire break path around the plantation, which is constantly mowed. An edge effect results when both biotic and abiotic conditions change along the boundary between two habitats, affecting the distribution, biodiversity and ecosystem functionality of species in both habitats [27,28]. A negative edge effect results when species found the edge of the habitat patch are exposed to an increased risk of parasitism, disease, increased predation, adverse microclimate conditions and increased competition from invasive species [27,29].

We suspect that this was also the case in this study and that this was due to the fact that the mites at the edge of the plantation would have been exposed to harsher environmental conditions, such as strong winds, exposure to rainfall and sunlight, as well as low relative humidity. Many species have been shown to suffer from edge effects [27,29–31], and in terms of arthropods, Lacasella et al. [32] showed that grassland species of spiders, centipedes and ground beetles were affected by the edge and this influence was evident up to 15 m from the habitat edge. Mites are delicate and are susceptible to these extreme conditions and thus avoid such habitats. Forest microclimate (patterns of temperature, moisture, wind and light) plays an important role in influencing insects and arthropod habitat selections [33,34]. Furthermore, relative humidity and exposure to sunlight have been shown to affect the distribution and diversity of mites within the tree canopy [35]. Croft et al. [36] showed that eggs and larvae of phytoseiid mites were sensitive to relative humidity and for all four species of phytoseiid mites subjected to humidity assays, egg and larvae mortality increased with decreasing humidity. To the best of our knowledge, this study is the first to show the impact of edge effects in mites.

5. Conclusions

Our results are from a single sampling period and do not show seasonal patterns of mite diversity. Even so, they indicate that, despite being adjacent to the indigenous forest, the coffee plantations at Beaver Creek were associated with a different suite of native species of predatory mites. This suggests that the coffee plantations are able to be successfully colonised by beneficial indigenous mites, but that migration of mite species between indigenous hosts and exotic crop species is limited. In addition, the coffee trees at the edge of the plantation were found to possess a lower diversity and slightly different suite of mites when compared to the plants from the middle of the plantation. These results highlight the importance of the mutualism between leaf domatia and mites in both commercial plants and indigenous species. This is important, as it implies that indigenous mites be effective at controlling pests on exotic crop species as an alternative to chemical control. Future studies that sample multiples times during the year and that compare mites found on leaves, branches and fruits will provide us with a better understanding of the diversity of mites found in coffee and the role of predatory mites in keeping plants healthy.

Author Contributions: Conceptualization, S.S. and N.P.B.; methodology, S.S.; formal analysis, S.S.; writing—original draft preparation, S.S.; review and editing, N.P.B. and S.V.; funding acquisition, N.P.B. and S.S. All authors have read and agreed to the published version of the manuscript.

Funding: This research was funded by the National Research Foundation (NRF) of South Africa and the Department of Science and Technology/National Research Foundation Centre of Excellence in Tree Health Biotechnology (CTHB) Grant UID 40943. The first author also acknowledges financial support from the Canon Collins Trust and GreenMatter.

Acknowledgments: We would like to thank Robby Cumming and Beaver Creek Coffee Estate for allowing us the opportunity to conduct sampling on their property. The assistance of friends and colleagues in the field is also gratefully acknowledged. Edward Ueckermann (Agricultural Research Council of South Africa, Biosystematics Division) is thanked for his assistance with mite identification.

Conflicts of Interest: The authors declare no conflicts of interest.

References

1. Vega, F.E.; Rosenquist, E.; Collins, W. Global project needed to tackle coffee crisis. *Nature* **2003**, *425*, 343. [CrossRef] [PubMed]
2. Vega, F.E.; Ochoa, R.; Astorga, C.; Walter, D.E. Mites (Arachnida: Acari) inhabiting coffee domatia: A short review and recent findings from Costa Rica. *Int. J. Acarol.* **2007**, *33*, 291–295. [CrossRef]
3. O'Dowd, D.J. Mite association with the leaf domatia of coffee (*Coffea arabica*) in North Queensland, Australia. *Bull. Ent. Res.* **1994**, *84*, 361–366. [CrossRef]
4. Agrawal, A. Do leaf domatia mediate a plant–mite mutualism? An experimental test of the effects on predators and herbivores. *Ecol. Entomol.* **1997**, *22*, 371–376. [CrossRef]
5. Norton, A.P.; English-Loeb, G.; Gadoury, D.; Seem, R.C. Mycophagous mites and foliar pathogens: Leaf domatia mediate tritrophic interactions in grapes. *Ecology* **2000**, *81*, 490–499. [CrossRef]
6. Onzo, A.; Hanna, R.; Zannou, I.; Sabelis, M.W.; Yaninek, J.S. Dynamics of refuge use: Diurnal, vertical migration by predatory and herbivorous mites within cassava plants. *Oikos* **2003**, *101*, 59–69. [CrossRef]
7. Vega, F.E.; Infante, F.; Castillo, A.; Jaramillo, J. The coffee berry borer, *Hypothenemus hampei* (Ferrari) (Coleoptera: Curculionidae): A short review, with recent findings and future research directions. *Terr. Arthrop. Rev.* **2009**, *2*, 129–147.

8. Oliveira, C.M.; Ferreira, J.A.M.; Olivereira, R.M.; Santo, F.O.; Pallini, A. *Ricoseius loxocheles*, a phytoseiid mite that feeds on coffee leaf rust. *Exp. Appl. Acarol.* **2014**, *64*, 223–233. [CrossRef]
9. English-loeb, G.; Norton, A.P.; Gadoury, D.M.; Seem, R.C.; Wilcox, W.F. Control of Powdery Mildew in Wild and Cultivated Grapes by a Tydeid Mite. *Bio. Con.* **1999**, *14*, 97–103. [CrossRef]
10. Ferreira, J.A.M.; Eshuis, B.; Janssen, A.; Sabelis, M.W. Domatia reduce larval cannibalism in predatory mites. *Ecol. Entomol.* **2008**, *33*, 374–379. [CrossRef]
11. Ferreira, J.A.M.; Cunha, D.F.S.; Pallini, A.; Sabelis, M.W.; Janssen, A. Leaf domatia reduce intraguild predation among predatory mites. *Ecol. Entomol.* **2011**, *36*, 435–441. [CrossRef]
12. Mineiro, J.L.D.; Sato, M.E.; Raga, A.; Arthur, V. Population dynamics of phytophagous and predaceous mites on coffee in Brazil, with emphasis on *Brevipalpus phoenicis* (Acari: Tenuipalpidae). *Exp. Appl. Acarol.* **2008**, *44*, 277–291. [CrossRef]
13. Fahl, J.I.; Queiroz-Voltan, R.B.; Carrelli, M.L.C.; Schiavinato, M.A.; Pradro, A.K.S.; Souza, J.C. Alterations in leaf anatomy and physiology caused by the red mite (*Oligonychus ilicis*) in plants of *Coffea arabica*. *Bra. J. Pl. Physiol.* **2007**, *19*, 61–86. [CrossRef]
14. Romero, G.Q.; Daud, R.D.; Salomão, A.T.; Martins, L.F.; Feres, R.J.F.; Benson, W.W. Mites and leaf domatia: No evidence of mutualism in *Coffea arabica* plants. *Bio. Neo.* **2011**, *11*, 27–33. [CrossRef]
15. Pemberton, R.W.; Turner, C.E. Occurrence of predatory and fungivorous mites in leaf domatia. *Amer. J. Bot.* **1989**, *76*, 105–112. [CrossRef]
16. Agricultural Reseach Council. Available online: http://www.arc.agric.za/arc-itsc/Pages/Coffee-Information.aspx (accessed on 20 November 2019).
17. Situngu, S. An Investigation of the Leaf Domatia—Mite Mutualism in South Africa: Insights from Ecological Studies. Ph.D. Thesis, Rhodes University, Grahamstown, South Africa, 2017.
18. Clarke, K.R.; Gorley, R.N. *PRIMER v6: User Manual/Tutorial*; PRIMER-E: Plymouth, UK, 2006.
19. Matos, C.H.C.; Pallini, A.; Chaves, F.F.; Schoereder, J.S.; Janssen, A. Do domatia mediate mutualistic interactions between coffee plants and predatory mites? *Ent. Exp. Et. App.* **2006**, *118*, 185–192. [CrossRef]
20. Mineiro, J.L.C.; Sato, M.E.; Berton, J.H.C.; Raga, A. Mites (Arachnida: Acari) on coffee plants in forest fragment and conventional plantation in Monte Alegre do Sul, Atate of São Paulo, Brazil. *Divulgação Scientífica* **2019**, *81*, 1–30.
21. Agrawel, A.A.; Karban, R.; Colfer, R.G. How leaf domatia and induced plant resistance affect herbivores, natural enemies and plant performance. *Oikos* **2000**, *89*, 70–80. [CrossRef]
22. O'Connell, D.M.; Lee, W.G.; Monks, A.; Dickinson, K.J.M. Does microhabitat structure affect foliar mite assemblages? *Ecol. Entomol.* **2010**, *35*, 317–328. [CrossRef]
23. English-Loeb, G.; Norton, A.; Walker, M.A. Behavioral and population consequences of acarodomatia in grapes on phytoseiid mites (Acari: Mesostigmata) and implications for plant breeding. *Ent. Exp. Et. Appl.* **2002**, *104*, 307–319. [CrossRef]
24. Melidossian, H.S.; Seem, R.C.; English-Loeb, G.; Wilcox, W.F.; Gadoury, D.M. Suppression of grapevine powdery mildew by a mycophagous mite. *Plant Dis.* **2005**, *89*, 1331–1338. [CrossRef] [PubMed]
25. Chagas, C.M.; Kitajima, E.W.; Rodrigues, J.C.V. Coffee ringspot virus vectored by *Brevipalpus phoenicis* (Acari: Tenuipalpidae) in coffee. *Exp. Appl. Acarol.* **2003**, *30*, 203–213. [CrossRef] [PubMed]
26. Duarte, M.E.; Navia, D.; dos Santos, L.R.; Rideiqui, P.J.S.; Silva, E.S. Mites associated with sugarcane crop and with nativetrees from adjacent Atlantic forest fragment in Brazil. *Exp. Appl. Acarol.* **2015**, *66*, 529–540. [CrossRef] [PubMed]
27. Demite, P.R.; Feres, R.J.F.; Lofego, A.C. Influence of agricultural environment on the plant mite community in forest fragments. *Braz. J. Biol.* **2015**, *75*, 396–404. [CrossRef] [PubMed]
28. Murcia, C. Edge effects in fragmented forests: Implications for conservation. *Trends Ecol. Evol.* **1995**, *10*, 58–62. [CrossRef]
29. Harper, K.A.; Macdonald, S.E.; Burton, P.J.; Chen, J.; Brosofske, K.D.; Saunders, S.C.; Euskirchen, E.S.; Roberts, D.; Jaiteh, M.S.; Esseen, P. Edge influence on forest structure and composition in fragmented landscapes. *Con. Biol.* **2005**, *19*, 768–782. [CrossRef]
30. Ries, L.; Sisk, T.D. A predictive model of edge effects. *Ecology* **2004**, *85*, 2917–2926. [CrossRef]
31. Tian, C.; Yang, X.; Liu, Y. Edge effect and its impacts on forest ecosystem: A review. *Chin. J. Appl. Ecol.* **2011**, *22*, 2184–2192.

32. Lacasella, F.; Gratton, C.; Felici, S.D.; Sbordoni, V. Asymmetrical responses of forest and "beyond edge" arthropod communities across a forest–grassland ecotone. *Bio. Con.* **2015**, *24*, 447–465. [CrossRef]
33. Chen, J.; Saunders, S.C.; Crow, T.R.; Naiman, R.J.; Brosofske, K.D.; Mroz, G.D.; Brookshire, L.; Franklin, J.F. Microclimate in Forest Ecosystem and Landscape Ecology: Variations in local climate can be used to monitor and compare the effects of different management regimes. *BioScience* **1999**, *49*, 288–297. [CrossRef]
34. Fukui, A. Indirect interactions mediated by leaf shelters in animal-plant communities. *Pop. Ecol.* **2001**, *43*, 31–40. [CrossRef]
35. Situngu, S.; Barker, N.P. Position, position, position: Mites occupying leaf domatia are not uniformly distributed in the tree canopy. *SA J. Bot.* **2016**, *108*, 23–28. [CrossRef]
36. Croft, B.A.; Messing, R.H.; Dunley, J.E.; Strong, W.B. Effects of humidity on eggs and immatures of *Neoseiulus fallacis*, *Amblyseieus andersoni*, *Metaseiulus occidentalis* and *Typhlodromus pyri (Phytoseiidae)*: Implications for biological control on apple, caneberry, strawberry and hop. *Exp. App. Acarol.* **1993**, *17*, 451–459. [CrossRef]

diversity

MDPI

Article

Water Mite Diversity (Acariformes: Prostigmata: Parasitengonina: Hydrachnidiae) from Karst Ecosystems in Southern of Mexico: A Barcoding Approach

Lucia Montes-Ortiz and Manuel Elías-Gutiérrez *

El Colegio de la Frontera Sur, Department of Aquatic Ecology and Systematics, Av. Centenario Km.5.5, Chetumal, 77014 Quintana Roo, Mexico; lumontes@ecosur.edu.mx

* Correspondence: melias@ecosur.mx

Received: 15 July 2020; Accepted: 21 August 2020; Published: 29 August 2020

check for updates

Abstract: Water mites represent the most diverse and abundant group of Arachnida in freshwater ecosystems, with about 6000 species described; however, it is estimated that this number represents only 30% of the total expected species. Despite having strong biotic interactions with their community and having the potential to be exceptional bioindicators, they are frequently excluded from studies of water quality or ecology, due to actual and perceived difficulties of taxonomic identification in this group. The objective of this study is to use the variations in the sequences of the mitochondrial cytochrome oxidase subunit I (COI), also known as the DNA barcodes region, as a tool to assess the diversity of water mites at 24 sites in the Yucatan Peninsula of Mexico. We found 77 genetic groups or putative species corresponding to 18 genera: *Arrenurus, Atractides, Centrolimnesia, Eylais, Geayia, Hydrodroma, Hydryphantes, Hygrobates, Koenikea, Krendowskia, Limnesia, Limnochares, Mamersellides, Mideopsis, Neumania, Piona, Torrenticola*, and *Unionicola*. This was significant, since there are only 35 species described for this region. Furthermore, this molecular information has allowed us to infer that there are characteristic assemblies per site. These data will facilitate the incorporation of water mites in different studies while the curatorial work continues to assign a Linnaean name.

Keywords: COI; Yucatan Peninsula; assemblages; richness; Acari

1. Introduction

Water mites belong to the Hydrachnidiae subcohort and represent the most important, abundant, and diverse group of the Arachnida in freshwater ecosystems [1,2]. There are about 6000 named species, with 1300 of them reported from the Neotropics. According to Goldschmidt [3], the neotropical water mite fauna is far from being completely described, and approximately 5440 species could reasonably be expected in this area.

Mexico is a mega-diversity country due to its position in a transition region between the Nearctic and Neotropical zones and its complex physiography [4]. As a result, it is the country in the world with the second highest number of ecosystems and the fourth in terms of biodiversity [4]. In relation to aquatic environments, we know only a small fraction of their biological diversity. Regarding water mites, 317 species have been described and some reported here in the last 40 years [5]. Only 35 were from the Yucatán Peninsula that comprises three Mexican states (Quintana Roo, Yucatan, and Campeche) [6–8].

The Yucatan region includes one of the world's largest karstic aquifers and that represents a mosaic of different geochemistry and hydrogeologic properties on its water ecosystems [9,10]. For example, the Cenote Azul, located in the southern part of the Yucatan Peninsula (18.647 N and 88.412 W, Datum WGS84), is a unique extreme environment, characterized by a high sulfate and strontium content water [10]. Lake Bacalar, also located in the south, hosts the largest living freshwater microbialites in the world [11,12] and has a rich mite fauna, which is still unknown [13].

According to Cook (1980), we were far from knowing all the local water mite diversity in the neotropics, and this situation has not improved significantly over the last 40 years. Other authors have observed that neotropical water mite fauna shows regional diversification, and a high degree of richness and endemism should be expected in this region [1,3].

The taxonomy of water mites is difficult, and systematics is constantly subject to changes [14–16], first, due to the complex life cycle composed of three active stages: parasitic larva, depredatory deutonymph, and adult and three resting stages, namely prelarva, protonymph, and tritonymph, plus the egg [2]. Some groups, such as adults arrenurids, also present a strong sexual dimorphism, where males and females are completely different morphologically. In other cases, this dimorphism is visible in the modification of the legs IV for males. Finally, the diagnostic characteristics, such as setaes, coxal groups, acetabular plate, glandularias, or palps are difficult to identify without taxonomic training. Due to these challenges, many synonyms, cryptic species, subspecies, and "forms" with questionable identity exist in the literature [6,14,17].

The application of molecular biology techniques adds new characters to taxonomy. A particular region of the mitochondrial COI (cytochrome c oxidase I) gene, one of the groups known as DNA barcode region, is the most common sequence used in water mite taxonomy research. Public databases, including the Barcode of Life Database (www.boldsystems.org) or GenBank (www.ncbi.nlm.nih.gov), and the use of new bioinformatic tools represent a breakthrough in species identification [18,19]. On the other hand, these molecular data allow us to understand, from another perspective, not only the identity of species, but also ecological relationships that exist between these animals. It is also another important character for new species descriptions [20,21] that can solve problems related to cryptic species complexes [14,15], and matching of different development stages from eggs to adult males and females despite their morphological differences [16,21–24].

Additionally, DNA barcoding and the BOLD database (boldsystems.org) can be used to obtain a preliminary approximation of distribution patterns, species assemblages, richness, and diversity among other analysis [25]. The Barcode Index Number (BIN) is a fast-computational algorithm based on differences of the COI fragment. It is a unique Operational Taxonomic Unit (OTU) that highlights a putative species, assigning an exclusive code composed of alphanumeric characters [26]. The BIN system provides information about specimens with their associated metadata (taxonomy, distribution, images, sequences, collector, identifier, and institution where the voucher/specimens is deposited) [26]. This system has been used with success in diverse invertebrate surveys, biodiversity assessments, and species delimitation [13,25,27,28]. Currently there are 77,666 Trombidiformes records in the database where Hydrachnidia is a subcohort.

The aim of this study was to assess water mite diversity in different water bodies from the Center to the Southern Yucatan Peninsula, using DNA barcoding and the subsequent BIN representing each OTU, and their correspondence with identified morphotaxa, as the main approach.

2. Materials and Methods

2.1. Collection of Samples

Data were mined from both BOLD corresponding to previous published studies by the authors [13,23] and unpublished data from a last sampling survey carried out in April and August corresponding to the dry and rainy seasons. By the end, all data represented 24 sites (Table 1) from Yucatan Peninsula (PY), Mexico (Figures 1 and 2). All the samples were collected according to the methods in earlier studies [23], with the exception of two systems: Acapulquito and Palmar, where the collection was carried out by using manual nets with a mesh of 100 μm.

Table 1. Collection locations and Barcode Index Numbers (BINs) associated.

Number	Site	Lat N	Long W	BINs
1	Acapulquito	18.4321	88.5312	11
2	El palmar	18.4407	−88.5273	6
3	Cenote Azul	18.651	−88.4098	14
4	Cenote Cocalitos	18.652	−88.408	21
5	Cenote Escuela Normal	18.651	−88.40g.	9
6	North Bacalar Lake	18.9176	−88.171	14
7	Cenote Pucte 1	19.079	−87.994	11
8	Cenote Pucte 2	19.091	−87.994	9
9	Cenote el Toro	19.098	−88.021	2
10	Ramonal	19.3921	−88.6242	10
11	Cenote Sijil Noh Ha	19.475	−88.052	3
12	Cenote Chancah Veracruz	19.486	−87.988	4
13	Cenote del Padre	19.604	−88.003	6
14	Minicenote	19.607	−87.989	2
15	Cenote Tres Reyes 1	19.668	−87.881	3
16	Cenote Tres Reyes 2	19.692	−87.877	6
17	Santa Teresa	19.723	−87.813	2
18	Chichancanab	19.924	−88.7708	7
19	Cueva de las serpientes	19.93	−88.806	1
20	Cenote km 48	19.943	−87.794	6
21	Chunyaxche Lagoon 1	20.042	−87.581	3
22	Chunyaxche Lagoon 2	20.06	−87.576	12
23	Muyil Lagoon 1	20.069	−87.594	8
24	Muyil Lagoon 2	20.075	−87.607	4

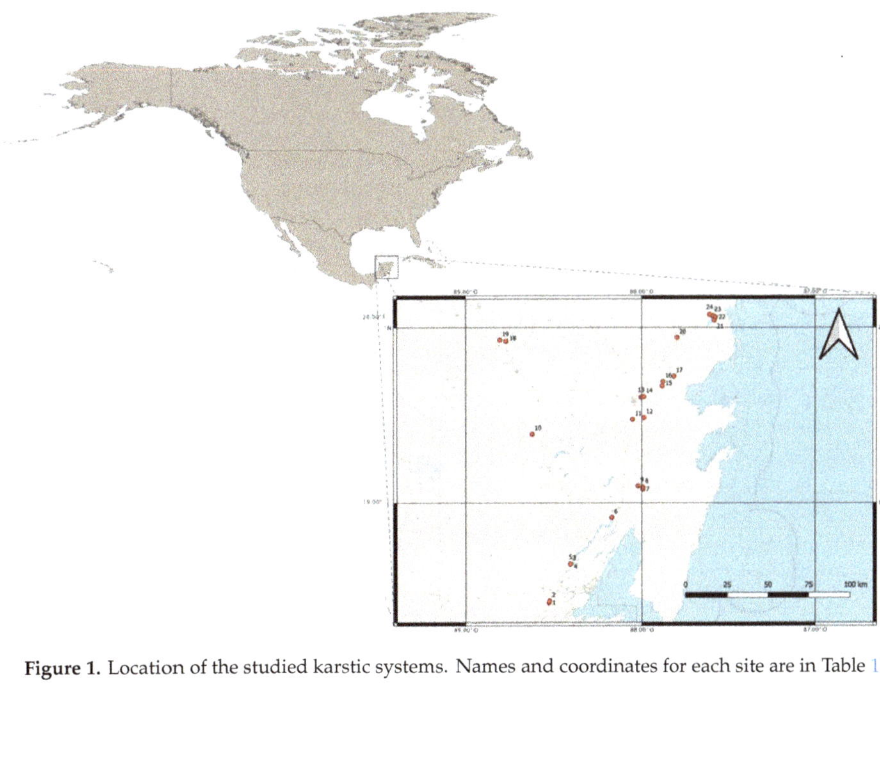

Figure 1. Location of the studied karstic systems. Names and coordinates for each site are in Table 1.

Figure 2. Examples of some sampled localities: (**a**) Cenote Azul, (**b**) Cenote Cocalitos, (**c**) Bacalar lake, and (**d**) large microbialites from Bacalar lake. Photos taken by ©HBahena/ECOSUR.

2.2. Specimen Preparational Analysis

In the laboratory, the fixed samples were viewed under a stereo microscope, and water mites were removed from each one. Representatives of each morphologically distinct group were separated and stored in 5 mL vials with 4 mL of 96° ethanol. All the water mites were identified to genus, using published keys [2,14,28]. All mites were photographed in a stereo microscope Zeizz Stereo Discovery with an Eos Rebel T3i camera.

2.3. DNA Extraction and Amplification

Whenever it was possible, five individuals of each genus were selected for genetic analysis. The whole water mite specimens were placed into 96-well plates, and DNA extraction was carried out by using a standard glass fiber method [29]. After the DNA extraction, the vouchers were recovered and preserved in Koenike's solution for future curatorial labor and deposited in the Reference Collection at El Colegio de la Frontera Sur, Unidad Chetumal (ECOCH-Z-10339-10364).

The PCR mixtures contained a final volume of 14 μL and were prepared as follows: 2 μL of Hyclone ultra-pure water, 6.25 μL of 10% trehalose (previously prepared: 5 g D-(+)-trehalose dehydrate, in 50 mL of total volume of molecular grade ddH2O), 1.25 μL of 10X PCR buffer, 0.625 μL of MgCl2 (50 mM), 0.0625 μL of dNTP (10 mM), 0.125 μL of each primer (10 μM), 0.06 μL of Platinum Taq DNA polymerase, and 3 μL of DNA template. All specimens were amplified with the zooplankton primers (ZplankF1_tl and ZplankR1_tl). The reactions were cycled at 94 °C for 1 min, followed by five cycles of 94 °C for 40 s, 45 °C for 40 s, and 72 °C for 1 min, followed by 35 cycles of 94 °C for 40 s, 51 °C for 40 s, and 72 °C for 1 min, with a final extension of 72 °C for 5 min. PCR products were visualized on 2% agarose gel (E-Gel 96 Invitrogen); finally, positive PCR products were selected for sequencing.

PCR products were sequenced, using a modified BigDye © Terminator v. 3.1 Cycle Sequencing Kit (Applied Biosystems, Inc., Foster City, CA, USA), and sequenced bidirectionally on an ABI 3730 capillary sequencer at Eurofins Scientific. Sequences were edited by using Codon Code v.3.0.1 (CodonCode Corporation, Dedham, MA, USA). Sequence data, trace files, collection data, and primer details for all specimens are available within the public dataset DS-YUCWM through the public data portal of the Barcode of Life Data Systems (www.boldsystems.org) and in GenBank (www.ncbi.nlm.nih.gov).

2.4. Sequencing and Data Analysis

All sequences that met minimal quality standards (≥500 bp, without ambiguous bases or stop codons) were assigned to a BIN [19,26]. These BINs are considered putative species or OTUs [13].

The analysis of all sequences with a BIN assignment was conducted by using MEGA v.6. We constructed Neighbor Joining trees for the most families with large numbers of BINs (Arrenuridae, Limnesiidae, Unionicolidae, and Hygrobatidae). The simplified trees were prepared by using Figtree v1 4.4.

A Jaccard index and a dendrogram were calculated with Excel software, to assess beta diversity and the similarity of water mites' BINs among the 24 locations.

3. Results

A total of 607 water mite sequences representing 77 BINs were obtained. These corresponded to 13 families: Anisitsiellidae, Arrenuridae, Eylaidae, Hydrodromidae, Hydryphantidae, Hygrobatidae, Krendowskiidae, Limnesiidae, Limnocharidae, Mideopsidae, Pionidae, Torrenticolidae, and Unionicolidae.

The number of BINs per site varied from one at Cueva de las serpientes to 21 at Cenote Cocalitos (Table 1).

We observed a correspondence between the BINs and the morphospecies for all the mite specimens. In Figure 3, we can see the correspondence between BINs and representatives of the Krendowskiidae family and *Limnesia* genera. In most cases, we matched molecularly and morphologically each BIN to a genus level, except for the following 15 that could only be assigned to families: Torrenticolidae, Limnesiidae, Hygrobatidae, Pionidae, Unionicolidae, and Eylaidae; and three BINs pertaining to Trombidiformes (Table 2).

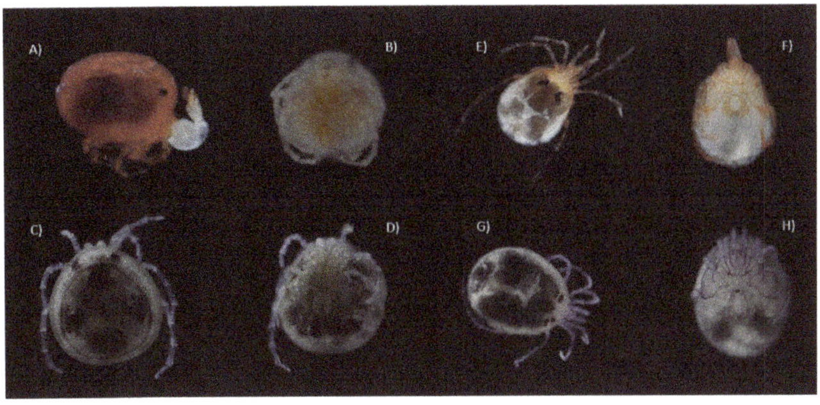

Figure 3. Members of Krendowskiidae family and *Limnesia* genus. (**A**,**B**) Lateral and ventral view of *Geayia* BIN ACT6195; (**C**,**D**) Dorsal and ventral view of *Krendowskia* BIN ACX8435; (**E**,**F**) dorsal and ventral view of *Limnesia* BIN ACY7380; and (**G**,**H**) dorsal and ventral view of *Limnesia* BIN AEA5595.

Table 2. Summary of taxa identified, BIN, and location.

Family	Genera	BIN	Location
Limnocharidae	*Limnochares*	ADI4862 *	3
		AEA4515 *	14
		AEB4511 *	24
		ACY6840	3, 6, 4
Hydrodromidae	*Hydrodroma*	ADF3732 *	3, 4, 23, 6, 18, 2, 20, 8.
Hydryphantidae	*Hydryphantes*	AEA5005 *	3
Torrenticolidae	*Torrenticola*	AEA7372 *	1
	Unknown genera	AEA4395 *	1
Limnesiidae	*Limnesia*	AEA5595 *	13, 6, 18, 7, 12.
		AEA6471 *	10
		ACX7759	5,4
		ACY7380	19, 5, 22, 4, 2, 6
	Centrolimnesia	AEA3914 *	9, 8, 16, 17
	Unknown genera	AEA4382 *	16, 9
Krendowskiidae	*Krendowskia*	ACX8435	20, 13, 5, 16, 24, 18, 6, 4
	Geayia	ACT6195	1
Mideopsidae	*Mideopsis*	AEA6512 *	1
		ACX8679	20, 13, 18, 5, 4, 24, 11, 23, 22, 8.
		ACY7169	7, 4, 22, 5.
	Unknown genera	AEB4633 *	12
Hygrobatidae	*Hygrobates*	AEA3689 *	1
		AEA3690 *	2
		AEA3924 *	1, 2
		ACX7887	3, 18
		ADO7098	6
	Atractides	ACX7786	5, 4
	Unknown genera	AEA4089 *	23
		AEA5236 *	21, 22, 23
Pionidae	*Piona*	AEB1670 *	6
		ACX8296	13, 12, 3, 6, 4, 23, 24, 7.
	Unknown genera	AEA4809 *	22
Unionicolidae	*Unionicola*	ACX8035 *	4
		AEB4634 *	8
		ACX8034	5, 4, 3, 7, 8, 14
		ACX9008	4, 5, 6
		ADM7936	21, 22, 23, 3
		ADP1665	4, 7, 22, 6
	Koenikea	ADI2928 *	3
		ACY7384	4, 5, 22
		ADI3114	2, 22, 6, 20, 3, 18, 8, 1.
	Neumania	AEA8101 *	20, 7, 10
		AEA5358 *	10
		ACY6829	6, 4
	Unknown genera	AEA4829 *	22, 8, 16
		AEA6062 *	23, 22.
		AEA6668 *	16
		AEA7951 *	16
		AEB1594 *	8
		ACY7381	4
		AEA3726 *	7
		AEA4514 *	1

Table 2. *Cont.*

Family	Genera	BIN	Location
Eylaidae	*Eylais*	ADD9174 *	4
	Unknown genera	AEA4696 *	15
		AEA5669 *	15
Arrenuridae	*Arrenurus*	ACX8462 *	4
		ACX8780 *	4, 2, 1
		ADI3752 *	3
		AEA3972 *	10
		AEA7182 *	7, 20
		AEA7842	10
		AEA7843 *	10
		AEA7844 *	1
		AEA8234 *	10
		ACL2418	4
		ACX8463	6, 4, 18, 23, 3, 12, 11, 13.
		ACX8464	4, 3, 13, 10, 6, 18.
		ACX8788	5
		ACY6809	7, 4, 21, 22, 24, 4,3
		AEB7095	1
		ADI4458 *	3
		AEA4828 *	17
Anisitsiellidae	*Mamersellides*	AEA6955 *	10
		AEA6956 *	10
Unknown		AEA4343 *	11
		AEA3823 *	15
		AEB1898	8

Localities are the same as Table 1. * Unique BINs in the Barcode of Life Database (BOLD) system.

3.1. Water Mite BINs Richness

Unionicolidae was the most diverse and abundant family, with 20 BINs and 230 sequences distributed among three genera, which were identified as *Unionicola, Koenikea,* and *Neumania*, and unidentified specimens. Fifty percent of the BINs of this family appear to have a restricted distribution inhabiting only one locality, while the other half was found in two to eight localities as *Koenikea* with the BIN ADI3114 (Figure 4 and Table 2).

The Arrenuridae was the second most diverse family, with 123 sequences and 17 BINs. All of them belonged to the genus *Arrenurus*. For nine BINs from this group, it was possible to correlate males and females and nymphs for three of them (Figure 5. Most of the BINs apparently inhabit only one location, and only three of them seem to have a wide distribution: ACX8463, ACX8464, and ACY6809 (Figure 6 and Table 2).

The Hygrobatidae and Limnesiidae families each had a moderate number of BINs. Hygrobatidae was represented by 48 sequences corresponding to eight BINs; two of them could be identified to genera *Hygrobates* and *Atractides*, and two more BINs could be identified only to family. Most of the Hygrobatidae occur only in one or two localities (Figure 7 and Table 2).

The Limnesiidae are represented by 68 sequences and six BINs, with four of them identified as *Limnesia*, one *Centrolimnesia*, and one unidentified genus. More than 80% of the limnesiids occurred in in two or more localities (Figure 8 and Table 2).

Other, less diverse families were the Limnocharidae, represented by nine sequences and four BINs, all of them *Limnochares*. Each BIN was found in a single locality, except for ACY6840, which was found in three close systems: Cenote Azul, Cenote Cocalitos, and North Bacalar Lake. Mideopsidae was represented by 36 sequences clustering in four BINs, with three of them from *Mideopsis* and the other one identified only at the family level; *Mideopsis* BIN ACX8679 seems to have a wide distribution, as it was found in ten localities (Table 2).

Pionidae and Eylaidae were composed of three BINs and were each represented by one genus, *Piona* and *Eylais*, respectively; however, in both families, there were BINs with no genus assignment. In the case of Eylaidae, each BIN inhabited one system, while *Piona* ACX8296 was found in eight localities (Table 2).

Krendowskiidae was represented by two genera, *Geayia* and *Krendowskia*, with 32 sequences and two BINs (Figure 3). Krendowskia ACX8435 was widely distributed. Hydrodromidae was represented by one BIN and 22 sequences belonging to *Hydrodroma* genus. This OTU is widely distributed in eight systems in the sampled area, and all the morphotypes corresponded with one putative species.

Hydryphantidae was a singleton of the genus *Hydryphantes*. Finally, there were five sequences represented by three BINs that belonged to the order Trombidiformes. These individuals were nymphs, which are not included in any taxonomic keys. They cannot be further identified until an adult can be sequenced, as for *Arrenurus* specimens (Figure 5).

From the 77 BINs, 51 were sequenced for the first time and appear as unique in the BOLD database (Table 2). Only four BINs had a wide distribution, from Neotropical Mexico to Eastern–Central Canada. These are the *Unionicola* ADP1665, *Arrenurus* ACL2418, *Geayia* ACT6195, and *Piona* ACX8296.

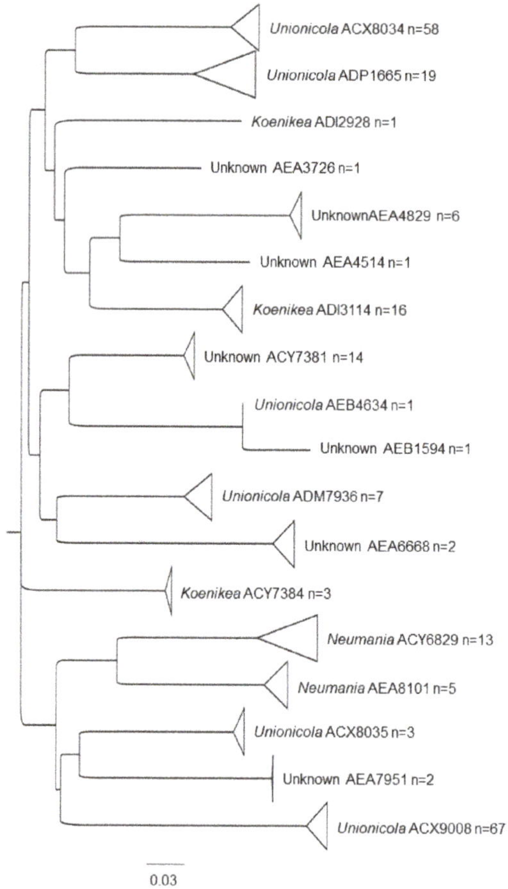

Figure 4. Neighbor Joining (NJ) tree for Unionicolidae family.

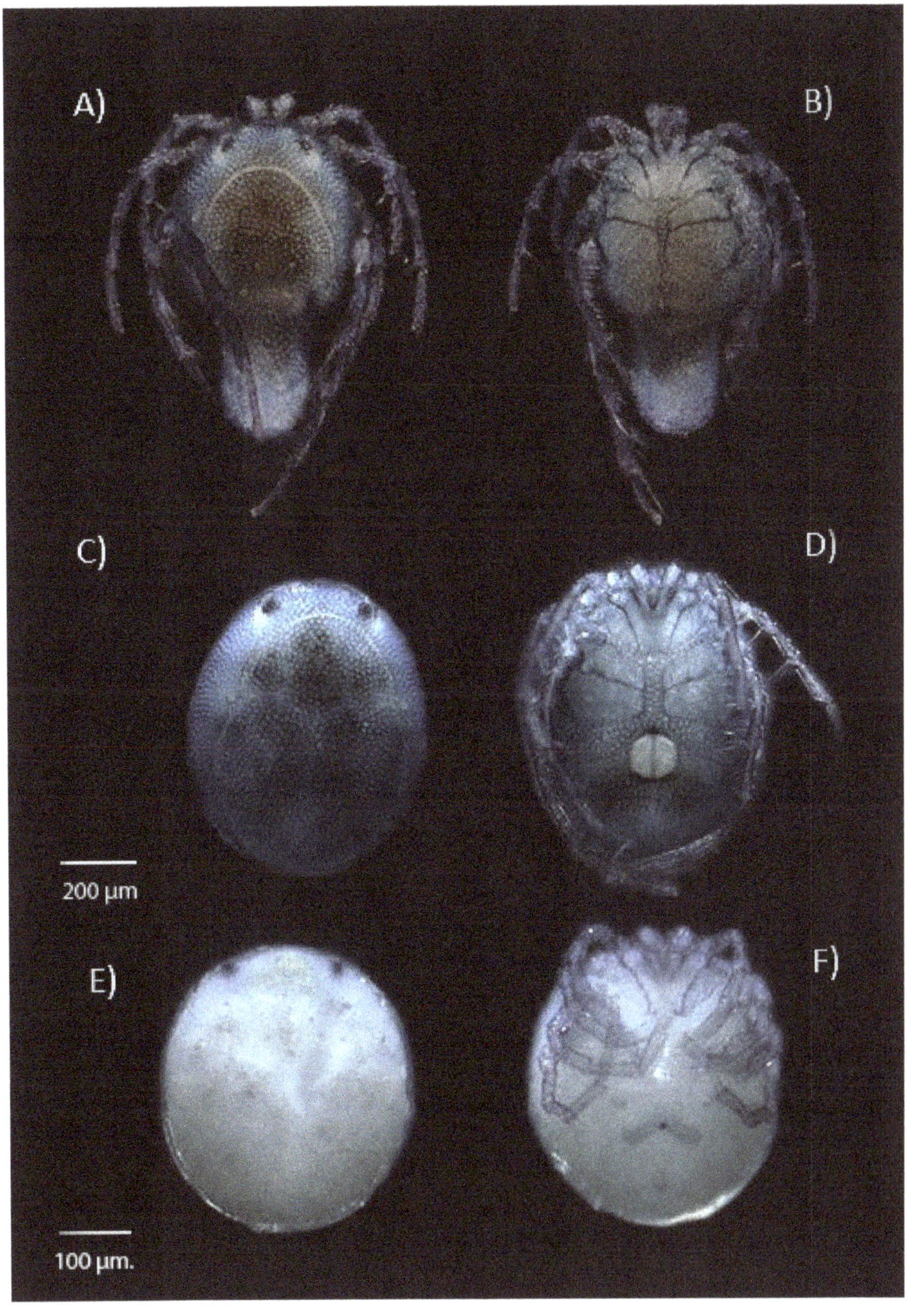

Figure 5. *Arrenurus* sp. BIN ACX8463: (**A**,**B**) dorsal and ventral view of male, (**C**,**D**) dorsal and ventral view of female, and (**E**,**F**) dorsal and ventral view of nymph.

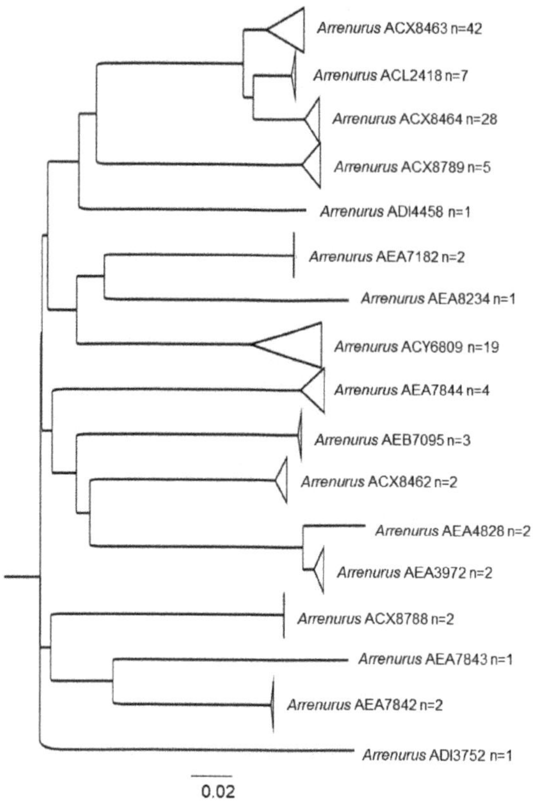

Figure 6. NJ tree for Arrenuridae family.

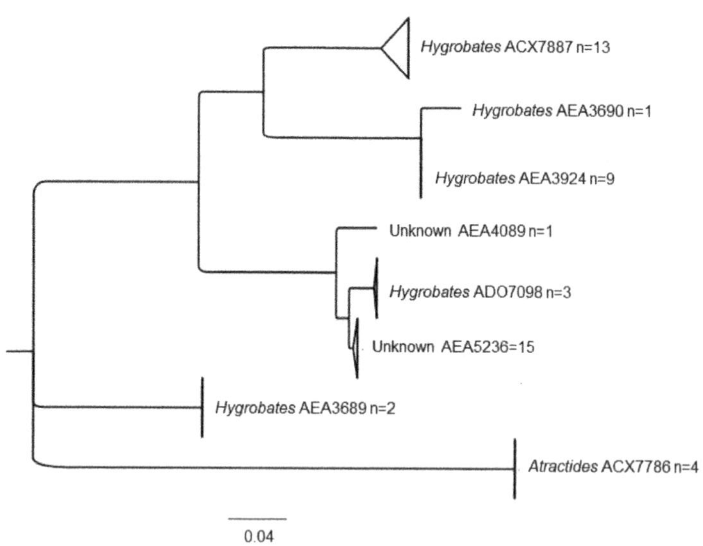

Figure 7. NJ tree for Hygrobatidae family.

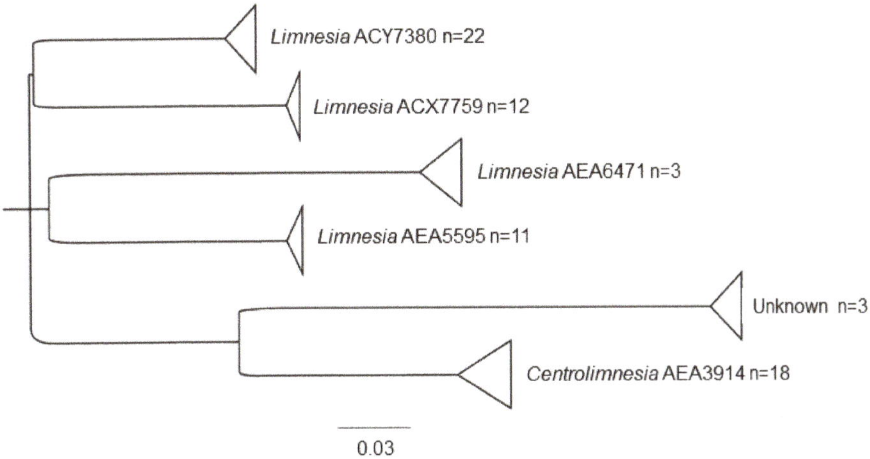

Figure 8. NJ tree for Limnesiidae family.

3.2. BIN Assemblies in the PY

From the total, 58 BINs were present in one or a maximum of three localities, possibly forming unique species assemblages (Tables 1 and 2). The Jaccard index value, in general, for all the localities, was zero or extremely low. However, some systems shared a percentage of their water mite fauna composition as follows: Chichancanab lagoon and Cenote El Padre (44%), Chichancanab lagoon and Cenote Km 48 (44%), Cenote El Toro and Cenote Santa Teresa (33%), Cenote Tres Reyes II and Cenote El Toro (33%), and Cenote Cocalitos and Cenote Escuela Normal (43%). The two latter are important because they are two different water systems inside the Bacalar Lagoon. Despite having such spatial relationship, each system seemed to have a different composition of water mites (Figures 9 and A1, Appendix A).

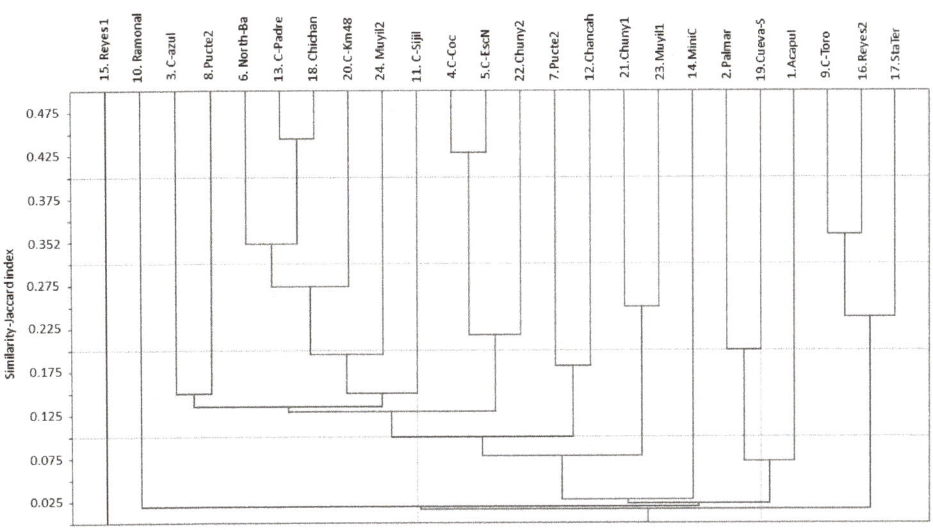

Figure 9. Similarity in water mites' composition between locations. Numbers in front of location name are the same as in Table 1.

4. Discussion

For the first time, a general analysis of the potential richness of water mite fauna in the central–southern part of the Yucatan Peninsula (Mexico), based on DNA barcodes was completed. Our results indicate an 11-fold increase in the number of species found previously in Quintana Roo state and twice the number of species registered in all the PY (in the three states, namely Campeche, Quintana Roo, and Yucatan) [6–8]. Out of 77 BINs, 58 are new in BOLD and seem to have a restricted distribution. This result indicates the presence of a unique set of environmental conditions and a particular water mite fauna composition of which most likely could be undescribed taxa. We need to study mite fauna in a wider geographic region to support this point; however, we have seen that most species are not widely distributed in our study area.

In the case of *Hydrodroma*, our results indicate the presence of only one morphospecies in eight sampling sites and has a proper correspondence with the unique BIN ADF3732. Previous research identified two species in the PY, *H. peregrina* Cook, 1980, and *H. despiciens* Marshall, 1936. However, for both species, Cook (1980) noticed distinctions from the type specimen. A recent study, using integrative taxonomy [30], compared sequences with the specimens collected from the Cenote Azul (Mexico) (*Hydrodroma* ADF3732), and the authors concluded that it was not *H. despiciens* [30] and probably not *H. peregrina*, due to the differences noticed by Cook [6]. Consequently, *Hydrodroma* BIN ADF3732 is probably a new species that needs to be formally described and is likely endemic to Southern Mexico.

Similarly, in the case of Unionicolidae, we registered 18 BINs (Figure 4 and Table 2). Previous records include ten species for the PY. Three of them correspond to descriptions of *Koenikea indistincta* Marshall, 1936; *Koenikea neopectinifera* Cook, 1980; and *Neumania cenotea* Marshall, 1936. All of them were apparently restricted to this region. The rest are described from other localities in Mexico or different regions. For example, *Unionicola gracilipalpis tenuis* Cook, 1980, was recorded in Campeche, Michigan, and Canada, but the type locality is in Haiti. Nevertheless, *U. gracilipalpis* was originally described from Europe. It is possible that this subspecies could be a full species, but we need to compare the type material to reach such a conclusion. *Unionicola (Pentatax) furculopsis* Cook, 1980, was described from Oaxaca state, but it was found in the Cenote Azul and Bacalar Lagoon by Otero-Colina [7]. Nevertheless, he noticed a similarity with *U. furcula* (Lundblad, 1935) and described some characteristics that the type species did not have, such as denticles in the gnathosoma base. These differences could be critical to identifying a different species, but more detailed research is required. *Neumania (Neumania) diversipalpa* Cook, 1980, was originally described from a single male in a river in Chiapas, based on an adult female, was recorded in the Cenote Azul by Otero-Colina [7]. The match male–female should be made from the same locality or at least after DNA barcodes have been obtained.

These are some examples of the taxonomic uncertainties that exist for water mites from the PY; however, our goal was not to discuss all previous identifications. These are just examples of the taxonomic impediment that still exists about "subspecies", "forms", and species recorded far away from the type locality or in an extremely different habitat from the original site. Some studies have revealed that species previously considered to be cosmopolitan are not really [30]. Many of them could be actually new species or species complexes. We consider that at least 15 OTUs of the Unionicolidae recognized by different BINs are possible new species.

Likewise, for Arrenuridae, there are seven species reported from the PY [8], six *Arrenurus* from the subgenus *Megaluracarus*, and one from the subgenus *Arrenurus*. Most of these reports are from Campeche and only one from Quintana Roo and Yucatan. We found 17 putative species of *Arrenurus* (Table 2 and Figure 6). Only three of these 17 BINs appear in multiple locations. The remaining 14 were found in only one or two close sampling sites (Table 2). After a comparison with the 135 BINs of arrenurids currently in BOLD, 94% of the BINs that we found appear juts in the south of Mexico. Other studies have previously documented the endemism of this family in other regions of the world [31–33]; however, this cannot be verified until a detailed morphological review of the specimens is made.

Another important achievement of this study is the pairing of males and females in nine BINs of this group that exhibit a high sexual dimorphism. Pairing the nymphal state in another three BINs will also allow us to make more complete formal descriptions if this species turns out to be undescribed (Figure 5).

The Hygrobatidae were the third richest group that we found (Table 2 and Figure 7). These are the first records for the PY. They were common in some locations (personal observation) that were previously surveyed [6,7]. This family seemed to be rare in the 1980s, when the previous studies took place. Some authors [2,34,35] suggest that several members of this family, *Hygrobates* included, are indicators of pollution and environmentally stressed water bodies. They were found in places like Cenote Cocalitos, Palmar, and Cenote Azul, with strong development of tourism (Tables 1 and 2). However, we must clarify the identity and habitat preferences of the species found in order to conclude if they indicate some level of environmental degradation. They may just be adapted to the extreme conditions of these places due to the presence of carbonates [10]. Nevertheless, previous surveys overlooked them.

The uniqueness of each aquatic system is clearly supported by the low values of the Jaccard index between the localities (Figure 9 and Appendix A Figure A1). For example, Cenote Azul and Cenote Cocalitos (Figures 1 and 2) are two localities with a distance of 160 m, but their similarity index is only 0.13 (Figure 9). This supports previous studies that found a difference in water quality and absence of communication between Cenote Azul and Bacalar [10]. Of the 14 BINS found in Cenote Azul and 21 found in Cocalitos, they only share four: *Limnochares* ACY6840, *Hydrodroma* ADF3732, *Unionicola* ACX8034, and *Arrenurus* ACX8463. These two systems have been extensively sampled, and their differences are also reflected in the composition of their planktonic communities [13,23]. Related studies have found that water mite assemblages are partially explained by environmental parameters such as temperature, conductivity, or pH and can almost be predicted by the potential prey groups, mainly cladocerans, copepods, and chironomids [36].

The PY ecosystems are characterized for being a mosaic of multiple habitats, with extreme differences in hydrogeochemistry conditions [9,10,12]. Their unique configuration that is structured after faults, underground and surface intermittent connections, and sinkholes (the most common surface water systems) suggests that they could be isolated. Therefore, they exhibit a distinctive diversity. Additionally, the distribution of water mites is known to be influenced by the substrate, type of vegetation, water flow, and depth. For example, El Palmar and Acapulquito present microhabitats with slow flow current combined with pools and submerged vegetation. As a result, we found a mixture of taxa with lotic environment preferences as *Torrenticola* and species with lentic preferences as *Arrenurus* or *Unionicola* [2].

Evidently there are still several unanswered questions in terms of the diversity of water mites in the PY. For example, are there specific assemblies for microhabitats? What causes differences in abundance between species? What are the phylogenetic relationships between them, or how is their evolutive history in the PY? Finally, we consider this analysis as a preliminary step toward the formal description of all the species that we found, including morphological details of the vouchers, in order to assign them a Linnaean name; once this step has been carried out, many of our hypotheses about restricted distributions and new endemic species could be fully tested.

Author Contributions: Conceptualization, L.M.-O. and M.E.-G.; methodology, M.E.-G.; software, L.M.-O.; validation, M.E.-G.; formal analysis, L.M.-O.; investigation, L.M.-O.; resources, M.E.-G.; data curation, L.M.-O.; writing—original draft preparation, L.M.-O.; writing—review and editing, M.E.-G. and L.M.-O.; visualization, L.M.-O. and M.E.-G.; supervision, M.E.-G.; project administration, M.E.-G.; funding acquisition, M.E.-G. All authors have read and agreed to the published version of the manuscript.

Funding: This study is a result of the PhD thesis of Lucia Montes Ortiz, at El Colegio de la Frontera Sur, supported by a scholarship by the National Council of Science and Technology (CONACYT). Some of the results presented here were obtained from the Project financed by Global Environment Fund through the Programa de Naciones Unidas para el Desarrollo (PNUD México), Comisión Nacional para el Conocimiento y Uso de la Biodiversidad (CONABIO) and Comisión Nacional de Áreas Naturales Protegidas (CONANP): Programa de detección temprana piloto de especies acuáticas invasoras a través de los métodos de código de barras de la vida y análisis de ADN ambiental en la Reserva de la Biosfera Sian Ka'an. Proyecto 00089333 "Aumentar las capacidades de México para manejar especies exóticas invasoras a través de la implementación de la Estrategia Nacional de Especies Invasoras" conducted by Martha Valdez-Moreno.

Acknowledgments: We thank Tom Goldschmidt for the support on morphological identifications, Alexei and Iurthitsi Elías Valdéz, and Jonas Goldschmidt for the field assistance, and all the team that worked in the Project 000089333, in particular Martha Valdez Moreno, José Angel Cohuo Colli, Adrian Emmanuel Uh Navarrete, Ivan Canul Palma, and Georgina Alexandra Prisco Pastrana. Alma Estrella García Morales from the Mexican Barcode of Life (MEXBOL), node Chetumal assisted with DNA extraction, PCR reactions, and sequence edition of all material presented here. Isaac Farraz Montes assisted with the map presented here, and Humberto Bahena provided the pictures from Cenote Azul, Cenote Cocalitos, Bacalar, and the Microbialites. Brianna Jacobson kindly performed a style review on the manuscript. ME-G wants to recall about the lack of interest of the Mexican Government on Biodiversity studies in a country that is the fourth place in the world. It's always been the same, but recently it's been worst. We will continue training human resources in this area as a commitment to Mexico and our planet.

Conflicts of Interest: The authors declare no conflict of interest. The funders had no role in the design of the study; in the collection, analyses, or interpretation of data; in the writing of the manuscript; or in the decision to publish the results.

Appendix A

	AC	PAL	CAZ	COC	CEN	BAN	CP1	CP2	CT	RAM	CSN	CCV	CP	MIC	CR1	CR2	CST	CHI	CS	K48	CH1	CH2	MU1	MU2
AC	-																							
PAL	0.14	-																						
CAZ	0.00	0.06	-																					
COC	0.03	0.08	0.13	-																				
CEN	0.00	0.08	0.05	0.43	-																			
BAN	0.00	0.12	0.22	0.35	0.15	-																		
CP1	0.00	0.00	0.10	0.15	0.13	0.15	-																	
CP2	0.00	0.08	0.15	0.11	0.13	0.10	0.06	-																
CT	0.00	0.00	0.00	0.00	0.00	0.00	0.00	0.10	-															
RAM	0.00	0.00	0.05	0.00	0.00	0.05	0.06	0.00	0.00	-														
CSN	0.00	0.00	0.06	0.09	0.09	0.06	0.00	0.09	0.00	0.00	-													
CCV	0.00	0.00	0.06	0.09	0.00	0.20	0.18	0.00	0.00	0.00	0.17	-												
CP	0.00	0.00	0.11	0.17	0.15	0.33	0.15	0.07	0.00	0.07	0.29	0.43	-											
MIC	0.00	0.00	0.07	0.05	0.10	0.00	0.10	0.10	0.00	0.00	0.00	0.00	0.00	-										
CR1	0.00	0.00	0.00	0.00	0.00	0.00	0.00	0.00	0.00	0.00	0.00	0.00	0.00	0.00	-									
CR2	0.00	0.00	0.00	0.04	0.07	0.05	0.00	0.15	0.33	0.00	0.00	0.00	0.09	0.00	0.00	-								
CST	0.00	0.00	0.00	0.00	0.00	0.00	0.00	0.10	0.33	0.00	0.00	0.00	0.00	0.00	0.00	0.14	-							
CHI	0.00	0.09	0.24	0.12	0.14	0.31	0.07	0.23	0.00	0.07	0.11	0.10	0.44	0.00	0.00	0.08	0.00	-						
CS	0.00	0.20	0.00	0.05	0.11	0.07	0.00	0.00	0.00	0.00	0.00	0.00	0.00	0.00	0.00	0.00	0.00	0.00	-					
K48	0.00	0.10	0.11	0.13	0.15	0.18	0.15	0.25	0.00	0.07	0.13	0.00	0.20	0.00	0.00	0.09	0.00	0.44	0.00	-				
CH1	0.00	0.00	0.13	0.00	0.00	0.00	0.09	0.00	0.00	0.00	0.00	0.00	0.00	0.00	0.00	0.00	0.00	0.00	0.00	0.00	-			
CH2	0.00	0.14	0.09	0.19	0.25	0.14	0.11	0.18	0.00	0.00	0.08	0.00	0.06	0.00	0.00	0.06	0.00	0.13	0.09	0.13	0.17	-		
MU1	0.00	0.00	0.17	0.12	0.00	0.17	0.07	0.07	0.00	0.00	0.11	0.22	0.18	0.00	0.00	0.00	0.00	0.08	0.00	0.08	0.25	0.20	-	
MU2	0.00	0.00	0.06	0.09	0.18	0.06	0.08	0.08	0.00	0.00	0.17	0.00	0.25	0.00	0.00	0.11	0.00	0.22	0.00	0.25	0.17	0.07	0.00	-

Figure A1. Matrix of Jaccard index values for all pairs of sampled locations, based on water mite BINs. AC = Acapulquito, PAL = Palmar, CAZ = Cenote Azul, COC = Cenote Cocalitos, CEN = Cenote Escuela Normal, BAN = Bacalar Norte, CP1 = Cenote Pucte 1, CP2 = Cenote Pucte 2, CT = Cenote El Toro, RAM = Ramonal, CSN = Cenote Sijil Noh Ha, CCV = Cenote Chancah Veracruz, CP = Cenote del Padre, MIC = Minicetonte, CR1 = Cenote Tres Reyes 1, CR2 = Cenote Tres Reyes 2, CST = Cenote Santa Teresa, CHI = Chichancanab, CS = Cueva de las serpientes, K48 = Cenote Km.48, CH1 = Chunyaxche 1, CH2 = Chunyaxche 2, MU1 = Muyil 1, and MU2 = Muyil 2.

References

1. Di Sabatino, A.; Smit, H.; Gerecke, R.; Goldschmidt, T.; Matsumoto, N.; Cicolani, B. Global diversity of water mites (Acari, Hydrachnidia; Arachnida) in freshwater. *Hydrobiologia* **2008**, *595*, 303–315. [CrossRef]
2. Proctor, H.C.; Smith, I.M.; Cook, D.R.; Smith, B.P. Subphylum Chelicerata, Class Arachnida. In *Thorp and Covich's Freshwater Invertebrades: Ecology and General Biology*; Thorp, J., Rogers, D.C., Eds.; Academic Press: Cambridge, MA, USA, 2015; pp. 599–660, ISBN 9780123850270.

3. Goldschmidt, T. The biodiversity of Neotropical water mites. In *Acarid Phylogeny and Evolution: Adaptations in Mites and Ticks;* Bernini, F., Nannelli, R., Nuzzaci, G., de Lillo, E., Eds.; Kluwer Academic Publisher: Dordrecht, The Netherlands, 2002; pp. 91–99.
4. Goldschmidt, T.; Ramírez-Sánchez, M.; Pedroza-Ramos, A.; Rico-Sánchez, A.E.; Ojeda, P.M.; Pérez-Munguia, R.M.; Pimiento-Ortega, M.; Carlos-Delgado, A.M.; Durán-Suáres, S. Primer registro de ácaros acuáticos (Acari: Hydrachnidia) del estado de Querétaro, México First records of water mites (Acari: Hydrachnidia) from Queretaro, Mexico. *Dugesiana* **2015**, *22*, 21–27.
5. Rivas, G.; Hoffmann, A. Los ácaros acuáticos de México. Estado actual de su conocimiento. *Mexicoa* **2000**, *2*, 33–39.
6. Cook, D.R. *Studies on Neotropical Water Mites;* The American Entomological Institute: Gainesville, FL, USA, 1980.
7. Otero-Colina, G. Datos originales sobre los ácaros acuáticos (Prostigmata:Parasitengona) del sureste de México. *Folia Entomol. Mex.* **1988**, *76*, 195–223.
8. Ramírez-Sánchez, M.M.; Rivas, G. New species of subgenus *Megaluracarus* (Acari: Hydrachnidiae: Arrenuridae: *Arrenurus*) from Mexico. *Zootaxa* **2013**, *3718*, 317–330. [CrossRef] [PubMed]
9. Gondwe, B.R.N.; Lerer, S.; Stisen, S.; Marín, L.; Rebolledo-Vieyra, M.; Merediz-Alonso, G.; Bauer-Gottwein, P. Hydrogeology of the south-eastern Yucatan Peninsula: New insights from water level measurements, geochemistry, geophysics and remote sensing. *J. Hydrol.* **2010**, *389*, 1–17. [CrossRef]
10. Perry, E.; Velazquez-Oliman, G.; Marin, L. The hydrogeochemistry of the karst aquifer system of the northern yucatan peninsula, Mexico. *Int. Geol. Rev.* **2002**, *44*, 191–221. [CrossRef]
11. Castro-Contreras, S.I.; Gingras, M.K.; Pecoits, E.; Aubet, N.R.; Petrash, D.A.; Castro-contreras, S.M.; Dick, G.; Planavsky, N.; Konhauser, K.O. Textural and Geochemical Features of Freshwater Microbialites. *Palaios* **2014**, *29*, 192–209. [CrossRef]
12. Gischler, E.; Gibson, M.A.; Oschmann, W. Giant holocene freshwater microbialites, Laguna Bacalar, Quintana Roo, Mexico. *Sedimentology* **2008**, *55*, 1293–1309. [CrossRef]
13. Elías-Gutiérrez, M.; Valdez-Moreno, M.; Topan, J.; Young, M.R.; Cohuo-Colli, J.A. Improved protocols to accelerate the assembly of DNA barcode reference libraries for freshwater zooplankton. *Ecol. Evol.* **2018**, *8*, 3002–3018. [CrossRef]
14. Pešić, V.; Smit, H. Evidence of cryptic and pseudocryptic speciation in *Brachypodopsis baumi* species complex (Acari, Hydrachnidia, Aturidae) from Borneo, with description of three new species. *Syst. Appl. Acarol.* **2016**, *21*, 1092–1106. [CrossRef]
15. Pešić, V.; Asadi, M.; Cimpean, M.; Dabert, M.; Esen, Y.; Gerecke, R.; Martin, P.; Savić, A.; Smit, H.; Stur, E. Six species in one: Evidence of cryptic speciation in the *Hygrobates fluviatilis* complex (Acariformes, Hydrachnidia, Hygrobatidae). *Syst. Appl. Acarol.* **2017**, *22*, 1327–1337. [CrossRef]
16. Pešić, V.; Broda, L.; Dabert, M.; Gerecke, R.; Martin, P.; Smit, H. Re-established after hundred years: Definition of *Hygrobates prosiliens* Koenike, 1915, based on molecular and morphological evidence, and redescription of H. longipalpis (Hermann, 1804) (Acariformes, Hydrachnidia). *Syst. Appl. Acarol.* **2019**, *24*, 1490–1511. [CrossRef]
17. Stålstedt, J.; Bergsten, J.; Ronquist, F. "Forms" of water mites (Acari: Hydrachnidia): Intraspecific variation or valid species? *Ecol. Evol.* **2013**, *3*, 3415–3435. [CrossRef] [PubMed]
18. Hebert, P.D.N.; Cywinska, A.; Ball, S.L.; DeWaard, J.R. Biological identifications through DNA barcodes. *Proc. R. Soc. B Biol. Sci.* **2003**, *270*, 313–321. [CrossRef]
19. DeWaard, J.R.; Ratnasingham, S.; Zakharov, E.V.; Borisenko, A.V.; Steinke, D.; Telfer, A.C.; Perez, K.H.J.; Sones, J.E.; Young, M.R.; Levesque-Beaudin, V.; et al. A reference library for the identification of Canadian invertebrates: 1.5 million DNA barcodes, voucher specimens, and genomic samples. *Sci. Data* **2019**, *6*, 308. [CrossRef]
20. Stryjecki, R.; Bańkowska, A.; Gryzińska, M.; Sarnacka, E.; Rutkowska, M.; Zawal, A. The use of molecular techniques in the taxonomy of water mites (Hydrachnidia, Acari). *Acta Biol.* **2016**, *23*, 117–126. [CrossRef]
21. Pešić, V.; Smit, H. A second Palaearctic species of the genus *Wettina* Piersig, 1892 based on morphological and molecular data (Acari, Hydrachnidia: Wettinidae). *Syst. Appl. Acarol.* **2018**, *23*, 724–732. [CrossRef]
22. Blattner, L.; Gerecke, R.; von Fumetti, S. Hidden biodiversity revealed by integrated morphology and genetic species delimitation of spring dwelling water mite species (Acari, Parasitengona: Hydrachnidia). *Parasites Vectors* **2019**, *12*, 1–13. [CrossRef]

23. Montes-Ortiz, L.; Elías-Gutiérrez, M. Faunistic survey of the zooplankton community in an oligotrophic sinkhole, cenote azul (Quintana roo, Mexico), using different sampling methods, and documented with DNA barcodes. *J. Limnol.* **2018**, *77*, 428–440. [CrossRef]
24. Martin, P.; Dabert, M.; Dabert, J. Molecular evidence for species separation in the water mite *Hygrobates nigromaculatus* Lebert, 1879 (Acari, Hydrachnidia): Evolutionary consequences of the loss of larval parasitism. *Aquat. Sci.* **2010**, *72*, 347–360. [CrossRef]
25. Young, M.R.; Proctor, H.C.; DeWaard, J.R.; Hebert, P.D.N. DNA barcodes expose unexpected diversity in canadian mites. *Mol. Ecol.* **2019**, *28*, 5347–5359. [CrossRef]
26. Ratnasingham, S.; Hebert, P.D.N. A DNA-based registry for all animal species: The barcode index number (BIN) system. *PLoS ONE* **2013**, *8*, e66213. [CrossRef] [PubMed]
27. Montes-Ortiz, L.; Goldschmidt, T.; Elías-Gutiérrez, M. First evidence of parasitation of a bosmina (Cladocera) by a water mite larva in a karst sinkhole, in Quintana Roo (Yucatán Peninsula, México). *Acarologia* **2019**, *59*, 111–114. [CrossRef]
28. Gutiérrez-Aguirre, M.A.; Cervantes-Martínez, A.; Elías-Gutiérrez, M.; Lugo-Vázquez, A. Remarks on mastigodiaptomus (Calanoida: Diaptomidae) from Mexico using integrative taxonomy, with a key of identification and three new species. *PeerJ* **2020**, *8*, e8416. [CrossRef] [PubMed]
29. Ivanova, N.V.; Dewaard, J.R.; Hebert, P.D.N. An inexpensive, automation-friendly protocol for recovering high-quality DNA. *Mol. Ecol. Notes* **2006**, *6*, 998–1002. [CrossRef]
30. Więcek, M.; Szydło, W.; Dabert, J.; Proctor, H. Delimiting species of water mites of the genus *Hydrodroma* (Acari: Hydrachnidiae: Hydrodromidae) from North America and Europe: Integrative evidence of species status from COI sequences and morphology. *Zool. Anz.* **2020**, *284*, 16–29. [CrossRef]
31. Esen, Y.; Erman, O. A new species of the genus *Arrenurus* Dugès, 1834 (Acari: Hydrachnidia: Arrenuridae) for the Turkish fauna: *Arrenurus* (*Truncaturus*) corsicus (E. Angelier, 1951). *Turk. J. Zool.* **2013**, *37*, 372–375. [CrossRef]
32. Erman, O.; Arslan, B.; Gülle, P. Türkiye ve Bazı Palearktik Bölge Ülkelerinin *Arrenurus* Duges (Arrenuridae: Hydrachnidia: Acari) Türlerinin Karşılaştırılması. *Mehmet Akif Ersoy Üniversitesi Fen Bilim. Enstitüsü Derg.* **2020**, *11*, 22–33. [CrossRef]
33. Wiles, P.R. The water mites (Acari: Hydrachnidia) of New Guinea. *Raffles Bull. Zool.* **1997**, *45*, 375–418.
34. Walter, D.E.; Proctor, H.C. *Mites: Ecology, Evolution & Behaviour: Life at a Microscale*, 2nd ed.; Springer: Dordrecht, The Netherlands, 2013; ISBN 9789400771642.
35. Goldschmidt, T. Water mites (Acari, Hydrachnidia): Powerful but widely neglected bioindicators—a review. *Neotrop. Biodivers.* **2016**, *2*, 12–25. [CrossRef]
36. Pozojević, I.; Juršić, L.; Vučković, N.; Dorić, V.; Gottstein, S.; Ternjej, I.; Mihaljević, Z. Is the spatial distribution of lentic water mite assemblages (Acari: Hydrachnidia) governed by prey availability? *Exp. Appl. Acarol.* **2019**, *77*, 487–510. [CrossRef] [PubMed]

diversity

MDPI

Article

Crenal Habitats: Sources of Water Mite (Acari: Hydrachnidia) Diversity

Ivana Pozojević [1], Vladimir Pešić [2], Tom Goldschmidt [3] and Sanja Gottstein [1,*]

[1] Department of Biology, Faculty of Science, University of Zagreb, Rooseveltov trg 6, 10000 Zagreb, Croatia; ivana.pozojevic@biol.pmf.hr

[2] Department of Biology, University of Montenegro, Cetinjski put b.b., 81000 Podgorica, Montenegro; vladopesic@gmail.com

[3] Zoologische Staatssammlung, Münchhausenstraße 21, D-81247 München, Germany; tomgoldschmidt@web.de

* Correspondence: sanja.gottstein@biol.pmf.hr

check for updates

Received: 29 July 2020; Accepted: 17 August 2020; Published: 20 August 2020

Abstract: Many studies emphasized the role that water mites play within the invertebrate communities of spring ecosystems, regarding species diversity and its significance within the crenal food web, as well as the specific preferences water mites exhibit towards spring typology. In pristine natural springs with permanent flow, water mites are nearly always present and usually display high diversity. This study aimed to determine whether significant differences in water mite assemblages between rheocrene (river-forming springs with dominant riffle habitats) and limnocrene (lake-forming springs with dominant pool habitats) karst springs could be detected in terms of species richness, diversity and abundance, but also in different ratios of specific synecological groups: crenobiont (exclusively found in springs), crenophilous (associated with springs) and stygophilous (associated with groundwater) water mite taxa. Our research was carried out on four limnocrenes and four rheocrenes in the Dinaric karst region of Croatia. Seasonal samples (20 sub-samples per sampling) were taken at each spring with a 200-μm net, taking into consideration all microhabitat types with coverage of at least 5%. Water mite abundance was found not to differ between morphological spring types. Significantly higher values of species richness and diversity indices were found in rheocrenes compared to limnocrenes, like those usually reported for this type of springs. However, unlike the studies previously reported, in this case, the higher shares of crenophilous and crenobiont water mite individuals were found in limnocrenes. The differences between stygophilous water mite taxa ratios among spring morphotypes were not statistically significant, indicating that the degree of the groundwater/surface water interaction (and water mite interaction therein) does not seem to be directly influenced by spring morphotype. Within this research, 40% of identified water mite species (eight out of twenty) were recorded for the first time in Croatia, thus highlighting again a huge gap in water mite knowledge of karst springs.

Keywords: spring ecology; crenal diversity; crenal water mites; spring morphology

1. Introduction

Springs are unique and complex ecosystems that house an array of different synecological groups of invertebrate species: from euryvalent generalist species, to highly specialized stenovalent crenobiont species (exclusively found in springs), as well as species that cohabitate between groundwater/surface water and aquatic/terrestrial habitats [1–4]. The eucrenal zone is frequently inhabited by stygophilous species (associated with groundwater), who depend on surface water habitats mostly for feeding and other biotic purposes and interactions [5]. Karst springs differ from other spring types usually by a higher degree of discharge heterogeneity and water level fluctuations [6,7] that are a result of the permeability of karstic geological layers [8]. Based on the classical Steinmann–Thienemann limnological spring classification, springs are usually categorized by ecomorphotype, as it is considered to be the key defining factor in determining crenal communities. Three main morphotypes include limnocrenes (lake-forming springs), rheocrenes (river-forming springs) and helocrenes (swamp-forming spring; [9]). In this research, we focused on limnocrene and rheocrene karst springs of Croatia. These springs mainly differ in water retention time and/or flow velocity that consequently influence an array of environmental parameters [7].

Most water mite species show clear preferences for specific habitats [10]; in fact, water mites are considered to be the invertebrate group with the highest ratio of crenophilous and crenobiont taxa (approximately 20%, [11,12]). Two are major crenobiont groups suggested by Gerecke et al. [11] that are called "paleocrenobiont" and "neocrenobiont" water mite species. Paleocrenobiont species include water mites from the superfamily Hydryphantoidea and represent evolutionary "older" species, developed from terrestrial ancestors, that inhabited the freshwater systems with a relatively large and stable transition zone between terrestrial and aquatic habitats-ponds and springs [13]. Neocrenobiont species are derived from rhithrobiont species that have secondarily inhabited spring ecosystems by upstream migration. This group of species is usually dominant in rheocrenes, where we usually find the mentioned rheocrenobionts, other crenobionts not specific to a morphotype, but also rhithrobiont species [14]. Limnocrenes are often inhabited by non-specific crenobionts because in Europe it is commonly believed that no true limnocrenobiont water mite species are documented, although a recent study suggests that large karstic limnocrenes on the Balkan peninsula (Ecoregion 5 and 6 by Illies [15]) may be inhabited by true limnocrenobionts [16]. Nevertheless, because of the aforementioned reasons, rheocrenes usually display higher water mite diversity.

The main focus of this study was to analyse water mite diversity in spring habitats with four specific goals in mind: (1) to analyse differences in environmental parameters between limnocrenes and rheocrenes and their influence on water mite assemblage composition; (2) to determine differences in water mite species richness, abundance and diversity indices between the two spring ecomorphotypes; (3) to analyse the differences in ratios of specific synecological groups (stygophilous and crenophilous/crenobiont taxa) between the ecomorphotypes and (4) to determine which water mite species contribute most to the similarity within a spring ecomorphotype.

2. Materials and Methods

2.1. Study Area

The study area is located in the Dinaric Western Balkan ecoregion [15] of Croatia. Within the study area, four river-forming springs—rheocrenes—and four lake-forming springs—limnocrenes [9]—were selected: rheocrenes Čabranka, Zrmanja, Krčić, Žrnovnica and limnocrenes Kupa, Kamačnik, Una, Cetina (Figure 1).

Each spring was sampled four times, once per season in 2015/2016. On every sampling occasion, 20 subsamples were taken with regard to microhabitat composition. Only microhabitat types that made up at least 5% of the total substrate mosaic were sampled [17]. A total of 640 benthic macroinvertebrate (sub)samples were collected (80 per spring, 320 from each spring morphotype). Averaged abundances from 20 subsamples are shown in Appendix A in site per season display of values. All benthic samples of macroinvertebrate fauna were collected with a kick-sampler (200 μm mesh size; from approximately 0.0625 m^2 surface area) and preserved in alcohol. Adult water mite individuals were determined to species or genus level using [18–20], deutonymphs were determined using [21]. All water mite specimens are stored at Zoological collection of the Faculty of Science, Department of Biology in Zagreb, Croatia.

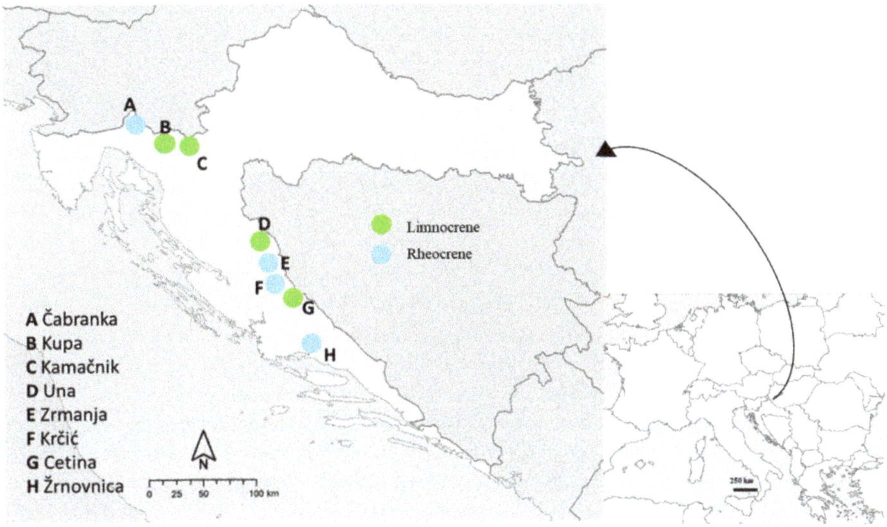

Figure 1. Map of the study area with the locations of the eight karst springs.

2.2. Environmental Variables

The following variables were chosen as environmental descriptors of differences between spring ecomorphotypes: water temperature, oxygen concentration and saturation (WTW optical oxygen sensor: FDO 925), chemical oxygen demand (COD; potassium permanganate oxidation by Standard Methods-APHA 2005), conductivity (WTW electrode conductivity measuring cell: TetraCon 925), concentration of calcium carbonate ($CaCO_3$ mg/L) and discharge (data obtained from Croatian meteorological and hydrological service).

2.3. Data Analysis

Canonical correspondence analysis (CCA) was used to assess the response of limnocrene vs. rheocrene water mite assemblages, as well as specific water mite species to environmental water properties. Only water mite taxa determined to species level were included in the analysis. A Monte Carlo test using 999 permutations ($p < 0.05$) was performed to test the significance of the analysis using the CANOCO package v.5.0, Ithaca, NY, USA [22]. These analyses were done in order to determine possible differences in environmental parameters between limnocrenes and rheocrenes and their influence on water mite assemblage composition.

Water mite local diversity was assessed by calculating the Shannon diversity index in PRIMER v.6.0, Plymouth, UK [23]. The environmental variables, water mite species richness, abundance and local diversity were tested for significant differences between spring morphotypes using the Student's t-test. Prior to this testing, the data were tested for normal distribution using the Shapiro–Wilk's test. These analyses were done in Statistica, 13.0; Tulsa, OK, USA [24] in order to determine differences in environmental factors, and water mite species' richness, abundance and diversity indices between the two spring ecomorphotypes.

Water mite species were divided into two synecological groups: (1) crenophilous/crenobiont and (2) stygophilous taxa. The ecological preferences were determined using literature data from [18–20]. The differences in ratios of the synecological groups was tested between morphotypes using the Mann–Whitney U test in Statistica, 13.0 [24].

Water mite assemblages from rheocrenes represented Group 1 and from limocrenes represented Group 2 in the SIMPER analysis of the (Bray–Curtis) similarity between water mite assemblages.

This was done in order to determine how water mite assemblages differ between ecomorphotypes in terms of species composition and abundance contribution. The analyses were done in PRIMER v.6.0 [23].

3. Results

A total of 20 water mite species was determined in this study, eight of which are recoded for the first time in Croatia: *Atractides latipalpis*, *A. walteri*, *Lethaxona cavifrons*, *Ljania macilenta*, *Partnunia steinmanni*, *Sperchon hibernicus*, *S. thienemanni* and *S. vaginosus*.

3.1. Environmental Conditions in Spring Ecomorphotypes

In the CCA analysis, the seven evaluated environmental parameters explained 33.6% of the total variation of water mite assemblages. The eigenvalues of the first two axes were 0.61 and 0.54 (Figure 2). A Monte-Carlo permutation test showed that the ordination was statistically significant ($F = 2.4$, $p = 0.002$). Limnocrenes were, in general, associated with higher levels of oxygen concentration and discharge, whereas rheocrenes were associated with higher values in conductivity and dissolved calcium carbonate. *Atractides pennatus* and *Partnunia angusta* were positively associated with higher water temperatures, whereas *Sperchon thienemanni* was negatively associated to higher temperatures. *Woolastookia rotundifrons* was positively associated with higher discharge levels. *Protzia squamousa* was found to positively associate with sites of higher levels of dissolved calcium carbonate and conductivity. *Partnunia steinmanni* was found to negatively associate with sites of higher COD values. Other species were positioned centrally in the analysis, or (as is the case with species like *Pseudotorrenticola rynchota* and *Sperchon vaginosus*) had multiple factors positioning them away from the centre, but were not significantly linked to any.

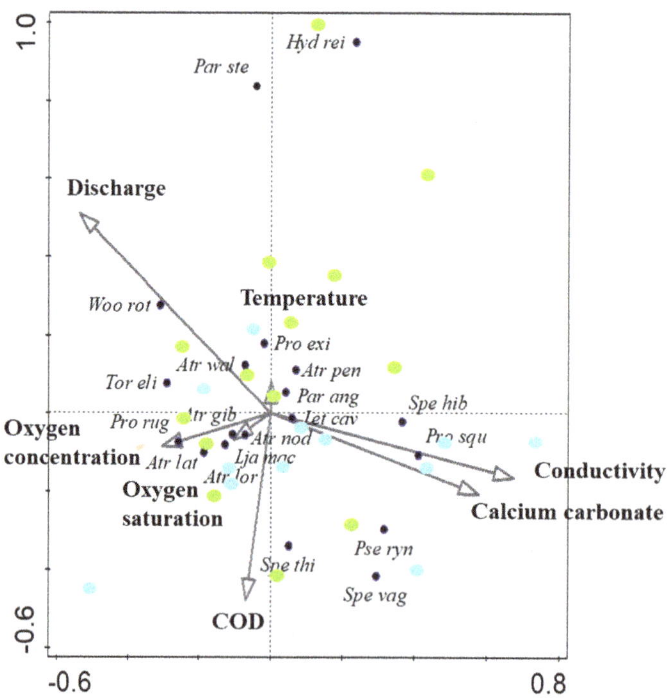

Figure 2. Canonical correspondence analysis (CCA) ordination of water mite samples, species and environmental water properties. Water mite taxa are marked with black dots, samples from limnocrenes

with green dots, samples from rheocrenes with blue dots and environmental water properties with arrows. Arrow length on the ordination indicates the relative importance of the explanatory variables (physico-chemical water properties), and their direction relative to each other and to the sites indicates positive or negative correlations. Abbreviations: COD = Chemical oxygen demand; *Atractides gibberipalpis* = Atr gib; *A. latipalpis* = Atr lat; *A. loricatus* = Atr lor; *A. nodipalpis* = Atr nod; *A. pennatus* = Atr pen; *A. walteri* = Atr wal; *Hydrodroma reinhardi* = Hyd rei; *Lethaxona cavifrons* = Let cav; *Ljania macilenta* = Lja mac; *Partnunia angusta* = Par ang; *Partnunia steinmanni* = Par ste; *Protzia eximia* = Pro exi; *Protzia squamosa* = Pro squ; *Protzia rugosa* = Pro rug; *Pseudotorrenticola rhynchota* = Pse ryn; *Sperchon hibernicus* = Spe hib; *S. vaginosus* = Spe vag; *S. thienemanni* = Spe thi; *Torrenticola elliptica* = Tor eli; *Woolastookia rotundifrons* = Woo rot.

Oxygen concentration values (*t*-test, F = 1.18; $p < 0.01$) and discharge levels (F = 7.52; $p < 0.001$) were significantly higher in limnocrenes, whereas conductivity (F = 3.26; $p < 0.05$) and values of dissolved $CaCO_3$ (F = 5.14; $p < 0.01$) were significantly higher in rheocrenes.

3.2. Water Mite Diversity in Spring Morphotypes

Water mite abundances were found not to differ significantly between spring morphotypes (Figure 3). However, water mite species richness (F = 1.517; $p < 0.05$) and local diversity (Shannon index; F = 1.257; $p < 0.05$) were found to be significantly higher in rheocrenes when compared to limnocrenes.

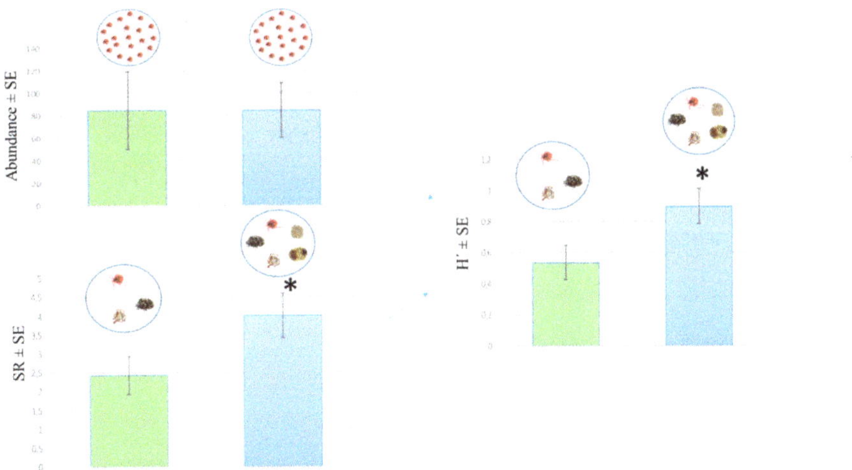

Figure 3. Illustration of the differences in water mite diversity (Shannon index, H') between limnocrenes (with green bars) and rheocrenes (with blue bars) as a result of water mite abundance and species richness (SR). Columns represent average values, whereas error bars represent the calculated standard errors (SE). The asterisk * indicates statistically significant ($p < 0.05$) differences between spring morphotypes (*t*-test).

Water mite species Atractides walteri, Hydrodroma reinhardi, Partnunia steinmanni, *Sperchon vaginosus*, *S. thienemanni*, *Torrenticola elliptica* and *Woolastookia rotundifrons* were found only in limnocrenes, whereas *Atractides latipalpis*, *A. loricatus*, *Lethaxona cavifrons*, *Ljania macilenta*, *Partnunia angusta* and *Sperchon hibernicus* were found exclusively in rheocrenes.

3.3. Synecological Water Mite Groups in Spring Morphotypes

Water mite species were divided into two synecological groups with regard to crenal and groundwater habitat preferences (Table 1). Two species, *Partnunia angusta* and *Torrenticola elliptica*, showed preferences to both habitat types. The ratio of crenophilous and crenobiont water mite taxa in the water mite assemblages varied greatly among sites, ranging from 0 to 100% of this synecological group within a local assemblage (details in appendix). The shares of stygophilous taxa ranged from 0 to 53% within a local assemblage. Stygophilous taxa, on average, made up around 4.4% ± 13.2% of water mite assemblages in limnocrenes and around 5.7% ± 11.1% in rheocrenes, but both ratios showed great variability, with standard deviations exceeding the average. The slightly higher ratios of stygophilous taxa in rheocrenes compared to limnocrenes were not statistically significant (Mann–Whitney *U*-test, $Z = 0.73$; $p > 0.05$). Crenobiont and crenophilous taxa, on average, made up around 19.6% ± 28.5% of water mite assemblages in limnocrenes and around 15.6% ± 27.7% in rheocrenes (again very high variability present shown in high standard deviations). The higher ratios of crenobiont and crenophilous taxa in limnocrenes compared to rheocrenes were statistically significant ($Z = 2.85$; $p < 0.05$).

Table 1. List of water mite species with determined ecological preferences (synecological groups) to crenal and groundwater habitats. Ecological preferences were determined using literature data from [18–20].

Water Mite Species	Preferences to Crenal Habitats	Preferences to Groundwater Habitats
Atractides latipalpis		Stygophilous
Atractides pennatus	Crenobiont	
Atractides walteri	Crenobiont	
Lethaxona cavifrons		Stygophilous
Ljania macilenta		Stygophilous
Partnunia angusta	Crenophilous	Stygophilous
Partnunia steinmanni	Crenobiont	
Protzia eximia	Crenophilous	
Protzia squamosa	Crenobiont	
Pseudotorrenticola rhynchota		Stygophilous
Sperchon thienemanni	Crenophilous	
Torrenticola elliptica	Crenophilous	Stygophilous

3.4. Water Mite Assemblage Similarity within Spring Morphotypes

The SIMPER group similarity analysis (only portraying taxa that contributed 4% to group similarity or more, Table 2) showed that both groups of sites (Group R- Rheocrenes and Group L-Limnocrenes) had a relatively high presence of *Atractides nodipalpis* and *A. gibberipalpis* individuals. *Protzia* species: *P. squamosa* and *P. rugosa* were associated with rheocrene sites whereas; *A. pennatus* individuals were associated with limnocrenes. Juvenile instars (nymphs) of numerous taxa also contributed to the similarity within spring groups.

Table 2. Results of the SIMPER analysis based on water mite assemblages from sites of different spring morphology: R-Rheocrenes and L-Limnocrenes.

Species	Average Abundance (Individuals/m^2)	Contribution to Group Similarity (%)
Group R-Rheocrenes		
Average similarity: 20.12		
Atractides gibberipalpis	1.73	20.63
Lebertia sp.	1.07	20.31
Atractides nodipalpis	1.22	10.15
Torrenticola sp.	1.1	9.44
Atractides sp.	1.15	8.87
Protzia squamosa	1.02	6.28
Sperchon sp.	0.9	5.01
Protzia rugosa	0.68	4.05
Group L-Limnocrenes		
Average similarity: 19.44		
Sperchon sp.	1.77	34.46
Atractides sp.	1.21	27.98
Atractides nodipalpis	0.82	7.31
Atractides gibberipalpis	0.9	5.23
Atractides pennatus	0.46	4.21

4. Discussion

Significant correlations with water temperature were determined for three water mite species, interestingly, all determined as crenobiont species. However, these significant relationships were not all showing the same trends. *A. pennatus* and *P. angusta* showed positive associations to higher water temperature, whereas *S. thienemanni* had the opposite trend. The idea that crenobiont species are cold stenothermic is questioned once more, and perhaps a more suitable angle of the problem, as proposed by Gerecke et al. [11], would suggest that the crenobiont water mite species are, in fact, "warm stenothermic". [11] proposed that such a vast speciation rate of crenal water mites might be due to the mild and stable environmental conditions in springs that are most obvious and pronounced in winter, thus taking the cold stenothermal theory and turning it upside-down. However, the constant temperature alone is most certainly not the only factor that contributes to the water mite diversity in springs, as crenobiont species are even found in tropical regions, where the temperature remains unchanged between the crenal area and lower river reaches [25,26]. The temperature arrow length in the CCA (indicates the relative importance of explanatory variable) was by far the smallest in the analysis (Figure 2). This is surely a result of the narrow temperature gradient present in springs, but also may indicate that, in fact, this factor is not the driving force of water mite diversity in springs. In this study, the environmental variables explained a substantial 33% of the total water mite variability in the studied springs, indicating that abiotic conditions are undoubtedly important contributors to water mite diversity. However, a more detailed portrait of the environmental effects would perhaps be shown if nutrient content in waters and available food sources were analyzed as well, but this was not the main focus of this paper. Here, a much broader approach regarding the influence of environmental conditions was taken when morphotypes were examined as the key drivers of water mite diversity in springs included in this research (as ecomorphotypes are hypothesized as one of the main drivers of environmental conditions within springs). A whole set of environmental variables (oxygen concentration, discharge levels, conductivity and dissolved $CaCO_3$) were found to significantly differ between ecomorphotypes, thus validating the approach.

Higher species richness and local diversity were recorded in rheocrenes when compared to limnocrenes, as expected when it comes to springs in Europe.

The differences between stygophilous water mite taxa ratios among spring ecomorphotypes were not statistically significant, indicating that the degree of the ground water/surface water interaction (and water mite interaction therein) does not seem to be directly influenced by spring ecomorphotype. Nevertheless, it is important to note that several/some stygophilous species—*A. latipapis, L. cavifrons, L. macilenta* and *P. angusta*—were found exclusively in rheocrenes.

Species that were exclusively found in limnocrenes fall into very different categories when it comes to habitat preferences: crenobiont (*A. walteri, P. steinmanni*), crenophilous (*S. thienemanni, T. elliptica*) to rheophilous species (*H. reinhardi, S. vaginosus, W. rotundifrons*). Species *A. walteri* and *P. steinmanni*, although crenobionts, are not considered highly specialized when it comes to spring ecomorphotype and environmental conditions. *S. thienemanni* is considered as one of the most widespread and abundant crenophilous taxa, and also as non-specific when it comes to ecomorphotype [27]. The distribution of many species in a large karstic limnocrenes (such as the ones discussed in this paper) depends on microhabitat/substrate composition, which provides conditions for different synecological groups; for example, the deeper part of limnocrenes provides conditions for good swimmers, while the spring brooks, often characterized by gravel/sandy substrate, can provide a habitat for crawlers such as *T. elliptica* and *W. rotundifrons*, which are usually associated with more riffle type habitats. *A. latipalpis* (stygophilous), *A. loricatus* (rheophilous), *L. cavifrons* (stygophilous), *L. macilenta* (stygophilous), *P. angusta* (crenophilous/stygophilous) and *S. hibernicus* (rheophilous) were found exclusively in rheocrenes.

5. Conclusions

With all the aforementioned taxa found exclusively in one spring ecomorphotype making up for more than 50% of the total taxa found in this research, we can conclude that spring ecomorphotype is indeed a crucial driver of water mite diversity. Furthermore, within this research, 40% of identified water mite species (eight out of twenty) were recorded for the first time in Croatia [28] (and quotes therein), thus highlighting once more the vast and undisclosed water mite diversity in spring ecosystems as well as the importance of spring habitats for the overall regional diversity of water mites. There is no ecology without knowing the fundamentals—species richness and diversity—and this is especially true in the case of research on one of the most diverse group of animals—water mites—in some of the most fragile ecosystems-freshwater springs.

Author Contributions: Conceptualization and supervision, I.P. and S.G.; methodology, S.G. and I.P.; formal analysis, I.P., S.G., T.G. and V.P.; resources, S.G. and I.P.; data curation, I.P., S.G, T.G. and V.P.; visualization, I.P.; funding acquisition, S.G.; project administration, S.G.; writing—original draft preparation, I.P.; writing—review and editing, I.P., S.G., T.G. and V.P. All authors have read and agreed to the published version of the manuscript.

Funding: This research was funded by The Environmental Protection and Energy Efficiency Fund (EPEEF) of Croatia, as part of the MULTISEC project (MULTImetric System of Evaluating Crenobiocoenosis) study conducted from September 2013 to December 2016. Project manager: Sanja Gottstein.

Acknowledgments: We kindly thank Ivica Barač for assistance in filed work.

Conflicts of Interest: The authors declare no conflict of interest.

Appendix A

Table A1. Seasonal values of measured environmental variables in the studied limnocrenes and rheocrenes. (T = water temperature, con. O_2 = oxygen concentration, sat. O_2 = oxygen saturation, σ = conductivity, $CaCO_3$ = calcium carbonate concentration, COD = chemical oxygen demand).

Spring (Morphotype)	Season	T (°C)	Con. O_2 (mg/L)	Sat. O_2 (%)	σ (µS/cm)	$CaCO_3$ (mg/L)	COD (mgO_2/L)
Čabranka	Fall	8.75	11.39	105.73	436.00	232.50	2.01
(Rheocrene)	Summer	8.80	11.12	101.60	420.33	206.67	0.68
	Spring	8.67	13.08	119.50	409.67	205.83	0.93
	Winter	8.38	11.09	101.50	378.75	192.50	1.81
Cetina	Fall	8.40	12.25	104.80	331.00	160.00	1.18
(Limnocrene)	Summer	9.10	13.02	114.83	294.00	145.00	1.10
	Spring	8.90	14.16	128.13	310.00	165.00	0.83
	Winter	8.90	12.81	111.38	338.00	175.00	0.24
Kamačnik	Fall	7.40	11.74	102.10	336.00	180.00	2.36
(Limnocrene)	Summer	9.23	12.57	114.70	303.25	166.25	0.58
	Spring	9.30	12.30	111.13	296.00	138.75	0.63
	Winter	7.20	11.95	103.70	322.00	162.50	1.02
Krčić	Fall	9.10	10.29	97.25	395.00	200.00	1.49
(Rheocrene)	Summer	9.70	10.63	101.43	395.00	202.50	0.94
	Spring	9.00	10.03	95.68	323.00	165.00	1.34
	Winter	8.90	9.90	94.30	406.00	195.00	1.49
Kupa	Fall	7.73	10.73	96.10	283.25	153.13	1.79
(Limnocrene)	Summer	9.40	11.50	103.00	266.75	135.63	0.85
	Spring	8.40	10.78	97.90	236.00	124.17	0.89
	Winter	7.57	11.49	103.80	256.33	125.83	1.55
Una	Fall	9.50	11.80	104.20	414.00	227.50	1.57
(Limnocrene)	Summer	11.80	11.72	102.70	420.00	215.00	1.18
	Spring	10.20	11.33	104.10	394.00	205.00	0.71
	Winter	9.30	11.35	101.10	440.00	247.50	1.41
Zrmanja	Fall	8.90	10.50	95.30	359.00	195.00	1.34
(Rheocrene)	Summer	9.80	11.84	108.70	347.00	170.00	1.10
	Spring	9.50	10.85	98.90	319.00	175.00	1.81
	Winter	8.40	12.08	107.50	363.00	185.00	1.81
Žrnovnica	Fall	12.52	11.26	102.30	413.80	190.00	1.20
(Rheocrene)	Summer	12.60	11.63	106.80	393.00	185.83	1.26
	Spring	12.56	11.14	107.90	369.57	185.00	0.79
	Winter	12.50	11.37	103.40	373.33	189.17	1.09

Table A2. Seasonal microhabitat ratios in the studied limnocrenes and rheocrenes.

Spring	Latitude	Longitude	Microhabitat Ratio (%)			
			Spring	Summer	Fall	Winter
Čabranka (Rheocrene)	45° 36′ 02.6″	14° 38′ 27.3″	Microlithal + Mesolithal (40%), Phytal (60%)	Microlithal + Mesolithal (60%), Phytal (40%)	Microlithal + Mesolithal (55%), Phytal (45%)	Mesolithal (70%), Phytal (30%)
Cetina (Limnocrene)	43° 58′ 50.9″	16° 25′ 81.4″	Akal (15%), Macrolithal (50%), Phytal (35%)	Microlithal + Mesolithal + Macrolithal (30%), Phytal (30%), Psammal (40%)	Akal (15%), Mesolithal (45%), Phytal (40%)	Microlithal + Mesolithal (55%), Phytal (45%)
Kamačnik (Limnocrene)	45° 20′ 49.3″	15° 03′ 38″	Ksilal + Microlithal + Mesolithal (40%), Phytal (60%)	Macrolithal + Megalithal (10%), Phytal (90%)	Microlithal (10%), Phytal (90%)	Microlithal + Mesolithal (10%), Phytal (90%)
Krčić (Rheocrene)	44° 25′ 48″	16° 33′ 61.9″	Phytal (100%)	Mesolithal (40%), Phytal (60%)	Mesolithal (20%), Phytal (80%)	Phytal (100%)
Kupa (Limnocrene)	45° 29′ 27.2″	14° 41′ 28.5″	Megalithal (15%), Mesolithal + Macrolithal (30%), Phytal (50%) Psammal (5%)	Microlithal (20%), Mesolithal (40%), Phytal (40%)	Microlithal + Mesolithal (80%), Phytal (10%), Psammal (10%)	Phytal (50%), Mesolithal (50%)
Una (Limnocrene)	44° 24′ 12.9″	16° 06′ 41.6″	Phytal (100 %)	Akal + Microlithal (10%), Mesolithal + Macrolithal (20%), Phytal (70%)	Microlithal + Mesolithal (80%), Phytal (20%)	Akal + Mesolithal (30%), Macrolithal + Mesolithal (50%), Phytal (20%)
Zrmanja (Rheocrene)	44° 11′ 79.2″	16° 03′ 38.1″	Microlithal + Mesolithal + Microlithal (55%), Phytal (45%)	Microlithal + Mesolithal (50%), Phytal (50%)	Microlithal + Mesolithal (40%), Phytal (60%)	Akal + Microlithal + Mesolithal (40%), Phytal (60%)
Žrnovnica (Rheocrene)	43° 31′ 24.8″	16° 34′ 29.2″	Mesolithal + Megalithal (60%), Phytal (40%)	Microlithal + Mesolithal (30%), Phytal (70%)	Microlithal + Mesolithal (40%), Phytal (60%)	Phytal (100%)

Table A3. Seasonal water mite taxa occurrence in the studied limnocrenes and rheocrenes.

Spring	Season	*Atractides* Sp.	*Atractides Gibberipapis*	*Atractides Latipapis*	*Atractides Loricatus*	*Atractides Nodipapis*	*Atractides Penatus*	*Atractides Walteri*	*Aturus* Sp.	*Feltria* Sp.	*Hydrodroma Reinhardii*	*Lebertia* Sp.	*Lethaxona Cavifrons*	*Ljania Macilenta*	*Partnunia Angusta*
Čabranka (Rheocrene)	Fall										+				
	Summer														
	Spring														
	Winter										+				
Cetina (Limnocrene)	Fall		+						+	+		+			
	Summer		+			+						+			
	Spring	+				+						+			+
	Winter	+										+			
Kamačnik (Limnocrene)	Fall						+								
	Summer														
	Spring	+								+					
	Winter	+					+								
Krčić (Rheocrene)	Fall														
	Summer	+										+			
	Spring														
	Winter														

Table A3. *Cont.*

Kupa (Limnocrene)	Fall	+	+												
	Summer	+	+			+			+	+		+			
	Spring	+	+			+			+	+		+			
	Winter	+	+			+			+	+		+			
Una (Limnocrene)	Fall	+	+			+		+	+						
	Summer						+		+			+			
	Spring						+		+						
	Winter					+									
Zrmanja (Rheocrene)	Fall	+	+	+	+	+			+	+		+		+	
	Summer	+	+			+			+	+		+	+	+	
	Spring					+									
	Winter	+	+			+			+						
Žrnovnica (Rheocrene)	Fall						+					+			
	Summer		+												+
	Spring		+												+
	Winter						+								+

Spring	Season	*Partnunia Steinmani*	*Protzia Eximia*	*Protzia Squamosa*	*Protzia Rugosa*	*Pseudotorrent-Icola Rhynchota*	*Sperchon Hibernicus*	*Sperchon* Deticulatus Group	*Sperchon Vaginousus*	*Sperchon* Sp.	*Sperchon Thienemanni*	*Torrenticola Eliptica*	*Torrenticola* Sp.	*Woolastookia Rotundifrons*	Hydrachnidia (Larvae)
Čabranka (Rheocrene)	Fall														
	Summer	+													
	Spring														
	Winter	+													
Cetina (Limnocrene)	Fall			+		+			+				+		
	Summer			+				+					+		
	Spring			+		+	+	+					+		+
	Winter			+				+					+		
Kamačnik (Limnocrene)	Fall										+				
	Summer														
	Spring							+							
	Winter									+	+				
Krčić (Rheocrene)	Fall														
	Summer														+
	Spring														
	Winter														
Kupa (Limnocrene)	Fall	+								+					
	Summer				+					+		+	+		+
	Spring				+					+		+	+		+
	Winter				+					+		+	+	+	+
Una (Limnocrene)	Fall		+							+			+		
	Summer		+							+					+
	Spring		+							+					
	Winter														+
Zrmanja (Rheocrene)	Fall				+					+			+		+
	Summer				+					+			+		+
	Spring				+					+					
	Winter				+					+			+		+
Žrnovnica (Rheocrene)	Fall														
	Summer									+			+		+
	Spring														
	Winter														

Table A4. Seasonal water mite taxa richness, abundance, Shannon diversity values and ratios of synecological (*** indicates no water mites present, whereas 0% indicates no taxa from synecological group present) water mite groups in the studied limnocrenes and rheocrenes.

Spring (Morphotype)	Season	Taxa Richness	Abundance (Individuals/m^2)	Shannon Diversity Index Value	Ratio of Stygophilous Taxa	Ratio of Crenobiont and Crenophilous Taxa
Čabranka (Rheocrene)	Fall	9	1280	1.446	1.82%	7.27%
	Summer	7	309	1.459	0.00%	37.50%
	Spring	9	451	1.447	5.92%	56.21%
	Winter	5	53	1.359	0.00%	79.17%
Cetina (Limnocrene)	Fall	2	57	0.693	0.00%	0.00%
	Summer	2	8	0.693	0.00%	100.00%
	Spring	0	0	0	***	***
	Winter	3	90	0.822	0.00%	92.86%
Kamačnik (Limnocrene)	Fall	3	7	1	0.00%	100.00%
	Summer	0	0	0	***	***
	Spring	3	144	0.958	0.00%	0.00%
	Winter	5	43	1.355	0.00%	45.83%
Krčić (Rheocrene)	Fall	0	0	0	***	***
	Summer	3	32	0.856	0.00%	0.00%
	Spring	0	0	0	***	***
	Winter	0	0	0	***	***
Kupa (Limnocrene)	Fall	4	14	1.091	0.00%	14.29%
	Summer	8	220	1.528	7.27%	7.27%
	Spring	10	1667	1.583	44.44%	44.44%
	Winter	12	856	1.479	53.08%	53.08%
Una (Limnocrene)	Fall	8	360	1.485	0.00%	27.27%
	Summer	5	162	1.044	0.00%	76.54%
	Spring	5	22	1.221	0.00%	35.71%
	Winter	2	32	0.693	0.00%	0.00%
Zrmanja (Rheocrene)	Fall	13	900	1.711	13.39%	0.00%
	Summer	11	630	1.552	1.52%	0.00%
	Spring	4	9	1.242	0.00%	0.00%
	Winter	8	288	1.444	0.00%	0.00%
Žrnovnica (Rheocrene)	Fall	3	5	1.04	0.00%	50.00%
	Summer	5	25	1.274	36.36%	36.36%
	Spring	3	12	1.011	33.33%	33.33%
	Winter	3	35	0.986	27.27%	100.00%

References

1. Gerecke, R.; Meisch, C.; Stoch, F.; Acri, F.; Franz, H. Eucrenon-hypocrenon ecotone and spring typology in the Alps of Berchtesgaden (Upper Bavaria, Germany). A study of microcrustacea (Crustacea: Copepoda, Ostracoda) and water mites (Acari: Halacaridae, Hydrachnellae). In *Studies in Crenobiology*; The Biology of Springs and Springbrooks; Botosaneanu, L., Ed.; Backhuys Publisher: Leiden, The Netherlands, 1998; pp. 167–182.
2. Knight, R.L.; Notestein, S.K. Effects of nutrients on spring ecosystems. In *Summary and Synthesis of the Available Literature on the Effects of Nutrients on Spring Organisms and Systems*; Florida Department of Environmental Protection: Tallahassee, FL, USA, 2008; pp. 271–304.
3. Stoch, F.; Gerecke, R.; Pieri, V.; Rossetti, G.; Sambugar, B. Exploring species distribution of spring meiofauna (Annelida, Acari, Crustacea) in the south-eastern Alps. *J. Limnol.* **2011**, *70* (Suppl. 1), 65–76. [CrossRef]
4. Reiss, M.; Chifflard, P. Hydromorphology and Biodiversity in headwaters: An Eco-faunistic substrate preference assessment in forest springs of the German subdued mountains. In *Biodiversity in Ecosystems—Linking Structure and Function*; Lo, Y.-H., Blanco, J.A., Roy, S., Eds.; InTech Open: Rijeka, Croatia, 2015; pp. 223–258.
5. Gottstein, M.S.; Bakran-Petricioli, T.; Bedek, J.; Bukovec, D.; Buzjak, S.; Franičević, M.; Tvrtković, N. An overview of the cave and interstitial biota in Croatia. *Nat. Croat.* **2002**, *11* (Suppl. 1), 1–112.
6. Bonacci, O. Karst springs hydrographs as indicators of karst aquifers. *Hydrol. Sci. J.* **1993**, *38*, 51–62. [CrossRef]
7. Smart, C.; Worthington, S.R.H. Springs. In *Encyclopedia of Caves and Karst Science*; Gunn, J., Ed.; Taylor & Francis: London, UK, 2004; pp. 1495–1505.
8. Prelovšek, M. Hydrology. In *Introduction to the Dinaric Karst*; Mihevc, A., Prelovšek, M., Zupan, H.N., Eds.; Karst Research Institute: Postojna, Slovenija, 2010; pp. 14–19.
9. Glazier, D.S. Springs. In *Likens GE (ur.) Encyclopedia of Inland Waters*; Elsevier: Oxford, UK, 2009; Volume 1, pp. 734–755.
10. Goldschmidt, T. Water mites (Acari, Hydrachnidia): Powerful but widely neglected bioindicators—A review. *Neotrop. Biodivers.* **2016**, *2*, 12–25. [CrossRef]
11. Gerecke, R.; Martin, P.; Gledhill, T. Water mites (Acari: Parasitengona: Hydrachnidia) as inhabitants of groundwater-influenced habitats—Considerations following an update of Limnofauna Europaea. *Limnologica* **2018**, *69*, 81–93. [CrossRef]
12. Pešić, V.; Savić, A.; Jabłońska, A.; Michoński, G.; Grabowski, M.; Bańkowska, A.; Zawal, A. Environmental factors affecting water mite assemblages along eucrenon-hypocrenon gradients in Mediterranean karstic springs. *Exp. Appl. Acarol.* **2019**, *77*, 71–486. [CrossRef] [PubMed]
13. Smith, I. Water mites (Acari: Parasitengona: Hydrachnida) of spring habitats in Canada. *Mem. Entomol. Soc. Can.* **1991**, *155*, 141–167. [CrossRef]
14. Gerecke, R.; Di Sabatino, A. Water mites (Acari, Hydrachnellae) and spring typology in Sicily. *Crunoecia* **1996**, *5*, 35–41.
15. Illies, J. *Limnofauna Europaea*; Gustav Fischer: Stuttgart, Germany; New York, NY, USA, 1978; pp. 1–532.
16. Pešić, V.; Asadi, M.; Cimpean, M.; Dabert, M.; Esen, Y.; Gerecke, R.; Martin, P.; Savic, A.; Smit, H.; Stur, E. Six species in one: Evidence of cryptic speciation in the Hygrobates fluviatilis complex (Acariformes, Hydrachnidia, Hygrobatidae). *Syst. Appl. Acarol.* **2017**, *22*, 1327–1377. [CrossRef]
17. AQEM Consortium. Manual for the application of the AQEM system. In *A Comprehensive Method to Assess European Streams Using Benthic Macroinvertebrates Developed for the Purpose of the Water Framework Directive*; Version 1.0; European Commission: Duisburg, Germany, 2020.
18. Davids, C.; Di Sabatino, A.; Gerecke, R. Chelicerata: Araneae, Acari I. In *Chelicerata: Araneae, Acari I. Süßwasserfauna von Mitteleuropa 7/2-1*; Bartsch, I., Davids, C., Deichsel, R., Eds.; Spektrum Akademischer: Heidelberg, Germany, 2007; pp. 241–333.
19. Di Sabatino, A.; Gerecke, R.; Gledhill, T.; Smit, H. Chelicerata: Acari II. In *Süßwasserfauna von Mitteleuropa 7/2-2*; Gerecke, R., Ed.; Spektrum: Heidelberg, Germany, 2010; pp. 1–216.
20. Gerecke, R.; Gledhill, T.; Pešić, V.; Smit, H. Chelicerata: Acari III. In *Süßwasserfauna von Mitteleuropa 7/2-3*; Spektrum: Heidelberg, Germany, 2016; pp. 1–417.
21. Tuzovskij, P. *Key to Deutonymphs of Water Mites*; Akademia Nauka Moscow, UdSSR: Moscow, Russia, 2000.

22. Ter Braak, C.J.F.; Šmilauer, P. *CANOCO Reference Manual and User's Guide: Software for Ordination (Version 5.0)*; Microcomputer Power: Ithaca, NY, USA, 2012.
23. Clarke, K.R.; Gorley, R.N. PRIMER V6: User manual/tutorial. In *Plymouth: Primer-E*; Plymouth Marine Laboratory: Plymouth, UK, 2006; pp. 1–192.
24. TIBCO Software Inc. Statistica (Data Analysis Software System), Version 13. Available online: http://statistica.io (accessed on 7 May 2019).
25. Goldschmidt, T. Environmental parameters determining the water mite communities in Costa Rican freshwater habitats. *Exp. Appl. Acarol.* **2004**, *34*, 171–197. [CrossRef] [PubMed]
26. Goldschmidt, T. Water mites (Acari, Hydrachnidia) in tropical springs—Diversity, specificity, monitoring possibilities. *Verh. Des. Int. Ver. Limnol.* **2009**, *30*, 669–672. [CrossRef]
27. Martin, P.; Stur, E. Parasite-host associations and life cycles of spring-living water mites (Hydrachnidia, Acari) from Luxembourg. *Hydrobiologia* **2006**, *573*, 17–37. [CrossRef]
28. Pozojević, I.; Pešić, V.; Gottstein, S. Two water mite species (Acari: Hydrachnidia) from karst springs new for the fauna of Croatia with notes on distribution and environmental preferences. *Nat. Croat.* **2019**, *28*, 417–424. [CrossRef]

diversity

MDPI

Review

The Biodiversity of Water Mites That Prey on and Parasitize Mosquitoes

Adrian A. Vasquez [1,2,*,†], Bana A. Kabalan [3,†], Jeffrey L. Ram [4] and Carol J. Miller [1]

1 Healthy Urban Waters, Department of Civil and Environmental Engineering, Wayne State University, Detroit, MI 48202, USA; ab1421@wayne.edu
2 Cooperative Institute for Great Lakes Research, School for Environment and Sustainability, University of Michigan, 440 Church Street, Ann Arbor, MI 48109, USA
3 Fisheries and Aquatic Sciences Program, School of Forest Resources and Conservation, University of Florida, Gainesville, FL, 32611, USA; bana.kabalan@ufl.edu
4 Department of Physiology, School of Medicine Wayne State University, Detroit, MI 48201, USA; jeffram@gmail.com
* Correspondence: avasquez@wayne.edu
† These authors contributed equally to this work.

Received: 2 May 2020; Accepted: 4 June 2020; Published: 6 June 2020

Abstract: Water mites form one of the most biodiverse groups within the aquatic arachnid class. These freshwater macroinvertebrates are predators and parasites of the equally diverse nematocerous Dipterans, such as mosquitoes, and water mites are believed to have diversified as a result of these predatory and parasitic relationships. Through these two major biotic interactions, water mites have been found to greatly impact a variety of mosquito species. Although these predatory and parasitic interactions are important in aquatic ecology, very little is known about the diversity of water mites that interact with mosquitoes. In this paper, we review and update the past literature on the predatory and parasitic mite–mosquito relationships, update past records, discuss the biogeographic range of these interactions, and add our own recent findings on this topic conducted in habitats around the Laurentian Great Lakes. The possible impact on human health, along with the importance of water mite predator–prey dynamics in aquatic food webs, motivates an increase in future research on this aquatic predator and parasite and may reveal novel ecological functions that these parasitic and predator–prey relationships mediate.

Keywords: *Arrenurus*; *Lebertia quinquemaculosa*; Lake St. Clair Metropark; Belle Isle; Detroit; phoresy; mesocosm; Diptera; freshwater ecology

---

1. Introduction

Water mites are both abundant and ubiquitous aquatic arachnids that are found globally in freshwater habitats, except in Antarctica. Water mites have high species richness and biomass and can easily be collected in the many habitats they occupy. Reports of over 600 specimens that represent up to 13 genera can be collected in under three hours in typical freshwater habitats in the Great Lakes region by using a basic dip net to collect aquatic debris, which, when placed on a white enamel pan, allows the mites to be easily siphoned by using a pipette, as they scurry out of the debris [1]. More recently, we reported 17 genera occupying one location in this region within the Detroit River [2]. A species accumulation curve from the Palearctic shows that a plateau in the curve has not been reached, indicating that many more species remain to be discovered [3]. Previous studies report that perhaps only half of water mite species in North America have been named, constituting around 6000 water mite species with potentially 10,000 or more species found globally [3]. In other regions of the world, such as the Neotropics, water mite biodiversity is largely unexplored, with species counts expected to be four times what is currently known [4]. Molecular barcoding has helped improve the knowledge of water mite diversity in North America, as is evident in our work in the Detroit River and Western Lake Erie, where we contributed several previously unknown molecular barcodes for multiple genera of water mites [2,5]. The high biodiversity attributed to water mites is thought to have been a result of repeated instances of rapid diversification that enabled exploitation of host insects such as Dipterans [6]. Some groups, such as *Lebertia*, are thought to have co-evolved with nematocerous Dipterans, such as chironomids, which are closely related to mosquitoes [6]. Subsequent sections in this review focus on the biodiversity of water mites that prey on mosquito eggs/larvae and parasitize emerging adults, with an emphasis on updating and correcting the known biodiversity, summarizing the biogeography, and identifying future research avenues with discussion of our recent findings.

Beyond the lack of knowledge about water mite biodiversity, studies on their life history strategies are also lacking, with virtually all previous reports of water mite diets based on laboratory observations. Water mites have a complex life cycle that has co-evolved with important freshwater insect groups, especially Diptera, including mosquitoes and midges [6], which have frequently, though not exclusively, been identified as water mite prey. Proctor and Pritchard [7] reviewed prey consumed by water mites, and their list included copepods, mosquito larvae, chironomid larvae, *Daphnia*, and ostracods [8–10]. This illustrates their importance as predators in aquatic habitats due to their widespread presence, voracious appetite, and high biomass [7]. Water mites are also important constituents of aquatic habitats due to their usefulness as bioindicator species of the habitats in which they are found [11]. Although this is a new area of investigation, studies in Central America and Europe have already shown the benefits of using water mites as bioindicators [12,13]. These studies underscore the importance of water mites in aquatic ecology and suggest the need for more investigation in water mite life history.

Water mite life history begins with a fertilized egg from which a larva hatches. Water mite larvae often develop into ectoparasites that parasitize aquatic insect adult hosts, such as mosquitoes, as the hosts eclose from their pupal case and enter an aerial environment. The host is used by the larval water mite for nutritional value and dispersal to a suitable habitat for post-larval development [6]. The effects of water mite parasitism on the host includes morphological damage and reduced survival and fecundity, therefore negatively affecting population sizes of host species if infection rates are high [6]. Almost two-thirds of host order species have been found to be in the order Diptera, which has been the main focus in studies of the effects of water mite parasitism [14]. After the parasitic larval stage, water mites detach from the host and develop into the deutonymph stage. During the deutonymph stage, water mites rapidly grow in body size mainly through predation on insect larvae, such as mosquito larvae, and other macroinvertebrates. Some water mite species, such as *Parathyas barbigera* and *P. stolli* have been found to prey on mosquito eggs in laboratory studies [15] and will be discussed later. Water mites subsequently develop into quiescent trytonymphs, and finally into predatory adult water mites. The presence of both parasitic and predatory behaviors and the combination of aquatic and terrestrial/aerial stages of water mites suggest that they may be important model species for understanding population dynamics of macroinvertebrate species that have a mix of aquatic and semi-aquatic life histories.

Mosquitoes have been more intensively studied than water mites, resulting in a more comprehensive understanding of their global biodiversity, comprising about 3500 species in at least 42 genera [16]. Mosquitoes have many predators and are considered an important food source for many aquatic organisms [16]. Mosquitoes have a semi-aquatic holometabolous life cycle that consists of four different stages—egg, larva, pupa, and adult [16]. The immature stages can exist in many types of aquatic habitats, allowing mosquitoes to have high species richness and biomass. Beyond inhabiting all types of permanent and ephemeral lentic and lotic freshwater habitats, mosquitoes are found to colonize rock holes, tree holes, parts of vegetation, and artificial containers, such as buckets, tires, flower vases, bird feeders, and more. The life cycle begins when fertilized eggs hatch into an aquatic larval stage and the larva typically hangs suspended from the water surface. The larva molts and sheds its skin, lasting for one to three weeks, depending on species type, water temperature, and food availability. The mosquito larval stage suffers the greatest threat of predation from aquatic species such as water mites. The pupa is a resting stage that is solely aquatic, with no feeding, and lasts from one to three days, during which the pupa metamorphoses into a flying adult that lives its life in both a terrestrial and aerial environment. The switching of adult mosquito hosts from animal to human can occur seasonally, enabling zoonotic disease transmission [16].

During the larval aquatic stage in the life history of mosquitoes, they are preyed upon by water mites. Water mites are true aquatic organisms, but many species (which are reviewed here) have an ectoparasitic larval stage that parasitizes organisms that may become airborne, such as mosquitoes. Biotic interactions, such as those of water mites and mosquitoes, contribute to functional biodiversity, which might be critical in sustaining ecosystems [17]. The impact of contemporary global biodiversity decline has prompted the United Nations (UN) to declare 2011 to 2020 as the "UN decade of biodiversity" [18]. Freshwater biodiversity is the most threatened form of biodiversity, and experts implore an increased investment in research and documentation of freshwater biodiversity [19].

Through the highly complex web of interactions among species, such as parasitism, ecosystem functioning and biodiversity can be altered [20]. One example of an ecosystem service provided by water mites includes lowering mosquito fecundity, and thus reducing mosquito prevalence [21–23]. These positive effects can be attributed to the presence of high diversity, because the likelihood of selection effects, facilitation from long-term coexistence, and niche complementarity are greater as diversity increases [20]. The ecosystem-level consequences from biodiversity loss are significant, being of the same magnitude as the effects on environments from other anthropogenic global-change stressors [24]. This reinforces the urgency of the conservation and restoration of biodiversity worldwide [25]. The importance of biodiversity can be demonstrated through several theories that link higher diversity to increased productivity, ecosystem stability, and resistance to invasion from exotic species [25].

Through the loss of biodiversity, we not only lose the species themselves, but we harm the direct and indirect community and ecosystem-level biotic interactions that they are embedded in, as well. Beyond just consumer diversity, the role of parasite diversity on ecosystem functions has rarely been regarded. Parasites are ubiquitous organisms that are capable of regulating host abundance and community assemblages, which in turn can impact host biodiversity and the ecosystem processes those hosts influence [20]. For example, a parasite that uses a herbivore host can reduce herbivore abundance, which can have a trophic cascade that increases plant primary productivity through reduced grazing pressure [26]. Parasites are capable of increasing or decreasing biodiversity through facilitating or removing novel traits, as well as increasing or decreasing trait diversity [20].

Water mites are globally diverse aquatic arachnids and increase the complexity of trophic networks by being both predators and parasites. This review updates our current knowledge of the diversity of water mites that interact with mosquitoes and updates past records. We summarize the literature on biogeography and discuss possible life history strategies of water mites. This work aims to advance water mite research by exploring new avenues of research revealed by preliminary data from mesocosm experiments regarding water mite predation on mosquito larvae in urban parks. Our mesocosm experiments reported here can be used to identify other water mite mosquito predators and could also be modified to study parasitism. This review also provides a platform to advance important aquatic ecological topics, such as predator–prey interactions and parasitism. The research on water mite parasitism and predation on mosquitoes remains a relevant area of investigation, given the many unknowns of the diversity of these biotic interactions and the continuing and ever-expanding threat from mosquitoes.

2. Materials and Methods

2.1. Literature Review

A literature review of all water mite associations with mosquitoes as predators and parasites, using the Wayne State University Web of Science© portal (Clarivate Analytics), resulted in 186 records from a total of 148,858,601 records. The search terms were "water mites parasite mosquito". A second search of "water mites predator mosquito" returned 24 records, of which none was relevant to the present review. A Google Scholar Publication search identified 41 records, comprising 2 books and 39 articles. Other articles of interest were obtained from primary authors themselves.

2.2. Field Experiments to Identify Water Mite Mosquito Predators

Mosquito-attracting mesocosms were deployed at 6 sites in Lake St. Clair Metropark (LSCMP), located in Harrison Township (42.5818° N, 82.8093° W), adjacent to Lake St. Clair, and at 6 sites in Belle Isle State Park (BI) (42.3433° N, 82.9743° W), a 400 acre urban island park in Detroit, MI, in the Detroit River, which forms the border between the United States and Canada (Figure 1). The mesocosms, consisting of buckets with a volume of 5 L suspended from wooden frames (Figure 2), were set up in wet-mesic flatwoods forest and marsh wetland habitats and monitored approximately every two weeks, from April through November 2018. The buckets filled naturally with rainwater, to varying depths, and by July, mosquitoes laid eggs in the buckets, and mosquito larvae were observed through October. After mosquito larvae were detected in the mesocosms, water mites (*Lebertia quinquemaculosa*, *Hydrachna*, or *Arrenurus*) were added to some mesocosms, at various intervals, while noting the presence or absence of mosquito eggs, larvae, and pupae at each interval (see Figure 2).

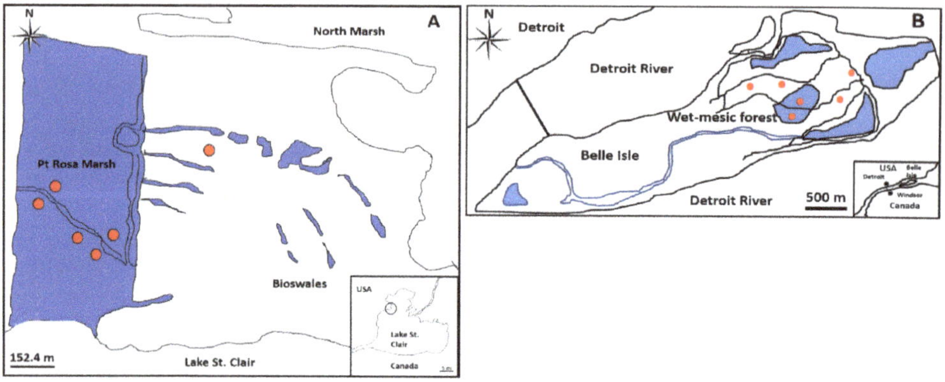

Figure 1. Maps of deployment sites (red markers) of mesocosm buckets. (**A**) Lake St. Clair Metropark; (**B**) Belle Isle State Park.

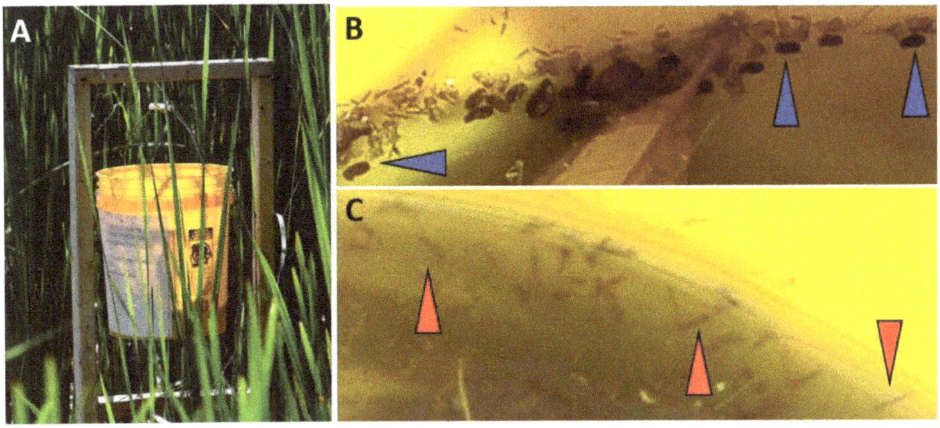

Figure 2. Mesocosms depicting mosquito larvae and pupae infestation. (**A**) Single-bucket mesocosm. (**B**) Mesocosm with mosquito pupae indicated by blue arrows. (**C**) Mesocosm with mosquito larvae indicated by red arrows.

3. Biodiversity of Water Mites and Mosquito Interactions

3.1. Predation

Since the 1700s, water mites have been studied by classical taxonomists such as Linnaeus and DeGeer. DeGeer (1778) reported water mite parasitism in his renowned work "Mémoires pour servir à l'histoire des insectes" (eight volumes, 1752–1778) [27]. However, observations of water mite predation on mosquito larvae were only reported much later, at the beginning of the 20th century. Water mites have been observed as predators of mosquito larvae under both natural conditions and controlled laboratory experiments. However, in comparison to the many studies of the diversity of water mites that parasitize mosquitoes, very few studies have been reported on the diversity of water mites as predators of mosquito larvae, eggs, or pupae. So far, there is only evidence of water mite predation on mosquito larvae and eggs, while there are no reports of predation on pupae. Smith [27] summarized the known material regarding mite predation of mosquitoes in his review, but there has been no update since then.

Here we discuss these previous observations and add six additional water mite records, since Smith [27], to the list of water mites that prey upon mosquito life stages (see Table 1). Mullen [15] observed *Thyas barbigera* and *T. stolli* preying on *Aedes* eggs in the laboratory. In that same work, he reported *Hydryphantes ruber* preying on *Aedes stimulans* larvae in the laboratory, and *Piona* feeding on mosquito larvae in woodland ponds. An earlier work by Laird observed *Limnesia jamurensis* feeding on *Culex pullus* and *Anopheles farauti* eggs and small larvae. They observed ponds devoid of mosquito larvae but filled with water mites and thus deduced that the mites might be the predators. They also conducted feeding experiments in the laboratory. Smith [27] also reported field observations of another water mite, *Piona*, feeding on mosquito larvae. Smith [27] suggested that adult *Arrenurus* mites feed on ostracods, while larval *Arrenurus* parasitize mosquitoes.

Table 1. Water mite impact on numbers of mosquito larvae in field-deployed mesocosms.

Mesocosm Identifier	Date Water Mite Added	Date Monitoring Mosquito Larvae
LSCMP#3	*Lebertia quinquemaculosa* 9 August 2018 Larvae present	No larvae observed 23 August 2018
LSCMP#2	*Hydrachna*, 31 August 2018 Larvae present	Larvae observed 7 September 2018
LSCMP#6	*Lebertia quinquemaculosa*, 11 October 2018 Larvae present	No larvae observed 19 October 2018
BI#6	*Lebertia quinquemaculosa* 15 September 18 and 17 October 2018, Pupae present, no larvae present	Pupae observed 22 October 2018
BI#3	*Arrenurus*, 15 September 2018 Larvae present	Periodically inspected no effect observed experiment ended 13 November 2018

Rajendran and Prasad [28] added a new taxon preying on mosquito larvae, *Encentridophorus similis* from the Unionicolidae family. Rajendran and Prasad [28] collected water mites of this species from adult mosquitoes and fed them *Aedes albopictus* larvae, which they preferred over copepods and ostracods. A subsequent study Rajendran and Prasad [29] provided the sole example of adult *Arrenurus* feeding on mosquito larvae. In experimental studies, *Arrenurus madaraszi* were fed larvae from *Aedes albopictus*, *A. hyrcanus*, and *A. vagus*. He noted that mosquito larvae became paralyzed when water mites attached themselves to the larvae. This suggests that water mites may be injecting venom that paralyzes the larvae, certainly a potential avenue for future research.

While Smith [27] mentions the work of Hearle [30], in which red water mites were observed to feed voraciously on mosquito "wrigglers" (larvae), known elsewhere as "wiggle waggles" (Pers comm. Belize colloquial use), he did not include it in his list. Perhaps this is because Hearle [30] did not identify the mite, although he wrote extensively on some life history characteristics where he kept mites and fed them. *Aedes vexans* larvae were provided upon which the mites then laid eggs that hatched after the season passed. Hearle [30] deduced that, in nature, the mites would lay eggs on leaves and debris and these would remain dormant until the following spring, at which point they would hatch. This observation has been confirmed in other mite species where seasonality is important in their life history. Smith [27] suggests that the mites observed by Hearle [30] were most likely *Piona*, which Smith [27] has also reported in his work as being predators of mosquito larvae. Bottger [31] had reported observations of *Teutonia cometes*, *Limnesia koenikei*, and *Hygrobates calliger* as preying on mosquito larvae, and although these were not included in the Smith [27] review of water mite predators of mosquitoes, these observations by Bottger were cited in the review by Proctor and Pritchard [7] on the scope of prey that water mites feed on.

Despite an estimated 57 families of water mites with 428 genera currently described [3], our current review observed only nine genera preying upon mosquito larvae or eggs (see Table 2). These observations include a new record, *Lebertia quinquemaculosa*, from our own study, described in this review (see Table 1). In this limited dataset, the addition of *Lebertia quinquemaculosa* to two mesocosms with mosquito larvae reduced the number of mosquito larvae to zero, whereas mosquito larvae continued to be present where *Hydrachna* and *Arrenurus* (see Table 1 and Figure 3) had been added. Interestingly, *L. quinquemaculosa* added to a mesocosm with only mosquito pupae present still had live pupae remaining when next inspected.

Table 2. The biodiversity of water mite predators of mosquitoes.

Family	Genus & Species	Mosquito Taxonomy and Prey Life Stage	Citation
Arrenuridae	*Arrenurus madaraszi*	*Anopheles* sp, *Armigerus* and *Aedes* sp larvae	[29]
Hydryphantidae	*Hydryphantes ruber*	*Aedes stimulans* larvae	[15]
Hydryphantidae	*Parathyas barbigera*	*Aedes* egg	[15]
Hydryphantidae	*Parathyas stolli*	*Aedes* egg	[15]
Hygrobatiidae	*Hygrobates calliger*	Unknown mosquito larvae	[31]
Lebertiidae	*Lebertia quinquemaculosa*	*Culex pipiens* larvae	This work
Limnesiidae	*Limnesia jamurensis*	*Anopheles farauti* and *Culex pullus* eggs and larvae	[32]
Limnesiidae	*Limnesia koenikei*	Unknown mosquito larvae	[31]
Pionidae	*Piona spp.*	*Aedes* larvae	[15,27]
Teutoniidae	*Teutonia cometes*	Unknown mosquito larvae	[31]
Unionicolidae	*Encentridophorus similis*	*Aedes albopictus* larvae	[28]

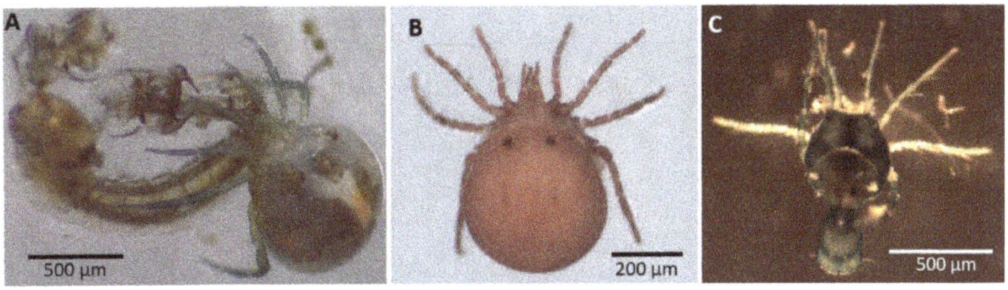

Figure 3. Micrographs of representative water mites used in mesocosms experiments: (**A**) *Lebertia quinquemaculosa* seen consuming a *Culex pipiens* larvae; (**B**) *Hydrachna* (ventral view), and (**C**) *Arrenurus* (ventral view).

3.2. Parasitism

Parasitism of mosquitoes by water mites was first recorded by DeGeer in 1778 and has since been an interesting focus of research for water mite investigators and others [27,33]. Comprehensive reviews of water mite parasitism on mosquitoes by Mullen [33] and Simmons and Hutchinson [34] revealed a global biodiversity of water mites that parasitize mosquitoes. The early work by Mullen [33] reported 15 genera of water mites that parasitize mosquitoes, but he disqualified five based on what he thinks were misidentifications or other inconsistencies. Smith [27] reported 10 genera mainly based on Mullen's work but not including the genera Mullen disqualified. Smith and Oliver [35] compiled an excellent review of parasitic hosts of larval water mites and their work agrees with the water mite–mosquito associations reported here. A newer study by Simmons and Hutchinson [34] reported seven genera and two families of water mites that parasitize mosquitoes. Water mites that parasitize mosquito adults overlap with six water mite genera that prey upon mosquito larvae and eggs (see Discussion). The mosquito hosts are similarly diverse, with 12 genera of mosquitoes having been identified by Simmons and Hutchinson [34]. In the present review, we add newer studies of water mites that parasitize mosquitoes; Table 3 lists the newer cases of water mite parasitism on mosquitoes that update the work since Simmons and Hutchinson [34].

Table 3. List of water mite–mosquito associations since Simmons and Hutchinson [34].

Parasitic Mite Taxa		Host Mosquito Taxa	Citation
Genus	**Species**		
Arrenurus	*acuminatus*	*Aedes pallidostriatus*, *Aedes pipersalatus*, *Anopheles barbirostris*, *Anopheles culicifacies*, *Anopheles minimus*, *Anopheles quinquefasciatis*, *Anopheles stephensi*, *Culex bitaeniorhynchus*, *Culex malayi*, *Culex nigropuntatus*, *Culex pipiens fatigans*	[36]
	danbyensis	*Culex infula*	[36]
	gibberifrons	*Aedes novalbopictus*	[36]
	kenki	*Aedes pallidostriatus*, *Aedes pipersalatus*,	[36]
		Anopheles quinquefasciatis, *Anopheles thomsoni*, *Culex malayi*, *Culex pipiens fatigans*, *Culex tritaeniorhynchus*, *Culex vishnui*, *Culex restuans*	[37]
	madaraszi	*Culex infula*	[36]
	spp.	*Aedes scapularis*, *Anopheles darlingi*, *Anopheles evansae*, *Psorophora ferox*, *Psorophora varipes*	[38]
Euthyas	spp.	*Culex restuans*	[37]
Hydrachna	spp.	*Aedes serratus*, *Mansonia wilsoni*,	[38]
		Psorophora varipes	[37]
Limnochares	spp.	*Aedes scapularis*, *Anopheles darlingi*	[37]
Parathyas	*barbigera*	*Aedes aegypti*, *Aedes albopictus*, *Aedes novalbopictus*,	[36]
		Aedes pallidostriatus, *Aedes pipersalatus*, *Aedes ramachandarai*, *Aedes vittatus*, *Anopheles barbarostris*, *Anopheles culicifacies*, *Anopheles minimus*, *Anopheles quinquefasciatis*, *Anopheles stephensi*, *Coquillettidia* spp., *Culex bitritaeniorhynchus*, *Culex infula*, *Culex malayi*, *Culex pipiens fatigans*, *Toxorhynchitis splendens*	[38]
	spp.	*Uranotaenia compestris* *Aedes albopictus*, *Aedes japonicus*, *Culex pipiens*, *Culex restuans*	[37]

Especially notable updates in Table 3 include observations by Atwa et al. [36], Manges et al. [37], and others. Atwa, Bilgrami and Al-Saggaf [36] reported new studies of water mites and their parasitic associations with mosquitoes, based on collections sourced in North India. Novel associations noted in this work included *Culex infula* associated with *Arrenurus danbyensis* and *Parathyas barbigera* [36]. Other *Culex* species, including *C. nigropuntatus*, *C. fatigans*, *C. malayi*, and *C. bitritaeniorhynchus* were reported associating with multiple *Arrenurus* genera, including *A. acuminatus*, *A. gibberifrons*, *A. madaraszi*, *A. kenki*, *A. danbyensis*, and *Parathyas barbigera* [36]. *Anopheles* genera also had new associations with *A. culcifactes*, *A. quinquefaciatis*, *A. stephensi*, *A. minimus*, and *A. barbarostris* with *Arrenurus* genera and *Parathyas barbigera* [36]. *Aedes* genera were also reported with new associations, including *A. albopictus*, *A. aegypti*, *A. pallidostriatus*, *A. pipersalatus*, *A. novalbopictus*, *A. vittatus*, and *A. ramachandarat* with *Arrenurus* genera. Likewise, *Parathyas barbigera* associated with *A. albopictus*, *A. aegypti*, *A. vittatus*, and *A. ramachandarat* [36]. Manges, Simmons, and Hutchinson [37] reported several new mosquito mite associations in North America, with mosquitoes that are considered invasive. *Aedes* genera, including *A. albopictus* and *A. japonicus*, were associated with *Parathyas* and *Culex restuands*, and *C. pipiens* were also associated with *Parathyas* [37]. Other interesting cases of parasitism, such as *Arrenurus* seen parasitizing a *Culex pipiens* pupae and *Unionicola* seen parasitizing a Cladoceran (*Bosmina tubicen*), are notable observations [38,39].

Our updated lists also include previously excluded data that should be considered, such as *Lebertia tauinsignata* reported by Marshall [40] but disqualified by Mullen [33]. We urge this reconsideration as we think *Lebertia tauinsignata* could possibly parasitize mosquitoes, as our own research shows *Lebertia* feeding on mosquito larvae and parasitizing chironomids, which are related to mosquitoes [10]. Our critical assessment of the work done by Marshall [40] did not find any reason to disqualify the observation. Another study that was disqualified by Mullen was the study by Mira [41] that identified *Unionicola* mites parasitizing *Anopheles* mosquitoes in what was Italian East Africa. Newer studies in the Arabian Peninsula adjacent to Ethiopia have identified *Unionicola* mites parasitizing mosquito pupae [38]. Other associations reported by Mullen [33] might need further assessment to determine why they were disqualified and if they should be considered again, given new research insights.

The *Arrenurus* genus commanded 61.67% of the parasitic associations, with 111 species of mosquitoes being parasitized (see Figure 4). The *Parathyas* genus was second highest, with 25.55% parasitic associations and 46 species of mosquitoes being parasitized. Further discussions on these two groups will be presented later, but it must be noted that the *Arrenurus* genera included several species of *Arrenurus*, but *Parathyas* was represented primarily by one species: *Parathyas barbigera*. In sum, the Arrenuridae water mite family (especially species of the genus *Arrenurus*) parasitized, by far, the greatest number of genera (11) and species (111) of mosquitoes (Figures 4 and 5). The water mite species and the number of mosquito species they parasitize are summarized in Figure 5. Within the *Arrenurus* genus, 27 species were found to parasitize mosquito larvae, with the most frequently observed species being *A. angustilimbatus*, *A. kenki*, and *A. madaraszi*. Since *Arrenurus* has the highest species richness of all water mite genera, and their larvae are generally difficult to identify at the species level, the diversity of *Arrenurus* species parasitizing mosquitoes may be even greater. Worldwide, 950 *Arrenurus* species have been documented [42], with 400 in the Nearctic region to date [6]. Newly assigned genera of water mites that parasitize mosquitoes included in the present review are *Lebertia* and *Unionicola*. These associations are based on our literature review, unpublished and published observations, and reassessment of previously rejected literature observations. For brevity, we did not include associations where the water mites could only be identified to family or subfamily taxonomic level, which included Euthyasinae and Thyadinae [34].

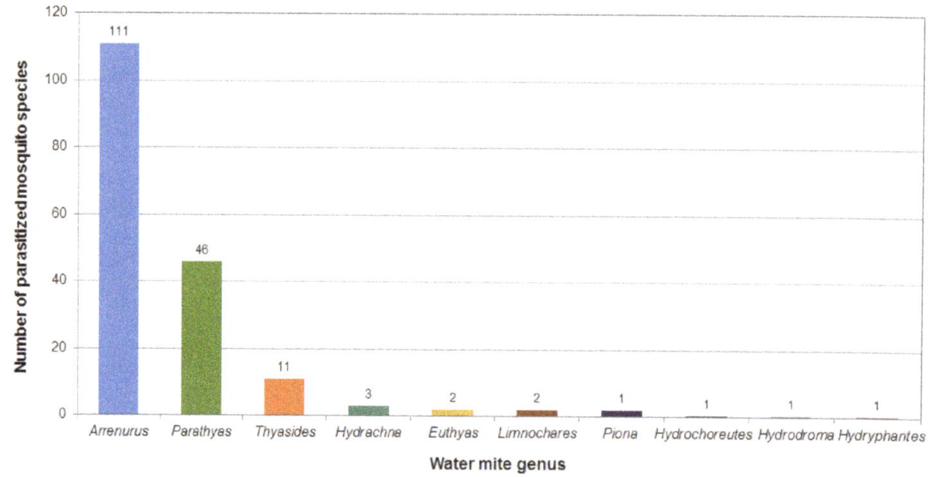

Figure 4. Number of parasitized mosquito species that are parasitized by each water mite genus, as reviewed by Atwa, Bilgrami, and Al-Saggaf [36]; Leal dos Santos [43]; Manges, Simmons, and Hutchinson [37]; and Simmons and Hutchinson [34].

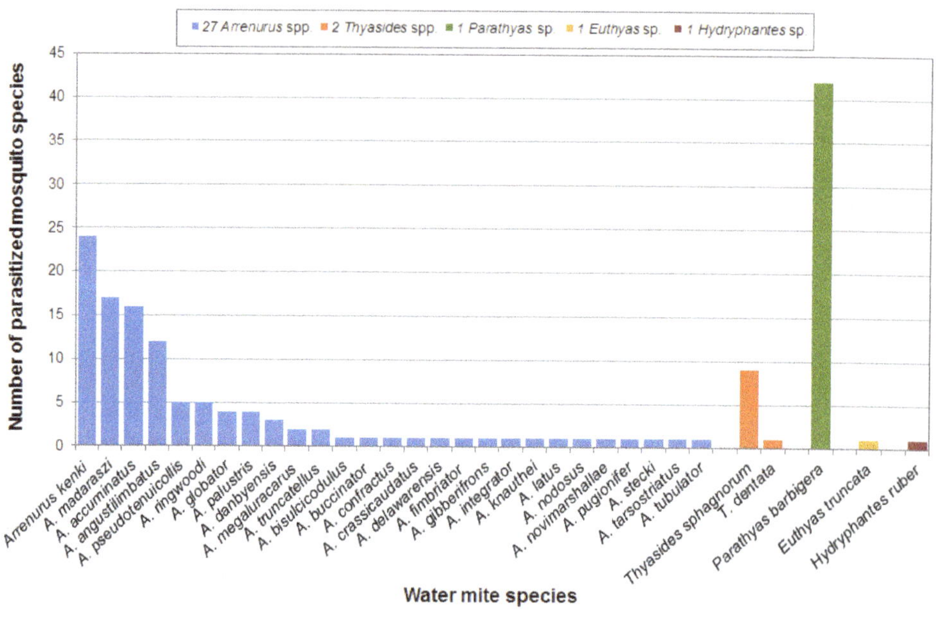

Figure 5. Water mites identified according to species and the number of mosquito species they parasitize, from Atwa, Bilgrami, and Al-Saggaf [36]; Leal dos Santos [43]; Manges, Simmons, and Hutchinson [37]; and Simmons and Hutchinson [34].

3.3. *Global Perspectives and Considerations of the Biodiversity of Water Mite Predation and Parasitism of Mosquitoes*

Despite Hydrachnidia (water mites) being the most biodiverse taxonomic group of the Arachnids, our analysis suggests that only about 3.5% of the total known water mite genera preys on and parasitize mosquitoes (see Table 4). However, some of the genera that have been shown to parasitize and prey on mosquitoes are believed to be some of the most specious, with *Piona* having potentially more than 100 species and *Arrenurus* up to 400 species [6].

Table 4. Overlap of water mite parasites and predators of mosquito adult and larvae, respectively.

Mite Genus	Mosquito Parasite	Mosquito Predator
Arrenurus	X	X
Encentridophorus		X
Euthyas	X	
Hydrachna	X	
Hydrochoreutes	X	
Hydrodroma	X	
Hydryphantes	X	X
Hygrobates		X
Lebertia	X [1]	X
Limnesia	X [1]	X
Limnochares	X	
Parathyas	X	X
Piona	X	X
Teutonia		X
Thyasides	X	

[1] Disqualified by Mullen, 1975 [33].

The aforementioned observations were based on both field and laboratory studies, comprising observations from the United States, Canada, Germany, Sweden, France, Denmark, New Zealand, Australia, Panama, Brazil, China, Japan, Uganda, Gambia, Madagascar, Nigeria, Angola, Indonesia, Malaysia, India, Sri Lanka, and Saudi Arabia (see Figure 6). While this demonstrates a broad global distribution of diverse water mite parasitism on mosquitoes, most reports originated from the United States and India, with over 100 records each.

Additionally, many biogeographical regions are left to be studied for water mite–mosquito associations, such as the Afrotropical and Neotropical regions, which are consequently known for mosquitoes and the diseases they cause (see Figure 6).

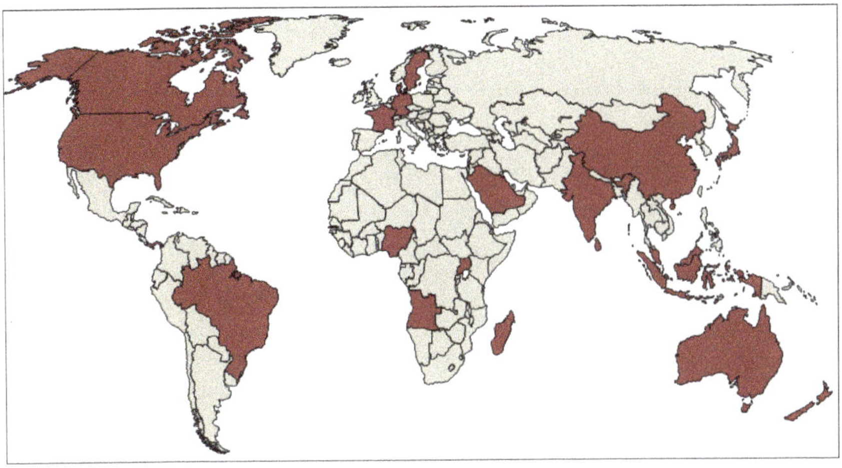

Figure 6. Map depicting locations (countries in red) where reports of water mite predatory and parasitic associations were obtained for this review.

4. Discussion

Water mites are known to be predators of a variety of aquatic invertebrates, including copepods, cladocerans, ostracods, and Dipteran larvae, and the larvae of water mites are also known to parasitize a diverse selection of invertebrates, such as dragonflies, mayflies, mosquitoes, and water beetles [7,35,44]. Given this diversity of biotic interactions they have as both predators and parasites, we consider their contribution to aquatic ecosystems to be very significant. The biodiversity of biotic interactions and the effect of parasitism on biodiversity are areas of research that are gaining renewed interest [17,20]. The specific aims of our work are to (i) update the known predatory and parasitic associations of water mites and mosquitoes (adults and larvae); (ii) update past records of water mite–mosquito associations; (iii) identify specific water mite genera and the biodiversity of their biotic interactions; and (iv) suggest future directions for studies with water mites, to increase our understanding of predatory biotic interactions in nature.

4.1. Identification of Major Water Mite Mosquito Parasites

Water mites are considered hyperparasites of mosquitoes, as they are a parasite that parasitizes another parasite, but despite these ecologically relevant biotic interactions, this area of research has been understudied [37,45]. Some mite species have been documented to parasitize multiple mosquito species throughout several genera, while other mite species have a very specific parasite–host association, with only a few mite species parasitizing a specific species of mosquito. Our review identified *Arrenurus kenki* as parasitizing 24 mosquito species of the genera *Aedes*, *Anopheles*, and *Culex* (see Table A1). Another water mite species that parasitizes a large number of mosquito species is *Parathyas barbigera*, which parasitizes 42 identified mosquito species in the genera *Aedes*, *Anopheles*, *Culex*, possibly *Coquillettidia* [46], *Psorophora*, *Toxorhynchitis*, and *Uranotaenia* (see Table A2).

The findings in our review suggest that the *Arrenurus* and *Parathyas* genus of water mites and their biotic interactions with mosquitos (as parasites of adult mosquitoes) might be an important area for future studies. *Arrenurus kenki* was listed as being "facultative tolerant" to organic wastes in a study that had a 0–5 range [47]. *Parathyas barbigera* (listed as *Thyas* in this reference) listed it as being facultative tolerant suggesting that both these species require an aquatic habitat that does not have excessive pollution [47]. This strengthens the idea that preserving biodiversity is important, especially in freshwater ecosystems that may contain these types of water mites which have a prominent role as mosquito parasites. It also emphasizes the loss of potential ecosystem services when the biodiversity of these biotic interactions is lost due to habitat loss or degradation. Work like this strengthens conservation efforts to improve freshwater habitats, since this is the habitat where biodiversity is disappearing at a faster rate than terrestrial systems [19].

4.2. The Potential Impact of Water Mite Life History Strategies on Their Biotic Interactions with Mosquitoes

Jalil and Mitchell [48] postulated that there are two types of water mites: the "thyasid-type", which belongs to the *Thyas* (=*Parathyas*) genus, and the "pionid-type", which includes those from the *Arrenurus* genus. In our review, the genera *Parathyas* and *Arrenurus* are those with the most significant parasitic associations with mosquito adult flies. The few studies focused on this topic have documented the possibility of water mite parasitism limiting the rate of survival and reproduction of mosquito hosts in natural environments to varying degrees [21,48–50]. The differing life history and behavior of thyasid-type and pionid-type mite larvae has previously been argued as playing a role in the attachment site and rate on mosquito hosts, and thus the intensity (as defined by the number of parasites on a host [51]) and severity of effects on mosquito survival and reproduction [48,49]. Thyasid-type mite larvae are believed to be closer relatives of terrestrial mites, from whom they evolved, than pionid-type larvae, because of their generalized, semi-aquatic life history [52]. Thyasid-type larvae are able to break through the water surface film immediately after hatching and "walk" on the water surface, having left the water altogether [48]. The thyasid-type larvae can only attach to adult mosquito hosts that return to the water surface, giving them only a few minutes of attack time to parasitize an adult mosquito host [49]. This may result in sexual discrimination by the thyasid-type mite larvae on female mosquitoes that exhibit a higher likelihood of returning to the water surface for oviposition of eggs than male mosquitoes [49]. In contrast, the pionid-type larvae are fully aquatic specialized swimmers that cannot leave the water until after they attach to the host. Therefore, pionid-type larvae can only seek a host during the mosquito's aquatic pupa stage, in which the mite larvae rest until mosquito ecdysis. After ecdysis of the mosquito pupa, the water mite remains on the adult mosquito, on which it initiates parasitism of the adult. This life history strategy may relate to the success in high intensity of mite parasitism with pionid-type larvae, specifically with *Arrenurus*, where *Arrenurus* mite load is commonly seen to be 30 or more mites per host [46,49]. Our review also identified *Arrenurus* as being the genus with the most species of *Arrenurus* parasitizing a wide diversity of mosquito hosts (see Figure 4 and Table A1).

Parathyas barbigera and *P. stolli* present an interesting case, as they were the second highest water mites having parasitic interactions with adult mosquitoes, with 42 different species of mosquitoes being parasitized by *P. barbigera* (see Figure 5 and Table A2). It was also found to prey on mosquito larvae (see Table 2). It was one of the few mites that had overlap with its larvae being parasites on mosquito adults and its adult form preying on mosquito larvae. This comparison can be seen in Table 4. *P. barbigera* and *P. stolli* belong to those water mites classified by Jalil and Mitchell [48] as "thyasid-type" mites, but despite this, they are very successful as predators and parasites of mosquitoes.

4.3. Water Mite Parasitism Reduces Mosquito Fecundity and Survivorship

The impact of high mite loads is evident from the linear relationship of mite-induced mortality and decreased fecundity on mosquito hosts, with the slope relating to the ratio of mite weight to host weight and the mite load on the host [22,50]. The few laboratory and natural experiments that have focused on the impact of high mite loads have commonly used the mite genus *Arrenurus* with the mosquito genus *Anopheles*, where reduced survivorship and reproduction of mosquitoes by *Arrenurus* species have been documented [21,23,50]. An experimental study by Lanciani and Boyt found that unparasitized female *Anopheles crucians* had a survival time of 23.32 days, while heavily parasitized females with *Arrenurus pseudotenuicollis* (around 17–32 mites) had a survival time of 6.25 days [21]. They also found that the number of eggs produced by *A. pseudotenuicollis* significantly decreased as *Arrenurus* mite load increased for both field-engorged mosquitoes and laboratory-fed mosquitoes, regardless of mosquito blood meal size [21]. Another experiment by Smith and McIver discovered that, when not accounting for blood meal size of *Coquillettidia perturbans*, the parasitism of *Arrenurus danbyensis* greater than five mites decreased the egg production of *C. perturbans* by 3.5 eggs per additional mite [23].

In a natural experiment by Lanciani and Boyt, the female *Anopheles crucians* that are unengorged with their first blood meal were found to have the highest proportion of pionid-type parasites compared to engorged female mosquitoes [21]. Another laboratory experiment by Lanciani discovered the sexual preference of *Arrenurus novimarshallae* mite larvae toward female *Anopheles crucians* pupae hosts compared to male pupae of that species, even when females were reared to smaller sizes with reduced food levels [53]. Female mosquitoes require sufficient energy to conduct flights for their required first blood meal to survive, but if they are heavily parasitized by pionid-type larvae, then they are most likely unable to attain this crucial blood meal, which can severely reduce mosquito densities.

Therefore, the effects and rates of parasitism by larval water mites and predation by deutonymph and adult water mites have been found to have a severe effect on population sizes of host and prey species. This, therefore, emphasizes the importance of future studies on loss of biodiversity of water mite parasitism and predation and the effects on the population size of species that they can potentially regulate. Since mosquito larvae have been found to be a possible host and prey for some species of water mites, it is crucial to determine the specific species of water mites that parasitize and prey on mosquito larvae.

4.4. Potential Water Mite Adaptations: Speciation and Niche Partitioning

We have also reviewed the water mites that have been shown to prey on mosquito larvae. Although the *Arrenurus* genus is substantially found to dominate the type of water mites that were found parasitizing mosquito adults (see Figures 4 and 5), we could only find one instance in which an *Arrenurus* adult water mite was found to prey on mosquito larvae in laboratory experiments (see Table 2) [29]. This demonstrates the hidden complexity within the biotic interactions since *Arrenurus madaraszi* was the only *Arrenurus* adult water mite found preying on mosquito larvae, but is the primary genus reported as parasitic (see Table 2 and Figure 5) [29].

Of the 17 genera listed, only six (*Parathyas*, *Piona*, *Hydryphantes*, *Arrenurus*, *Lebertia*, and *Limnesia*) were shown to both prey on and parasitize mosquito larvae and adults (see Table 4). Water mites are known to prey on and parasitize a wide variety of invertebrates. However, not all water mites are parasites. Up to 29 species are thought to not have parasitic larval stages, and studies that compare both parasitic and non-parasitic are needed to clarify possible adaptation benefits of one over the other [54]. A study comparing two species of *Arrenurus*, *A. angustilimbatus* (which is a parasite of mosquitoes and mentioned in this review) and *A. rufopyriformis* (which does not have a parasitic stage in its life cycle), concluded that *A. angustilimbatus* could be considered to be "ecologically successful" due to its higher heterozygosity and wide geographic range [55]. Studies on other species of water mites have also suggested species' separation as a consequence of parasitism [56]. Additionally, parasitic water mites have been observed to partition on a single host [57]. Up to nine water mite species were observed partitioning Chironomid hosts, with some species demonstrating preferred specificity to the thorax, while others to the abdomen [57]. This type of "niche" partitioning along with the phoresy associated with the parasitism of adult flies could contribute to the highly successful biodiversity and prevalence of water mites. However, due to poor taxonomy of water mites, particularly at the parasitic stage (larval stage), there is still much work needed to be done to fully appreciate the ecological and evolutionary contributions that these life history traits provide. However, with the ability for more accurate genus and species identification of water mites and mosquitoes via genetic analysis, and increased research on the abovementioned life history traits, an expanded understanding of these relationships will be made possible.

4.5. Biogeography of Water Mite–Mosquito Biotic Interactions

Because of the high species richness of both mites and mosquitoes and the extensive diversity of species-specific interactions between taxa, further investigation of the abundance of host exploitation and the effects of mite parasitism on mosquitoes seem likely to reveal a range of functional interactions. Even at the species level, these characteristics (attachment rate, mite load, and effects) of mite parasitism on mosquitoes have been previously found to vary depending on the species of both organisms involved in the parasitic interaction. Additionally, the biogeography related to these biotic interactions is also in need of clarification, since our overview of the biogeography of the groups discussed in this review could only be described at the family or subfamily level (see Table 5). Some of the genera have broad distribution, which may imply widespread impact through the diverse biotic interactions with mosquitoes. The cosmopolitan genus *Arrenurus* has published records of parasitism from widely distant regions, such as Japan and Canada, covering the Nearctic, Neotropical, Palearctic, Oriental, Australasian, and Afrotropical (see Table A3). More work is needed to understand the biodiversity, biogeography, and specificity of these biotic interactions, to document the extensive parasite–host association combinations that are present at a global scale.

Table 5. Biogeography of water mite genera.

Genus	Biogeography	Citation
Arrenurus	Cosmopolitan	[54]
Encentridophorus	Australasia, Asia, Africa	[54]
Euthyas	North America, Europe	[55,56]
Hydrachna	Cosmopolitan	[3]
Hydrochoreutes	Cosmopolitan	[54]
Hydrodroma	Cosmopolitan	[54]
Hydryphantes	Cosmopolitan, except New Zealand	[54]
Hygrobates	Cosmopolitan, except New Zealand	[54]
Lebertia	Cosmopolitan, except Australasian	[3]
Limnesia	Cosmopolitan	[3]
Limnochares	Cosmopolitan	[54]
Parathyas	Cosmopolitan family	[3]
Piona	Cosmopolitan	[54]
Teutonia	Holoarctic	[55]
Thyasides	Cosmopolitan family	[3]

4.6. Studies on Water Mite Predation of Mosquito Eggs, Larvae, and Pupae Are Needed

Similarly, with respect to predatory impacts of water mites on mosquitoes, further investigations are needed to determine ecological significance and, given the high health impact of mosquito-borne diseases, to determine if mite predation on mosquito larvae can be exploited to reduce these disease burdens. Commenting on the voraciousness of *Piona* spp. on mosquito larvae, Smith [27] noted that although *Piona* spp. are able to consume a large number of mosquito larvae, "quantitative studies on the ecology and feeding behavior of these mites are lacking", a statement that is still true almost 40 years later, as we write this review. A similar comment is made by Esteva et al. [58], who created a mathematical model of the roles of parasitism and predation in controlling the population dynamics of water mites and mosquitoes. The modelers observed in their model that predation had a more significant effect than parasitism in controlling the dynamics of mosquito and water mite populations. Indeed, in their model, populations of adult mosquitoes plummet to near zero as the water mite predation rate increases; the range of effective population-reducing predation occurs at a level of <0.9 mosquito larvae consumed per day per mite, which is a modest level compared to laboratory observed rates of six to eight mosquito larvae per mite by *Limnesia jamurensis* [32]. However, the modelers noted that "systematic studies about the extent of the impact of water mites on mosquito populations that could be used as a basis for a control program are scarce and fragmentary" [58].

Going forward, much new data needs to be collected on the intensity of water mite predation on mosquito larvae and about their impacts on mosquito populations. Our studies, reported elsewhere, applied high throughput sequencing to determine if mosquito DNA can be detected from the molecular gut contents of water mites freshly collected from the field [10]. The molecular gut contents from *Lebertia quinquemaculosa*, a second species of *Lebertia* with a novel COI barcode (tentatively named *Lebertia davidcooki*), and unidentified species of *Arrenurus* and *Limnesia*, was amplified with COI primers designed to amplify insects but not arachnids [10]. While DNA of many of the expected prey was present in these specimens (e.g., most sequences in *Lebertia* were from a multitude of chironomid species; *Arrenurus* had DNA from the ostracod *Podocopida*), sequences from mosquitoes were also present [10]. *Culex pipiens* sequences were observed in 20% of *L. quinquemaculosa* and 7% of *L. davidcooki* specimens, and neither of the other species [10]. We plan to apply these techniques to other water mite species, including *Piona*, which has figured prominently in this review of water mite–mosquito predation, to determine which, if any, species of water mites might utilize mosquito larvae as a predominant part of their diets. In addition, we have initiated mesocosm studies, reported here, to study water mite impacts on naturally recruited mosquito larvae (see Section 3.1).

These studies looked at naturally recruited mosquito larvae in aquatic mesocosms (see Figure 2) to which we have experimentally added water mites at various intervals. Our gut DNA studies in

Vasquez [10] identified a potentially novel water mite mosquito predator, *Lebertia quinquemaculosa*, which we added to the mesocosms reported here. Mimicking "natural" artifacts, such as ponds and streams, vernal ponds, puddles, and other damp areas such as plant phytolemata [59], and also man-made water-retaining structures, such as cisterns, rain gutters, and buckets, we deployed bucket mesocosms in urban parks (see Figures 1 and 2), all of which provide an extensive range of mosquito-breeding habitats. Examples of such features in an urban area have been documented in Detroit [10] and are generally found in urban areas elsewhere [60,61]. We showed, in our results, that the addition of *L. quinquemaculosa* reduced mosquito larvae population in the mesocosms (see Table 1). We have thus added *L quinquemaculosa* to the list of water mite mosquito predators (see Table 2).

To our knowledge, these observations of water mite and mosquito larvae in bucket mesocosms constitute the first field test of its kind that investigated water mite impacts on mosquito populations in a naturalistic environment. On a preliminary basis, at least, *L. quinquemaculosa* seems to have a greater effect on mosquito larvae recruitment or survival than do two other genera of mites (*Arrenurus* and *Hydrachna*) which we added to our mesocosms (see Table 1 and Figure 3), a result supported by our molecular diet research on *Lebertia* and consistent with previous diet preference research on the other species [6,7,10]. As noted above, a systematic investigation of diverse water mite species, including those suggested to have predatory associations with mosquito larvae or eggs, would be warranted, especially to provide data for mathematical models of water mite–mosquito interactions and ultimately to determine whether water mite predation could be exploited to control mosquitoes. These mesocosm studies could also be enhanced with cameras and other observational methods that would clarify the mechanisms by which water mites impacted the recruitment, growth, and/or survival of the mosquito larvae in mesocosms.

5. Final Considerations

This knowledge of the functional biodiversity of water mites that feed on and parasitize mosquitoes could be of great importance in understanding predator–prey dynamics [20] and developing new methods for controlling mosquitoes. Several different types of diseases, such as West Nile virus, eastern equine encephalitis, dengue, malaria, Zika, yellow fever, and chikungunya, are caused by mosquitoes. The human morbidity due to mosquitoes is estimated at 725,000 worldwide, making it potentially the deadliest animal on earth (https://www.cdc.gov/globalhealth/stories/world-deadliest-animal.html). While our current research is in temperate regions where eastern equine encephalitis and West Nile virus are especially of concern, most mosquito-borne diseases of human pathological importance are primarily found in tropical regions, where, ironically, water mite biodiversity is least understood [4]. DNA barcoding could potentially assist in improving the knowledge of water mite diversity, since water mite adult and larvae DNA barcodes could be matched, thereby greatly facilitating research on mite–mosquito interactions [34].

Climate change and increased international travel provides an additional motivation for understanding water mite–mosquito interactions, as rising temperatures may allow organisms—such as mosquitoes—of the tropics to invade more temperate regions, posing new threats to human health [62]. Such changes may increase the financial burden for cities trying to control mosquito populations. As an example, expenditures for mosquito control in Miami-Dade County were recently at ten million dollars annually, five times its proposed budget (http://www.wlrn.org/post/miami-dade-county-faces-10-million-tab-mosquito-control). The Environmental Protection Agency (EPA) and the Centers for Disease Control (CDC) recommend an Integrated Pest Management (IPM) approach for the control of mosquitoes (https://www.epa.gov/mosquitocontrol/joint-statement-mosquito-control-united-states) that emphasizes natural control, with minimal chemical intervention when possible. Among these natural control methods may be the application of diverse species of water mites, as more natural biocontrol agents for mosquitoes, to reduce their human disease burden.

Author Contributions: Conceptualization, A.A.V.; methodology, A.A.V. and B.A.K.; validation, A.A.V., B.A.K., J.L.R, and C.J.M.; formal analysis, A.A.V. and B.A.K.; investigation, A.A.V. and B.A.K.; resources, A.A.V.;

data curation, A.A.V. and B.A.K.; writing—original draft preparation, A.A.V. and B.A.K.; writing—review and editing, A.A.V., B.A.K., J.L.R., and C.J.M.; visualization, A.A.V.; supervision, C.J.M.; project administration, A.A.V.; funding acquisition, A.A.V., J.L.R., and C.J.M. All authors have read and agreed to the published version of the manuscript.

Funding: This research was funded by a National Institutes of Health grant from the National Institutes of Health Common Fund and Office of Scientific Workforce Diversity under three linked awards, RL5GM118981, TL4GM118983, and 1UL1GM118982, administered by the National Institute of General Medical Sciences and by the Fred A. and Barbara M. Erb Family Foundation to A.A.V.; B.K. was supported by Healthy Urban Waters with support from the Fred A. and Barbara M. Erb Family Foundation. J.L.R.; and C.M. partially covered the cost of open-access publication of this manuscript.

Acknowledgments: We are grateful to Jorge Cruz for building the mesocosms and to Janiel Cruz, who assisted with monitoring of the mesocosms. We extend our thanks to the Lake St. Clair Metropark (LSCMP) and Belle Isle State Park, where the preliminary studies on water mites and their impact on mosquitoes were carried out. We are especially thankful to Julie Champion and Dale Smart, who have assisted us at the LSCMP. These projects were carried out under Scientific permit 2018 and 2019 issued by LSCMP to A.A.V. and Permit PRD-SU-2018-016 and PRD-SU-2019-036 issued by the Michigan Department of Natural Resources to A.A.V. for work at Belle Isle. We apologize for anyone who assisted us and was not mentioned in this acknowledgement. We also extend our appreciation to three anonymous reviewers whose suggestions greatly improved the manuscript.

Conflicts of Interest: The authors declare no conflict of interest.

Appendix A

Table A1. *Arrenurus* water mite species and the mosquito species they parasitize [34,36,37,43].

Parasitic Mite Taxa		Host Mosquito Taxa	
Genus	**Species**	**Genus**	**Species**
Arrenurus	*acuminatus*	*Aedes*	*pallidostriatus, pipersalatus*
		Anopheles	*barbirostris, culicifacies, minimus, punctipennis, quadrimaculatus, quinquefasciatis, stephensi, walkeri*
		Culex	*bitaeniorhynchus, malayi, nigropuntatus, pipiens, pipiens fatigans*
		Culiseta	*melanura*
	angustilimbatus	*Aedes*	*abserratus, aurifer, cinereus, communis, diantaeus, excrucians, fitchii, provocans, punctor, stimulans*

Table A1. *Cont.*

Parasitic Mite Taxa		Host Mosquito Taxa	
Genus	**Species**	**Genus**	**Species**
		Culex	*restuans*
		Culiseta	*morsitans*
	bisulcicodulus	*Anopheles*	*maculipennis*
	buccinator	*Anopheles*	*maculipennis*
	confractus	*Culex*	*restuans*
	crassicaudatus	*Anopheles*	*maculipennis*
	danbyensis	*Aedes*	*canadensis*
		Coquillettidia	*perturbans*
		Culex	*infula*
	delawarensis	*Coquillettidia*	*perturbans*
	fimbriator	*Anopheles*	*maculipennis*
	gibberifrons	*Aedes*	*novalbopictus*
	globator	*Aedes*	*excrucians*
		Anopheles	*claviger, maculipennis*
		Culex	*pipiens*
	integrator	*Anopheles*	*maculipennis*
	kenki	*Aedes*	*abserratus, canadensis, communis, excrucians, fitchii, japonicus, pallidostriatus, pipersalatus, provocans, punctor, stimulans, trivittatus, vexans*
		Anopheles	*quinquefasciatis, thomsoni, walker*
		Culex	*malayi, pipiens fatigans, restuans, restuans, salinarius, territans, tritaeniorhynchus, vishnui*

Table A1. *Cont.*

Parasitic Mite Taxa		Host Mosquito Taxa	
Genus	**Species**	**Genus**	**Species**
	knauthei	*Anopheles*	*maculipennis*
	latus	*Anopheles*	*maculipennis*
	madaraszi	*Anopheles*	*annularis, culicifacies, hyrcanus, nigerrimus, pulcherrimus, sinensis, stephensi, subpictus, vagus*
		Culex	*epidesmus, fuscophala, infula, pipiens fatigans, pseudovishnui, tritaeniorhynchus*
	megaluracarus	*Mansonia*	*uniformis*
		Anopheles	*walker*
		Culex	*territans*
	nodosus	*Anopheles*	*maculipennis*
	novimarshallae	*Anopheles*	*crucians*
	palustris	*Anopheles*	*walker*
		Culex	*restuans, territans*
	pseudotenuicollis	*Culiseta*	*morsitans*
		Aedes	*triseriatus*
		Anopheles	*crucians, punctipennis, quadrimaculatus, walker*
	pugionifer	*Anopheles*	*maculipennis*
	ringwoodi	*Aedes*	*trivattatus*
		Anopheles	*punctipennis*
	stecki	*Culex*	*restuans, salinarius, territans*
		Anopheles	*maculipennis*
		Culex	*pipiens*
	tubulator	*Anopheles*	*maculipennis*

Table A1. *Cont.*

Parasitic Mite Taxa		Host Mosquito Taxa	
Genus	**Species**	**Genus**	**Species**
	spp.	*Anopheles*	*aconitus, annulipes, aquasalis, barbirostris, costalis, coustani, darlingi, earlei, evansae, fluviatilis, gambiae, jamesi, karwari, maculatus, maculatus willmori, pallidus, philippinensis, punctipennis, ramsayi, splendidus, squamoses, sundaicus, tessellatus*
		Coquillettidia	*bitaeniorhynchus, crassipes, richiardii, venezulensis*
		Culex	*bitaeniorhynchus, brevipalpis, cornutus, erraticus, gelidus, malayi, modestus, pipiens, quinquefasciatus, sinensis, tarsalis, vishnui, whitmorei*
		Culicidae	spp.
		Culiseta	*alaskaensis, annulate, impatiens, inornata*
		Deinocerites	*atlanticus, melanophylum*
		Ficalbia	*chamberlaini*
		Mansonia	*annulifera, indiana*
		Psorophora	*ferox, varipes*
		Uranotaenia	*maculipleura*

Table A2. *Parathyas* water mite species and the mosquito species they parasitize [34,36–39].

Parasitic Mite Taxa		Host Mosquito Taxa	
Genus	**Species**	**Genus**	**Species**
Parathyas	*barbigera*	*Aedes*	*abserratus, aegypti, albopictus, annulipes, canadensis, cantans, cantator, caspius, cataphylla, cinereus, communis, detritus, excrucians, fitchii, idahoensis, leucomelas, novalbopictus, pallidostriatus, pipersalatus, provocans, punctor, ramachandarai, sticticus, stimulans, trichurus, triseriatus, trivittatus, vexans, vittatus, zoosuphus*
		Anopheles	*barbarostris, culicifacies, minimus, quinquefasciatis, stephensi*
		Coquillettidia	*perturbans,* sp.
		Culex	*bitritaeniorhynchus, infula, malayi, pipiens fatigans*
		Psorophora	sp.
		Toxorhynchitis	*splendens*
		Uranotaenia	*compestris*
	spp.	*Aedes*	*albopictus, japonicus*
		Culex	*pipiens, restuans*

Table A3. *Arrenurus* species biogeography.

Zoogeographic Region	Country	*Arrenurus* Species	Source
Nearctic	Canada	*angustilimbatus*	[34]
		kenki	[34]
		megaluracarus	[34]
		palustris	[34]
	USA	*acuminatus*	[34]
		angustilimbatus	[34]
		confractus	[34]
		danbyensis	[34]
		delawarensis	[34]
		globator	[34]
		kenki	[34,37]
		megaluracarus	[34]
		novimarshallae	[34]
		palustris	[34]
		pseudotenuicollis	[34]
		ringwoodi	[34]
		tarsostriatus	[34]
Neotropical	Brazil	spp.	[34,43]
	Panama	spp.	[33]
Palearctic	France	spp.	[34]
	Germany	*bisulcicodulus*	[33]
		buccinator	[33,34]
		crassicaudatus	[33]
		fimbriator	[33]
		globator	[33,34]
		integrator	[33]
		knauthei	[33]
		latus	[33]
		nodosus	[33]
		pugionifer	[33]
		stecki	[33]
		truncatellus	[33,34]
		tubulator	[33]
	China	*madaraszi*	[34]
	Japan	*madaraszi*	[33]

Table A3. *Cont.*

Zoogeographic Region	Country	*Arrenurus* Species	Source
Oriental	India	*acuminatus*	[36]
		danbyensis	[36]
		gibberifrons	[36]
		kenki	[36]
		madaraszi	[34,36]
	Indonesia	spp.	[34]
	Japan	*madaraszi*	[33]
Australasian	Australia	spp.	[34]
Afrotropical	Angola	spp.	[33]
	Madagascar	spp.	[33]
	Nigeria	spp.	[33]
	Saudi Arabia	spp.	[38]

References

1. Cook, D.R.; Mitchell, R.D. Notes on collecting water-mites. *Turtox News* **1953**, *30*, 122–125.
2. Vasquez, A.A.; Carmona-Galindo, V.; Qazazi, M.S.; Walker, X.N.; Ram, J.L. Water mite assemblages reveal diverse genera, novel DNA barcodes and transitional periods of intermediate disturbance. *Exp. Appl. Acarol.* **2020**. [CrossRef] [PubMed]
3. Di Sabatino, A.; Smit, H.; Gerecke, R.; Goldschmidt, T.; Matsumoto, N.; Cicolani, B. Global diversity of water mites (Acari, Hydrachnidia; Arachnida) in freshwater. *Hydrobiologia* **2008**, *595*, 303–315. [CrossRef]
4. Goldschmidt, T. *The Biodiversity of Neotropical Water Mites*; Springer: Dordrecht, The Netherlands, 2002; pp. 91–99
5. Vasquez, A.A.; Qazazi, M.S.; Fisher, J.R.; Failla, A.J.; Rama, S.; Ram, J.L. New molecular barcodes of water mites (Trombidiformes: Hydrachnidiae) from the Toledo Harbor region of Western Lake Erie, USA, with first barcodes for *Krendowskia* (Krendowskiidae) and *Koenikea* (Unionicolidae). *Int. J. Acarol.* **2017**, *43*, 494–498. [CrossRef]
6. Smith, I.M.; Cook, D.R.; Smith, B.P. Water mites and other arachnids. In *Ecology and Classification of North American Freshwater Invertebrates*, 3rd ed.; Thorpe, J.H., Covich, A.P., Eds.; Academic Press: Waltham, MA, USA, 2010; pp. 485–586.
7. Proctor, H.; Pritchard, G. Neglected predators—water mites (acari, parasitengona, hydrachnellae) in fresh-water communities. *J. N. Am. Benthol. Soc.* **1989**, *8*, 100–111. [CrossRef]
8. Pozojevic, I.; Jursic, L.; Vuckovic, N.; Doric, V.; Gottstein, S.; Ternjej, I.; Mihaljevic, Z. Is the spatial distribution of lentic water mite assemblages (Acari: Hydrachnidia) governed by prey availability? *Exp. Appl. Acarol.* **2019**, *77*, 487–510. [CrossRef] [PubMed]
9. Matveev, V.F.; Martinez, C.C.; Frutos, S.M. Predatory—Prey Relationships in Sub-Tropical Zooplankton—Water Mite against Cladocerans in an Argentine Lake. *Oecologia* **1989**, *79*, 489–495. [CrossRef]
10. Vasquez, A.A. Digestive composition and physiology of water mites. Ph.D. Thesis, Wayne State University, Detroit, MI, USA, 2017.
11. Goldschmidt, T. Water mites (Acari, Hydrachnidia): Powerful but widely neglected bioindicators—A review. *Neotrop. Biodivers.* **2016**, *2*, 12–25. [CrossRef]
12. Goldschmidt, T.; Helson, J.E.; Williams, D.D. Ecology of water mite assemblages in Panama—First data on water mites (Acari, Hydrachnidia) as bioindicators in the assessment of biological integrity of neotropical streams. *Limnologica* **2016**, *59*, 63–77. [CrossRef]
13. Wiecek, M.; Martin, P.; Gabka, M. Distribution patterns and environmental correlates of water mites (Hydrachnidia, Acari) in peatland microhabitats. *Exp. Appl. Acarol.* **2013**, *61*, 147–160. [CrossRef]
14. Martin, P.; Gerecke, R. Diptera as hosts of water mite larvae—an interesting relationship with many open questions. *Lauterbornia* **2009**, *68*, 95–103.

15. Mullen, G.R. Predation by water mites (Acarina-Hydrachnellae) on immature stages of mosquitos. *Mosq. News* **1975**, *35*, 168–171.
16. Rueda, L.M. Global diversity of mosquitoes (Insecta:Diptera:Culicidae) in freshwater. *Hydrobiologia* **2008**, *595*, 477–487. [CrossRef]
17. Luna, P.; Corro, E.J.; Antoniazzi, R.; Dáttilo, W. Measuring and Linking the Missing Part of Biodiversity and Ecosystem Function: The Diversity of Biotic Interactions. *Diversity* **2020**, *12*, 86. [CrossRef]
18. United Nations Decade on Biodiversity. Available online: https://www.cbd.int/undb/goals/undb-unresolution.pdf (accessed on 2 May 2020).
19. Albert, J.S.; Destouni, G.; Duke-Sylvester, S.M.; Magurran, A.E.; Oberdorff, T.; Reis, R.E.; Winemiller, K.O.; Ripple, W.J. Scientists' warning to humanity on the freshwater biodiversity crisis. *Ambio* **2020**. [CrossRef]
20. Frainer, A.; McKie, B.G.; Amundsen, P.A.; Knudsen, R.; Lafferty, K.D. Parasitism and the Biodiversity-Functioning Relationship. *Trends Ecol. Evol.* **2018**, *33*, 260–268. [CrossRef]
21. Lanciani, C.; Boyt, A. Effect of a parasitic water mite, Arrenurus-Pseudotenuicollis (Acari-Hydrachnellae), on survival and reproduction of mosquito Anopheles-Crucians (Diptera-Culicidae). *J. Med. Entomol.* **1977**, *14*, 10–15. [CrossRef]
22. Smith, B. Host-parasite interaction and impact of larval water mites on insects. *Annu. Rev. Entomol.* **1988**, *33*, 487–507. [CrossRef]
23. Smith, B.; Mciver, S. The impact of *Arrenurus danbyensis* Mullen (Acari, prostigmata—Arrenuridae) on a population of *Coquillettidia perturbans* (Walker) (Diptera, Culicidae). *Can. J. Zool. Rev. Can. Zool.* **1984**, *62*, 1121–1134. [CrossRef]
24. Hooper, D.U.; Adair, E.C.; Cardinale, B.J.; Byrnes, J.E.K.; Hungate, B.A.; Matulich, K.L.; Gonzalez, A.; Duffy, J.E.; Gamfeldt, L.; O'Connor, M.I. A global synthesis reveals biodiversity loss as a major driver of ecosystem change. *Nature* **2012**, *486*, U105–U129. [CrossRef]
25. Tilman, D.; Isbell, F.; Cowles, J.M. Biodiversity and Ecosystem Functioning. In *Annual Review of Ecology, Evolution, and Systematics*; Futuyma, D.J., Ed.; Annual Reviews: Palo Alto, CA, USA, 2014; Volume 45, pp. 471–493.
26. Preston, D.L.; Mischler, J.A.; Townsend, A.R.; Johnson, P.T.J. Disease Ecology Meets Ecosystem Science. *Ecosystems* **2016**, *19*, 737–748. [CrossRef]
27. Smith, B.P. The potential of mites as biological control agents of mosquitoes. In *Research Needs for Development of Biological Control of Pests by Mites*; Hoy, M.A., Cunningham, G.L., Knutson, L., Eds.; University of California: Berkeley, CA, USA, 1983; pp. 79–85.
28. Rajendran, R.; Prasad, R.S. *Encentridophorus similis* (Acarina, Unionicolidae) an active predator of mosquito larvae. *Curr. Sci.* **1989**, *58*, 466–467.
29. Rajendran, R.; Prasad, R.S. A laboratory study on the life-cycle and feeding-behavior of Arrenurus madaraszi (Acari, Arrenuridae) parasitizing Anopheles mosquitos. *Ann. Trop. Med. Parasitol.* **1994**, *88*, 169–174. [CrossRef] [PubMed]
30. Hearle, E. The mosquitoes of the Lower Fraser Valley, British Columbia, and their control. *Nat. Res. Counc. Ott. Kept* **1926**, *17*, 94.
31. Bottger, K. Feeding of water mites Hydrachnellae acari. *Int. Rev. Gesamten Hydrobiol.* **1970**, *55*, 895–912. [CrossRef]
32. Laird, M. Some natural enemies of mosquitoes in the vicinity of Palmalmal, New Britain. *Trans. Roy. Soc. N. Z.* **1947**, *76*, 453–476.
33. Mullen, G.R. Acarine parasites of mosquitos 1. Critical review of all known records of mosquitos parasitized by mites. *J. Med. Entomol.* **1975**, *12*, 27–36. [CrossRef]
34. Simmons, T.W.; Hutchinson, M.L. A Critical Review of All Known Published Records for Water Mite (Acari: Hydrachnidiae) and Mosquito (Diptera: Culicidae) Parasitic Associations From 1975 to Present. *J. Med. Entomol.* **2016**, *53*, 737–752. [CrossRef]
35. Smith, I.M.; Oliver, D.R. Review of parasitic associations of larval water mites (Acari, parasitengona, Hydrachnida) with insect hosts. *Can. Entomol.* **1986**, *118*, 407–472. [CrossRef]

36. Atwa, A.A.; Bilgrami, A.L.; Al-Saggaf, A.I.M. Host-parasite interaction and impact of mite infection on mosquito population. *Rev. Bras. Entomol.* **2017**, *61*, 101–106. [CrossRef]
37. Manges, A.; Simmons, T.; Hutchinson, M. First Record of Aedes albopictus (Diptera: Culicidae) and Second Record of Aedes japonicus (Diptera: Culicidae) Parasitized by Water Mites (Acari: Hydrachnidiae) in North America. *J. Med. Entomol.* **2018**, *55*, 1617–1621. [CrossRef] [PubMed]
38. Shaalan, E.A.-S.; Bekhet, G.; Abdelmoaty, Z.; Ahmad, N.W. First Report on Mosquito Parasitic Mites in Saudi Arabia. *Pak. J. Zool.* **2016**, *48*, 1989–1992.
39. Montes-Ortiz, L.; Goldschmidt, T.; Elias-Gutierrez, M. First evidence of parasitation of a Bosmina (Cladocera) by a water mite larva in a karst sinkhole, in Quintana Roo (Yucatan Peninsula, Mexico). *Acarologia* **2019**, *59*, 111–114. [CrossRef]
40. Marshall, J.F. *The British Mosquitoes*; British Museum: London, UK, 1938.
41. Mira, G. Sulla presenza di forme larvali di un acaro acquatico parassita, della famiglia degli Hydracnidae, su alcune zanzare del genere Anopheles in A. O. I. *Boll Idrobiol Caccia E Pesca Afr. Orient. Ital* **1940**, *1*, 29–33.
42. Esen, Y.; Erman, O. A new species of the genus Arrenurus Duges, 1834 (Acari: Hydrachnidia: Arrenuridae) for the Turkish fauna: Arrenurus (Truncaturus) corsicus (E. Angelier, 1951). *Turk. J. Zool.* **2013**, *37*, 372–375. [CrossRef]
43. Leal dos Santos, F.; Thies, S.; Gonçalves, A.; Vasconcelos, K.; Ribeiro, M.; Damasceno, J.; Oliveira Dantas, E.; Leite Júnior, D. Aquatic Phoretic Mites (Acari: Hydrachnidia) Associated with Ectoparasitism of Mosquitoes (Diptera: Culicidae) in the Midwest Region of Brazil. *Adv. Entomol.* **2016**, *4*, 141–150. [CrossRef]
44. Nagel, L.; Zanuttig, M.; Forbes, M.R. Escape of parasitic water mites from dragonfly predators attacking their damselfly hosts. *Can. J. Zool.* **2011**, *89*, 213–218. [CrossRef]
45. Werblow, A.; Martin, P.; Dorge, D.; Koch, L.; Mehlhorn, H.; Melaun, C.; Klimpel, S. Hyperparasitism of mosquitoes by water mite larvae. *Parasitol. Res.* **2015**, *114*, 2757–2765. [CrossRef]
46. Kirkhoff, C.; Simmons, T.; Hutchinson, M. Adult mosquitoes parasitized by larval water mites in Pennsylvania. *J. Parasitol.* **2013**, *99*, 31–39. [CrossRef]
47. Klemm, D.J.; Lewis, P.A.; Fulk, F.; Lazorchak, J.M. *Macroinvertebrate Field and Laboratory Methods for Evaluating the Biological Integrity of Surface Waters*; Environmental Monitoring Systems Laboratory, U.S. Environmental Protection Agency: Cincinnati, OH, USA, 1990.
48. Jalil, M.; Mitchell, R. Parasitism of Mosquitos by Water Mites. *J. Med. Entomol.* **1972**, *9*, 305. [CrossRef]
49. Mullen, G. Acarine parasites of mosquitos 4. Taxonomy, life-history and behavior of Thyas barbigera and Thyasides sphagnorum (Hydrachnellae-thyasidae). *J. Med. Entomol.* **1977**, *13*, 475–485. [CrossRef] [PubMed]
50. Lanciani, C. Influence of parasitic water mites on the instantaneous death rate of their hosts. *Oecologia* **1979**, *44*, 60–62. [CrossRef] [PubMed]
51. Margolis, L.; Esch, G.; Holmes, J.; Kuris, A.; Schad, G. The use of ecological terms in parasitology (report of an ad hoc committee of the American-Society of Parasitologists). *J. Parasitol.* **1982**, *68*, 131–133. [CrossRef]
52. Mitchell, R. Major evolutionary lines in water mites. *Syst. Zool.* **1957**, *6*, 137–148. [CrossRef]
53. Lanciani, C. Sexual bias in host selection by parasitic mites of the mosquito anopheles-crucians (Diptera, Culicidae). *J. Parasitol.* **1988**, *74*, 768–773. [CrossRef]
54. Smith, B.P. Loss of larval parasitism in parasitengonine mites. *Exp. Appl. Acarol.* **1998**, *22*, 187–199. [CrossRef]
55. Bohonak, A.J.; Smith, B.P.; Thornton, M. Distributional, morphological and genetic consequences of dispersal for temporary pond water mites. *Freshw. Biol.* **2004**, *49*, 170–180. [CrossRef]
56. Martin, P.; Dabert, M.; Dabert, J. Molecular evidence for species separation in the water mite Hygrobates nigromaculatus Lebert, 1879 (Acari, Hydrachnidia): Evolutionary consequences of the loss of larval parasitism. *Aquat. Sci.* **2010**, *72*, 347–360. [CrossRef]
57. Martin, P. Specificity of attachment sites of larval water mites (Hydrachnidia, Acari) on their insect hosts (Chironomidae, Diptera)—Evidence from some stream-living species. *Exp. Appl. Acarol.* **2004**, *34*, 95–112. [CrossRef]
58. Esteva, L.; Rivas, G.; Yang, H.M. Modelling parasitism and predation of mosquitoes by water mites. *J. Math. Biol.* **2006**, *53*, 540–555. [CrossRef]
59. Kneitel, J.M.; Miller, T.E. Resource and top-predator regulation in the pitcher plant (Sarracenia purpurea) inquiline community. *Ecology* **2002**, *83*, 680–688. [CrossRef]

60. Wilke, A.B.B.; Chase, C.; Vasquez, C.; Carvajal, A.; Medina, J.; Petrie, W.D.; Beier, J.C. Urbanization creates diverse aquatic habitats for immature mosquitoes in urban areas. *Sci. Rep.* **2019**, *9*, 15335. [CrossRef] [PubMed]
61. Maimusa, A.H.; Ahmad, A.H.; Abu Kassim, N.F.; Ahmad, H.; Dieng, H.; Rahim, J. Contribution of public places in proliferation of dengue vectors in Penang Island, Malaysia. *Asian Pac. J. Trop. Biomed.* **2017**, *7*, 183–187. [CrossRef]
62. UN Intergovernmental Panel on Climate Change. Available online: http://www.ipcc.ch/report/sr15/ (accessed on 31 May 2020).

diversity

MDPI

Article

Cereal Straw Mulching in Strawberry—A Facilitator of Plant Visits by Edaphic Predatory Mites at Night?

Fernanda de Cássia Neves Esteca [1,*], Nina Trandem [2,3], Ingeborg Klingen [3], Jandir Cruz Santos [4], Italo Delalibera Júnior [1] and Gilberto José de Moraes [1]

1. Department of Entomology and Acarology, "Luiz de Queiroz" College of Agriculture (ESALQ), University of São Paulo (USP), Av. Pádua Dias, 11, Piracicaba SP 13418-900, Brazil; delalibera@usp.br (I.D.J.); moraesg@usp.br (G.J.d.M.)
2. Faculty of Environmental Sciences and Natural Resource Management, Norwegian University of Life Sciences (NMBU), P.O. Box 5003, 1432 Ås, Norway; nina.trandem@nibio.no
3. Biotechnology and Plant Health Division, Norwegian Institute of Bioeconomy Research (NIBIO), P.O. Box 115, 1431 Ås, Norway; ingeborg.klingen@nibio.no
4. Centre for Biodiversity Genomics, 50 Stone Road East, University of Guelph, Guelph, ON N1G2W1, Canada; jandir_jc@hotmail.com

* Correspondence: fernanda.esteca@usp.br

Received: 3 May 2020; Accepted: 27 May 2020; Published: 13 June 2020

check for updates

Abstract: In Norway, strawberry producers use cereal straw mulching to prevent berries from contacting the soil and to control weeds. We hypothesized that organic matter such as straw mulch also favors the maintenance of predatory mites which visit strawberry plants at nighttime. We compared mite diversity in cereal straw exposed for different periods in strawberry fields and evaluated their possible migration to plants in two experiments with potted plants in 2019. An 'Early season' experiment compared no mulching (T1), oat straw mulch exposed in field since 2018 (T2), or 2017 (T3), while a 'Mid-season' experiment compared no mulching (T1), barley straw mulch from 2018 (T2), or a mix from 2017 and 2018 (T3). To provide edaphic predatory mites with a potential source of food, all plants were infested with two-spotted spider mite (*Tetranychus urticae* Koch). Results suggested that straw mulch facilitates the prevalence of predatory mites in strawberry fields. Most predatory mite visits were at night, confirming our initial hypothesis. Predominant nocturnal mites on leaves belonged to Melicharidae (*Proctolaelaps* sp.) ('Early season', T2), Blattisociidae (*Lasioseius* sp.) ('Early and Mid-season', T3) and Phytoseiidae ('Mid-season', T2). *Parasitus consanguineus* Oudemans & Voigts was the predominant species ('Early season', T3) at the base of plants. Anystidae were diurnal visitors only ('Mid-season', T2). Future studies should evaluate the predation potential of *Proctolaelaps* sp. and *Lasioseius* sp. on two-spotted spider mite and other strawberry pests.

Keywords: oat straw mulch; barley straw mulch; biological control; two-spotted spider mite; edaphic mites

---

1. Introduction

The use of mulching (a plastic film or an organic material) is an important technique in strawberry (*Fragaria* × *ananassa* Duch.) cultivation, to regulate soil temperature, suppress weeds, prevent erosion, and prevent berries from contact with the soil [1].

In Norwegian strawberry production, cereal straw is commonly used as a mulching material. In roughly half of the strawberry production area, cereal straw is the only mulching material, whilst in another 25%, straw is combined with a black plastic film beneath it [2]. Strawberry is an important fruit crop in Norway, accounting for 70% of the yield of soft fruit production in the country [3]. In 2019, about 9000 metric tons of strawberries were produced by Norwegian farmers, with revenue of about NOK 455 million (USD 45.5 million) [4]. Most of the strawberry production is done in open fields, where the plants are harvested for 2–3 years. The most common pests in strawberry cultivation in Norway are: two phytophagous mites, the two-spotted spider mite, *Tetranychus urticae* Koch [5], and the strawberry tarsonemid mite, *Phytonemus pallidus* (Banks); two weevils, *Anthonomus rubi* Herbst and *Otiorhynchus sulcatus* F.; and a capsid bug, *Lygus rugulipennis* L. [6–8].

According to Kader et al. [9], there are many advantages of using organic material instead of plastic as mulching; for example, the former can improve soil structure throughout the degradation process, in addition to the more efficient balancing of soil temperature. Another advantage is the maintenance of higher densities of certain pest predators, such as beetles of the families Carabidae and Staphylinidae, and spiders of the families Linyphiidae and Lycosidae, providing them with a more suitable environment [10]. Further, Larentzaki et al. [11] reported reduction in thrips populations in onion crops whose soil was covered with oat straw.

Beneficial mites can also benefit from organic mulching. An increase in the abundance and diversity of Gamasina mites and/or a decrease in pest populations in different crops have been reported by several authors [12–15]. Gamasina is a cohort of the order Mesostigmata, which comprises a large group of mites found in different habitats, but especially in the soil. Although many of these can feed on fungi, they are well known as predators of small invertebrates, including mites, nematodes, and insects. For this reason, the members of this group have been studied as predators of edaphic pests [16].

Mites, as well as Collembola and some other Hexapoda, are part of the soil mesofauna (length 200 μm–2 mm), many of which fragmentize the organic matter available on the soil surface [17]. The macrofauna (>2 mm), in particular earthworms, also participate in the process of organic matter fragmentation [18]. The soil microbiota, which includes the microfauna and other small organisms (<100 μm), including bacteria, fungi and nematodes, is also a key element in the cycling of organic matter [19]. Thus, it is expected that throughout the process of decomposition of organic matter, such as straw mulch, the characteristics of the soil will vary, influencing the fauna of edaphic predatory mites.

Hence, soil with better coverage, naturally represented by dead plant structures on its surface or resulting from the purposeful introduction by growers, has a higher density of predatory mites, as extensively reported in the literature [20–23]. Consequently, in agricultural lands, the use of organic cover should benefit pest control by acting as a reservoir of edaphic predatory mites.

An aspect that has been little mentioned in the literature on predatory mites is the movement of soil mites from protected to exposed habitats, like plants, at night [24–26]. In Brazil, Esteca et al. [20] found that *Proctolaelaps pygmaeus* (Muller) (Melicharidae) and *Blattisocius dentriticus* (Berlese) (Blattisociidae) were present on strawberry leaflets mainly at night, indicating their possible daily migration from the edaphic environment to the plants. The reason for this behavior has not been properly studied. It could be casual or linked to the direct effect of light or other abiotic factors (such as temperature and humidity), or to biotic factors. Our previous experience has shown that representatives of edaphic Gamasina (*Lasioseius floridensis* Berlese (Blattisociidae), *Cosmolaelaps* sp. (Laelapidae), *Proctolaelaps bickleyi* (Bram) (Melicharidae), *Protogamasellopsis posnaniensis* Wisniewski & Hirschmann (Rhodacaridae) and *Stratiolaelaps scimitus* (Womersley) (Laelapidae)) will perish in a few hours when subjected to relative humidity levels below 70%. Due to the increase in air relative humidity at night, this period is probably the most suitable for predatory Gamasina mites to migrate temporarily onto plants in search of prey. The possible migration of *L. floridensis* from the soil to plants and vice versa was conceived by Britto et al. [27]. However, it is expected that the movement of edaphic Gamasina to plants varies from place to place, according to the faunistic composition and the intrinsic preference of the local mite fauna.

The hypotheses of the work we will present were: (a) Species richness and abundance of predatory mites (especially Gamasina) in organic material increases with the time the matter has been exposed on the soil surface in the strawberry field; and (b) mites present in the organic mulching at daytime climb to strawberry plants at nighttime in potting experiments in climatic chambers.

The objective of this study was therefore to compare the diversity and prevalence of mites in cereal straw subjected to different periods of exposure in the strawberry field, evaluating their possible temporary movement from the edaphic habitat to strawberry plant leaves at nighttime.

2. Materials and Methods

Two consecutive experiments, one 'Early season' and one 'Mid-season', were carried out during May–July 2019, in a climatic room at the Department of Biotechnology and Plant Health of Norwegian Institute of Bioeconomy Research (NIBIO), Ås, Norway. The room was maintained at 18 ± 2 °C, 70 ± 5% relative humidity and a daily photoperiod of 15 h, turning the lights on at 6:00 a.m. (500 µmol (quanta) $m^{-2}.s^{-1}$) and off at 9:00 p.m. These parameters simulated the average light conditions of the region during the study period.

2.1. Experimental Set-Up

In both experiments, potted strawberry plants received one of three treatments (T): In T1 no mulch was added (control treatment), in T2 cereal straw mulch present in a strawberry crop since last autumn (standard practice) was added, and in T3 cereal straw one year older than in T2 was added. Each treatment had 13 replicates per experiment, and each experiment lasted 5 weeks.

2.2. Preparing Experimental Pots and Plants

Pots (5 L) were filled to 80% capacity with a commercial organic substrate (Kompostert Plantejord®, Lillestrøm, Norway); composition: 50% peat, 50% garden waste, macro- and micronutrients). One strawberry plant of cv 'Korona', approximately 10 cm high and with about five trifoliate leaves, free of pests or diseases, was transplanted into each pot. Pots were placed 15 cm apart on four shelves in the climatic room. To avoid movement of mites from one pot to another, the upper edge of each pot was covered with entomological glue (Tangle trap, Biocontrol®, Grossdietwil, Switzerland).

To provide edaphic predatory mites with a potential source of food, two adult females of the two-spotted spider mite were transferred to each of six random leaflets per plant two days after transplanting. Infested leaflets were marked with a pen (3 mm in diameter), so that they could be examined for mite presence along the experiment. One week after the transplant, plants were fertilized with 8 g of Plantagen® (Uddevalla, Sweden) fertilizer (2N: 3P: 1K) per pot. The plants were irrigated manually to field capacity (about 10 mL per pot) once every two days.

2.3. Sampling Straw Mulch and Soil in the Field

For both experiments, cereal straw mulch of two ages (1 and 2 two years old), but otherwise as similar as possible, was collected on a strawberry farm near NIBIO (details given in Table 1). A frame of 55 × 25 cm was used to standardize sampling, each subsample taken in spots about 4 m apart, always next to bed margin, maximum 15 cm from a strawberry plant. Subsamples of ca. 2 L were collected and pooled into 40 L plastic bags and taken to the laboratory. The straw used in the strawberry field originated from cereal production at the same farm as the strawberry field, where no insecticides were used.

Soil samples from the same fields were also collected by the use of a standard probe (6 cm deep, 10 cm in diameter), transferring each sample to a plastic bag that was brought to the laboratory.

At the time of the 'Early season' sampling, the strawberry plants in the two fields were approximately 30 cm high and had about 15 leaves each (grade 55 according to BBCH-scale—phenological strawberry developmental stage [28]). They had been covered with Agryl fleece a few weeks before, to avoid frost damage to flowers, a common practice in the region. The fleece was removed to sample straw and soil. The cultivation practices of both fields were approximately the same, belonging to the same grower, and no pesticides had been used that growing season.

At the 'Mid-season' sampling, plants in the sampled field were approximately 30 cm high and had about 30 leaves each; with no flowers open BBCH-scale 57 [28]. The straw and soil collection process in the field was the same as in the 'Early season'. They had been covered with Agryl fleece a few weeks before, to avoid frost damage to flowers, a common practice in the region. The fleece was removed to sample straw and soil.

Table 1. Details of the different types of straw used in the study and the fields from which they were collected. All fields belonged to the same grower (County of Viken) in SE Norway. RH = Relative humidity.

Experiment Treatment (T) *	Date Collected	Temperature, RH, Precipitation **	Location	Type of Straw	Length of Exposure in the Field
'Early season' T2	May 6, 2019	3.3–4.0 °C; 77–82% RH; 6 mm	59°39′38″ N; 10°40′37″ E, Altitude 90 masl, Loam soil	Oat	Since autumn 2018
'Early season' T3	May 6, 2019	3.3–4.0 °C; 77–82% RH; 6 mm	59°39′30″ N; 10°41′13″ E, Altitude 100 masl, Silty loam soil	Oat	Since autumn 2017
'Mid-season' T2	June 23, 2019	11.4–13.0 °C; 75–83% RH; 29 mm	59°39′51″ N; 10°41′4″ E, Altitude 100 masl, Silty loam soil	Barley	Since autumn 2018
'Mid-season' T3	June 23, 2019	11.4–13.0 °C; 75–83% RH; 29 mm	59°39′51″ N; 10°41′4″ E, Altitude 100 masl, Silty loam soil	Barley	Mixture of autumn 2017 (lower layer) and 2018 (upper layer)

* Treatment 1 was without straw in both experiments. ** In the week preceding mulch collection.

2.4. Berlese Funnel Extraction of Mites

To get data on prevalence of predatory mites in the different straw treatments at the time of collection, straw samples were processed in the laboratory by placing a part of the samples (not the same as that used as mulching in the pots in the experiment of the climatic room) in modified Berlese funnels [29] at the beginning of each experiment. For each straw treatment, 13 Berlese funnels were employed. The corresponding soil sampled in the strawberry fields was also placed to Berlese funnels (13 samples) to evaluate the prevalence of predatory mites. The volume of sample in each Berlese funnel was 1 L. Mites dropping from each funnel were caught in a container with 70% ethanol. The extraction process lasted 7 days.

The commercial potting organic substrate used in experimental pots was investigated in the same way to reveal whether it contained mites that would affect the study.

At the end of each experiment, similar Berlese extractions were performed with the material from experimental pots (straw or potting substrate): For T1 (potting substrate only), a sample (1 L per pot) of the upper part of the potting substrate was taken from each pot. For T2 and T3, all the straw (1 L) covering the substrate of each pot was analyzed.

The material from each Berlese funnel extraction was screened in a stereomicroscope, and all mites were mounted in Hoyer's medium for identification. The identifications to family were carried out using taxonomic keys provided by Krantz and Walter [30], to genera by using unpublished keys provided by the Ohio Summer Program, Agricultural Acarology, Columbus, Ohio, USA, and to species by using published descriptions and redescriptions of the species of each family.

2.5. Observation and Sampling of Mites on Marked Leaflets during Experiments

Once a week for four consecutive weeks, the number of mites was counted on the six leaflets of each plant previously infested with two-spotted spider mite. Each mite was categorized as Mesostigmata, Prostigmata, or Oribatida. At each date, evaluations were carried out at 3, 7, and 11 a.m., and 3, 7, and 11 p.m. Evaluations were conducted with a hand-held lens (40×) on both leaflet surfaces. During the dark phase (3 a.m. and 11 p.m.), evaluations were conducted by the use of a hand-held lens (40×) illuminated with a headlight. All non-two-spotted spider mites found on the first evaluated leaflet of each plant in all second and fourth weekly evaluations were collected and mounted on slides using Hoyer's medium for later identification. The identifications were performed as described in the previous paragraph.

2.6. Extraction of Mites from the Whole Plant (End of Experiments)

Each experiment was ended after 5 weeks by a destructive sampling in which 6 of the plants of each treatment were cut at the plant base at 11 a.m., and the remaining 7 plants were cut at 11 p.m. Each plant was divided into three parts: (a) basal region (first two centimeters from the plant base); (b) young leaves, still folded; and (c) remainder of the plant (mature leaves and petioles). Each part was immediately placed in a plastic bag containing 70% ethanol and shaken vigorously. The liquid was subsequently poured through a 160 µm sieve. To extricate any remaining mites, the plant material was further rinsed using jets of 70% ethanol. The mites retained in the sieve were mounted for identification and quantification under a stereomicroscope (100×).

To investigate the possible occurrence of the fungus *Neozygites floridana* (Weiser & Muma) Remaudière & Keller (Entomophthorales: Neozygitaceae), a pathogen of the two-spotted spider mite, a sample of two-spotted spider mite was mounted in Hoyer's medium (maximum 10 mites/plant part/plant) for examination under the stereomicroscope (100×).

2.7. Statistical Analysis

Numbers of Mesostigmata and two-spotted spider mite in the different treatments were analyzed statistically as follows: For Mesostigmata, the means were contrasted (just between T2 and T3 in both experiments, given that Mesostigmata were not found in T1) by the F test, from the generalized linear model of the quasi-Poisson distribution (Anova, model.QuasiPoisson, test = "F"), in R [31]. For the two-spotted spider mites, the untransformed data were fitted to a quasi-Poisson distribution. Means were compared using the *glht* multicomp package, in R [31]. The predominant species of Mesostigmata were calculated as proposed by Pinzón and Spence [32].

3. Results

3.1. Mites Extracted from Commercial Potting Substrate, Soil and Mulching Straw

No mites were found in the commercial potting substrate, neither at the start nor at the end of the experiments. This strongly indicates that all mites eventually found on the strawberry plants of T2 and T3 originated from the straw used as mulch. Regarding the soil samples, the average number of mites per sample of the 'Early season' were low in both T2 and T3 (1.2 ± 0.4 and 1.7 ± 0.5 mites, respectively), none of them Gamasina. In the 'Mid-season', averages were also low (0.5 ± 0.2 and 0.7 ± 0.3 mites, respectively), and the only Gamasina found were *Amblygamasus* sp. and *Pergamasus longicornis* (Berlese) (both Parasitidae) and *Rhodacarellus epigynalis* Shels (Rhodacaridae).

In contrast, the total numbers of mites in straw samples to be used in T2 and T3 in the 'Early season', were relatively high, with the proportion of Gamasina higher in T3 than in T2 (18% and 6.0%, respectively) (Table 2). Consequently, the mean number of Gamasina per sample was significantly higher in T3 (7.8 ± 0.6) than in T2 (3.5 ± 0.7) (Df = 1; F = 16.5; $p < 0.001$). In the 'Mid-season', the total numbers of mites in T2 and T3 were lower at the start, of which 15.8% and 38.6% were Gamasina. As in the 'Early season', the mean number of Gamasina per sample was higher in the straw of T3 (11.9 ± 1.9) than of T2 (2.7 ± 0.7) (Df = 1; F = 26.9; $p < 0.001$).

Table 2. Mites extracted in Berlese funnels from samples of four types of cereal straw mulching used in two climate chamber experiments in 2019. The mulching had been present in strawberry fields for different periods of time, as indicated in column headings. B = Samples taken at the start of experiment; E = taken at the end of the experiment. n = 13 samples of 1 L per straw type and sampling occasion.

Experiments	'Early Season' Experiment				'Mid-Season' Experiment			
Treatments	**T2 (Oat Straw 2018)**		**T3 (Oat Straw 2017)**		**T2 (Barley Straw 2018)**		**T3 (Barley Straw 2017–2018)**	
Taxa/Collection	**B**	**E**	**B**	**E**	**B**	**E**	**B**	**E**
Sarcoptiformes, Oribatida, Astigmatina								
Acaridae								
Tyrophagus putrescentiae	48	92	130	35	17	22	15	7
Winterschimidtiidae	0	0	8	0	0	0	0	0
Sarcoptiformes, other Oribatida								
-	119	11	109	7	23	5	149	16
Trombidiformes, Prostigmata								
Anystidae								
Anystis sp.	0	0	0	0	32	27	2	0
Cunaxidae								
Cunaxoides croceus	0	0	3	0	0	0	0	0
Ereynetidae								
Ereynetes spp.	135	0	17	0	1	2	0	6
Eupodidae								
Eupodes spp.	239	0	46	0	64	50	31	66
Pygmephoridae								
Siteroptes sp.	1	0	2	0	0	0	1	0
Tydeidae								
Lorryia oregonensis	167	6	129	46	22	20	33	21
Parasitiformes, Mesostigmata, Gamasina								
Ascidae								
Gamasellodes bicolor	0	0	1	0	0	0	0	0
Neojordensia sinuata ♂	1	0	0	0	0	0	0	0
Blattisociidae								
Lasioseius sp.	14	12	3	1	5	15 *	1	14 *
Eviphididae								
Alliphis halleri	9 *	5	5	0	0	0	7	3
Alliphis sp. immature	1	0	0	0	0	0	0	0
Melicharidae								
Proctolaelaps sp.	14 *	102 *	0	0	0	0	0	0
Parasitidae								
Amblygamasus sp.	0	0	0	0	4	3	20	0
Amblygamasus sp. (immature)	0	0	0	0	0	1	16	0
Parasitus consanguineus	0	0	1	7	0	0	0	0
Parasitus sp. (deutonymph)	1	0	7	0	0	0	0	0
Pergamasus longicornis	0	0	3	2	14 *	4	42 *	2
Pergamasus septentrionalis	0	0	10	0	0	0	5	2
Pergamasus sp. (deutonymph)	0	0	26 *	2	1	4	17	0
Pergamasus sp. ♂	0	0	0	2	0	0	0	0
Porrhostaspis lunulata	0	0	0	0	0	8	4	7 *
Porrhostaspis sp. (deutonymph)	0	0	1	2	0	0	1	0
Immature	1	0	2	1	2	1	3	0
Phytoseiidae								
Neoseiulus alpinus	1	2	3	1	0	0	0	0
Neoseiulus cucumeris	0	0	0	0	2	2	0	1
Neoseiulus sp. immature	0	1	3	1	0	0	0	0
Proprioseiopsis okanagensis	0	0	3	0	1	2	0	0
Typhlodromips masseei	1	0	13	0	0	0	5	2
Immature	0	0	17	0	0	0	0	0
Rhodacaridae								
Rhodacarellus epigynalis	0	0	0	0	1	0	2	3
Rhodacarellus kreuzi	0	0	1	0	0	0	0	0
Rhodacarellus sp. immature	1	0	0	0	0	0	0	0
Veigaiidae								
Veigaia nemorensis	1	0	0	0	0	0	22 *	2
Veigaia sp. immature	0	0	2	0	0	0	0	0
Total	754	231	545	105	189	166	376	152

(*) Predominant Gamasina species; (-) Not identified family.

The predominant gamasine families in 'Early season' T2 were Blattisociidae (33.3%), Melicharidae (31.1%) and Eviphididae (22.2%), each of the others representing a maximum of 4.4% of the total Gamasina. In T3, the predominant gamasine families were Parasitidae (49.5%) and Phytoseiidae (38.6%), each of the others representing a maximum of 4.9%. At the species level in the 'Early season', the predominant ones were *Alliphis halleri* (G. & R. Canestrini) (Eviphididae) and *Proctolaelaps* sp. (Melicharidae) in T2, and *Pergamasus* sp. (Parasitidae) in T3. In the 'Mid-season', the predominant family was Parasitidae in both treatments with straw (T2 = 76.1% and T3 = 71.4%), other families representing a maximum of 4.5% and 14.3%, respectively. At the species level in this experiment, *P. longicornis*

(Parasitidae) and *Lasioseius* sp. (Blattisociidae) were predominant in T2 and T3, and *Veigaia nemorensis* (Koch) (Veigaiidae), in T3.

At the end of the 'Early season', the total numbers of mites found in the straw of T2 and T3 were 231 and 105, respectively, Gamasina representing 52.8% and 16.1% of these (Table 2). Consequently, the mean number of Gamasina per sample was significantly higher in T2 than in T3 (9.4 ± 1.3 and 1.3 ± 0.4, respectively) (Df = 1; F = 47.7; $p < 0.001$). At the end of the 'Mid-season', the total numbers of mites in T2 and T3 were similar (166 and 152, respectively), and the proportions of Gamasina were also similar, 24.1% and 23.6% respectively (Table 2), resulting in no statistical difference between their mean numbers per sample (3.1 ± 0.7 and 2.8 ± 0.7, respectively) (Df = 1; F = 0.1; $p = 0.75$).

At the end of experiments, the predominant gamasine families in 'Early season' were Melicharidae (83.6%) and Blattisociidae (9.8%) in T2, each of the other families representing a maximum of 4.1% of the total Gamasina. In T3, the predominant gamasine families were Parasitidae (84.2%) and Phytoseiidae (10.5%), each of the others representing 5.2%. In this experiment, the predominant species were *Proctolaelaps* sp. (Melicharidae) in T2, and *Pergamasus* sp. (Parasitidae) in T3. In the 'Mid-season', the predominant families in both treatments were Parasitidae (T2 = 52.5%; T3 = 30.5%) and Blattisociidae (T2 = 37.5%; T3 = 38.9%), other families representing a maximum of 10% and 8.3%, respectively. In this experiment, *Lasioseius* sp. (Blattisociidae) was the predominant species at the end of both T2 and T3 and *Porrhostaspis lunulata* Müller (Parasitidae), in T2.

3.2. Mites on Marked Leaflets

Two-spotted spider mite was the only mite species found on the T1 marked leaflets of both experiments, which was expected, given that no mites were observed in the commercial substrate used for all treatments and that this was the only material present in the pots of T1 (Table 3). On the other hand, the totals of Gamasina observed on T2 and T3 leaflets were low throughout both experiments, with a maximum of four mites in T2 (in the third week of evaluation of the 'Early season') and of five mites in T3 (in the fourth week of evaluation of the 'Mid-season'). This makes it difficult to compare treatments statistically. Nevertheless, gamasine mites could be seen moving on the straw surface at night.

Table 3. Total mites (except *Tetranychus urticae*) collected in the second and fourth evaluations at 3, 7, 11 a.m. and 3, 7, 11 p.m. on the six marked leaflets with *T. urticae*/plant ($n = 13$ plants/treatment) in two lab experiments for different periods of time: 'Early season' experiment (oat straw: T2, since 2018; T3, since 2017); 'Mid-season' experiment (barley straw: T2, since 2018; T3, lower half layer since 2017 and top half layer since 2018). Light was on from 6 a.m. to 9 p.m.

Experiments	'Early Season' Experiment		'Mid-Season' Experiment	
Taxa/Treatments	**T2 (Oat Straw 2018)**	**T3 (Oat Straw 2017)**	**T2 (Barley Straw 2018)**	**T3 (Barley Straw 2017–2018)**
Trombidiformes, Prostigmata				
Anystidae				
Anystis sp.	0	0	22	0
Parasitiformes, Mesostigmata, Gamasina				
Ascidae				
Neojordensia sinuata	1	0	0	0
Blattisociidae				
Lasioseius sp.	0	3	1	11 *
Melicharidae				
Proctolaelaps sp.	6 *	0	0	0
Parasitidae				
Parasitus sp. (deutonymph)	0	0	1	0
Pergamasus sp. (deutonymph)	0	0	0	1
Phytoseiidae				
Neoseiulus cucumeris	0	0	1	0
Typhlodromips masseei	1	0	3 *	0
Proprioseiopsis okanagensis	0	0	1	1
Total	8	3	29	13

(*) Predominant species.

In the 'Early season', the predominant Gamasina in T2 were Melicharidae, of which *Proctolaelaps* sp. was the most abundant species. In T3, the predominant family was Blattisociidae, of which *Lasioseius* sp. was the most abundant species. In the 'Mid-season', the predominant Gamasina in T2 were Phytoseiidae, with *Typhlodromips masseei* (Nesbitt) the most abundant species, whereas the predominant family in T3 was Blattisociidae, with *Lasioseius* sp. the most abundant species.

To summarize, in the 'Early season', the total numbers of non-two-spotted spider mite found on the marked leaflets of T2 and T3 were eight and three, of which only one mite was Phytoseiidae (in T2, *T. masseei*). In the 'Mid-season', the total number of non-two-spotted spider mite found on marked leaflets of T2 and T3 were 29 and 13, of which 22 on T2 plants were Anystis sp. (Anystidae); of Phytoseiidae there were five mites in T2 (one *Neoseiulus cucumeris* (Oudemans), three *T. masseei* and one *Proprioseiopsis okanagensis* (Chant)) and one in T3 (*P. okanagensis*). Despite the small numbers of edaphic Gamasina, they were seen on the plants only at night (Figure 1A–D), except for Phytoseiidae and Anystidae, seen only at daytime (Figure 1C). For two-spotted spider mite, the mean numbers observed per leaflet in the fourth week of the experiment were: 'Early season': T1 = 15.4 ± 2.3, T2 = 9.2 ± 3.8, and T3 = 10.4 ± 3.7; 'Mid-season': T1 = 20.4 ± 4.1, T2 = 14.3 ± 4.0, and T3 = 14.5 ± 4.4. Significant differences were observed in the 'Early season' between T1 and T2 (Df = 2; Z = 2.55; $p = 0.02$), and in the 'Mid-season' between T1 and T2 (Df = 2; Z = 2.4; $p = 0.03$) and T1 and T3 (Df = 2; Z = 2.3; $p = 0.03$) (Figure 2A,B). In addition, the number of two-spotted spider mites did not vary according to evaluation times, as expected, leading us to consider each evaluation as a replicate in the statistical analysis.

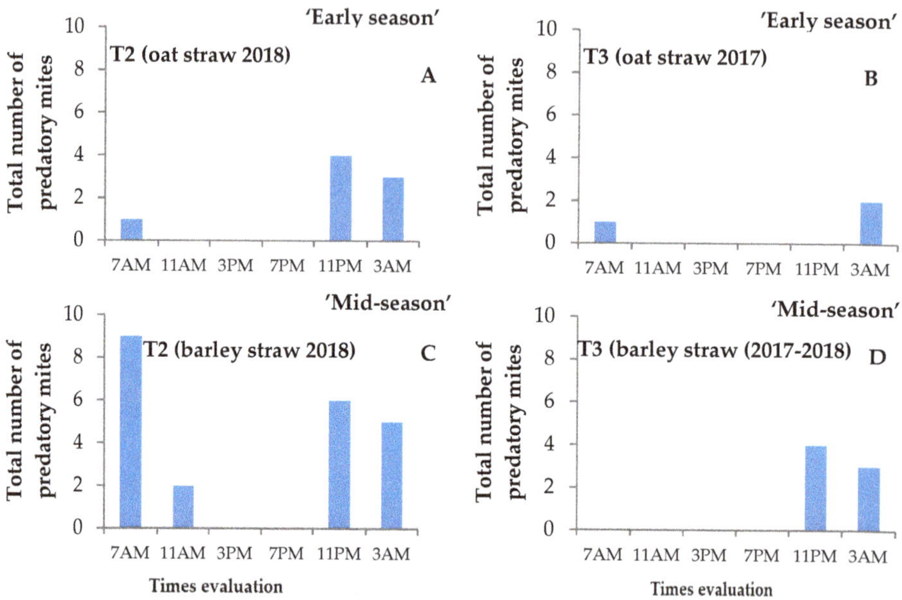

Figure 1. Weekly total numbers of Gamasina mites observed on six marked leaflets with *Tetranychus urticae* per plant ($n = 13$ plants per treatment). (**A**,**B**): 'Early season' experiment (T2, oat straw since 2018; T3, oat straw since 2017); (**C**,**D**): 'Mid-season' experiment (T2, barley straw since 2018; T3, barley straw lower half layer since 2017 and top half layer since 2018). Light was on from 6 a.m. to 9 p.m.

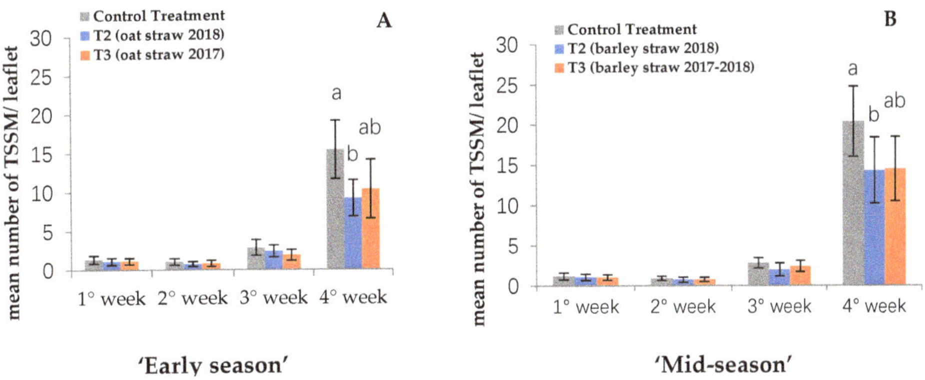

Figure 2. Weekly mean number of *Tetranychus urticae* observed on six marked leaflets per plant ($n = 13$ plants per treatment). (**A**): 'Early season' experiment (T2, oat straw since 2018; T3, oat straw since 2017); (**B**): 'Mid-season' experiment (T2, barley straw since 2018; T3, barley straw lower half layer since 2017 and top half layer since 2018. Light was on from 6 a.m. to 9 p.m.

3.3. Mites Extracted from Whole Plants at the End of Experiments

No Gamasina were found on T1 plants. Considering T2 and T3 together and the two evaluation times, the number of Gamasina found in the 'Early season' (107 mites) was higher than in the 'Mid-season' (76 mites) (Table 4), coinciding with the higher number of mites in the straw at the end in the 'Early season'. The largest number of Gamasina was found on plants of T2 in both experiments. In addition, considering T2 and T3 together, Gamasina were rarely found at 11 a.m. (total of five mites), with a much larger number found at 11 p.m. (total of 102 mites); this resulted in significantly higher density of Gamasina at 11 p.m. than at 11 a.m. in both treatments and in both experiments (Table 4).

In the 'Early season', at 11 a.m. no Gamasina were found on T2 or T3 plants, in contrast to the considerable number found at 11 p.m. (Table 5). The average number of Gamasina per plant was significantly different between night and day (between 11 a.m. and 11 p.m. in T2: Df = 1; F = 73.4; $p < 0.001$; between 11 a.m. and 11 p.m. in T3: Df = 1; F = 33.3; $p < 0.001$). In the 'Mid-season', the number of Gamasina on T2 plants at 11 p.m. was higher than at 11 a.m. (Df = 1, F = 10.6, $p < 0.01$), whilst in T3 there was no statistical difference between 11 a.m. and 11 p.m. (Df = 1, F = 2.1, $p= 0.15$).

As to the faunistic composition, in the 'Early season' at 11 p.m., the predominant families were Melicharidae, in T2, with *Proctolaelaps* sp. as the most abundant species, only 18.2% of which were located at the base of the plant or on the young leaflets. In T3, the predominant family was Parasitidae, with *Parasitus consanguineus* Oudemans & Voigts as the most abundant species; about 87.5% of these were found at the base of the plants. In the 'Mid-season', the predominant family in T2 was Blattisociidae, with *Lasioseius* sp. as the most abundant species. In T3, Gamasina always occurred at very low numbers. These results were similar to those of the evaluation of mites on leaflets only, that is, in both experiments, the largest number of predatory mites occurred in T2, both in the evaluations of only leaflets and on the whole plant.

The mean number of two-spotted spider mite per plant on leaves in T1, T2, and T3 were 176.3 ± 10.2, 143.6 ± 14.3, and 137.5 ± 11.2, respectively, in the 'Early season', and 223.8 ± 2.5, 138.4 ± 18.1, and 94.4 ± 18.5, respectively, in the 'Mid-season' experiment. Significant differences were observed in the 'Early season' between T1 and T3 (Df = 2; Z = −2.2; $p = 0.0$) and in the 'Mid-season' between T1 and T2 (Df = 2; Z = −3.2; $p = 0.003$) and T1 and T3 (Df = 2; Z = −5.1; $p < 0.001$). Infection of the two-spotted spider mite by the fungus *N. floridana* was not observed.

Table 4. Total number of mites (except *Tetranychus urticae*) of different groups extracted with 70% alcohol from whole strawberry plants. 'Early season' experiment (oat straw: T2, since 2018; T3, since 2017); 'Mid-season' experiment (barley straw: T2, since 2018; T3, lower half layer since 2017 and top half layer since 2018).

Treatments	T2				T3			
Experiments	**'Early Season'**		**'Mid-Season'**		**'Early Season'**		**'Mid-Season'**	
Taxa/Time	**11 a.m.**	**11 p.m.**	**11 a.m.**	**11 p.m.**	**11 a.m.**	**11 p.m.**	**11 a.m.**	**11 p.m.**
Trombidiformes, Prostigmata								
Anystidae								
Anystis sp.	0	0	11	0	0	0	0	0
Sarcoptiformes, Oribatida, Astigmatina								
Acaridae								
Tyrophagus putrescentiae	0	14	0	11	0	16	0	24
Parasitiformes, Mesostigmata, Gamasina								
Blattisociidae								
Lasioseius sp.	0	5	0	11 *	0	2	0	0
Lasioseius sp. immature	0	1	0	0	0	0	0	0
Eviphididae								
Alliphis halleri	0	11	0	0	0	0	0	1
Melicharidae								
Proctolaelaps sp.	0	33 *	0	0	0	0	0	0
Parasitidae								
Amblygamasus sp.	0	0	0	1	0	0	0	0
Amblygamasus (deutonymph)	0	0	0	1	0	0	0	0
Pergamasus sp. (deutonymph)	0	0	0	0	0	4	1	0
Porrhostaspis lunulata	0	0	2	1	0	0	0	0
Parasitus consanguineus	0	0	0	0	0	16 *	0	0
Pergamasus longicornis	0	0	0	0	0	1	0	4
Pergamasus septentrionalis	0	0	0	0	0	0	0	1
Immature	0	2	1	0	0	0	0	0
Phytoseiidae								
Neoseiulus cucumeris	0	1	0	2	0	0	0	0
Proprioseiopsis okanagensis	0	0	1	1	0	0	0	0
Typhlodromips masseei	0	0	0	1	0	1	0	0
Rhodacaridae								
Rhodacarellus sp. (immature)	0	0	0	1	0	0	0	0
Total	0	67	15	30	0	40	1	30

* Predominant species.

Table 5. Mean number (±SE) of Gamasina mites per plant at 11 a.m. and 11 p.m. ($n = 6$ plants) with straw mulching: 'Early season' (oat straw: T2, since 2018; T3, since 2017); 'Mid-season' (barley straw: T2, since 2018; T3, lower layer since 2017 and top layer since 2018).

Treatments	T2		T3	
Experiments/Time	**11 a.m.**	**11 p.m.**	**11 a.m.**	**11 p.m.**
'Early season' experiment	0.0 ± 0.0 a	4.1 ± 0.5 b	0.0 ± 0.0 a	1.8 ± 0.3 b
'Mid-season' experiment	0.3 ± 0.1 a	1.5 ± 0.4 b	0.2 ± 0.1 a	0.5 ± 0.2 a

In each row, different lower-case letters indicate significant statistical difference between sampling at 11 a.m. and at 11 p.m. within each treatment (Means contrasted by F test, generalized linear model of the quasi-Poisson type in R).

4. Discussion

Our study shows that Gamasina are present in cereal straw used as mulch in strawberry production. Both experiments confirmed our hypothesis that straw maintained longer in the field has the highest numbers and diversity of Gamasina, suggesting that organic mulch could serve as a reservoir for those organisms.

This is in accordance with the findings of Esteca et al. [20] for coffee husk maintained on the floor of a forest patch and then used as mulch in a Brazilian strawberry field. It should be expected, however, that there should be a limit to this effect over time, beyond which abundance and diversity would be reduced, due to the natural process of decomposition of the organic matter and changes in ecological conditions described by some authors [33–35].

The results also show that typically edaphic predators (except Anystidae and Phytoseiidae) move from cereal straw onto strawberry plants at night. This is in accordance with the findings of Britto et al. [27], who collected gerbera (*Gerbera* sp.) (Asteraceae) leaves from a commercial plantation Brazil. Further, it is in accordance with Esteca et al. [20] who found that the Gamasina *P. pygmaeus* (Melicharidae) and *B. dentriticus* (Blattisociidae) were present on strawberry leaflets mainly at night, indicating their possible daily migration from the edaphic environment to strawberry plants. Our results also show a lower prevalence of two-spotted spider mite in pots with straw than without, suggesting that the Gamasina moving from the straw to the plant at night might have preyed on two-spotted spider mite. However, other factors associated with the presence of straw should not be ruled out, and this would be an aspect for complementary investigation.

4.1. Mites from the Straw, Underlying Soil, and Commercial Potting Substrate

Although in this study our main interest centered on the Gamasina and the two-spotted spider mite, mites of other groups were also found in the straw. The diversity of mites found in the soil taken from the field was apparently low, with the occurrence of representatives of about 3–5 families (non-Astigmatina Oribatida were not identified to family) in each field. However, at least 13–14 families (also without taking into account families of non-Astigmatina Oribatida) were represented in the straw used in the experiments. Among the mites collected in this study, some Prostigmata (especially the Tydeidae) and some Astigmatina (especially Acaridae) have been considered important as alternative prey for several predatory mites. Some species of Astigmatina are extensively used as factitious food in the mass production of Gamasina for use as biological control agents [36]. In the present study, acarids were quite numerous in both experiments, and might have served as prey for some Gamasina. In fact, the acarid *T. putrescentiae* entered strawberry plants at night, in both the mulch treatments of both experiments. These were not present on plants at daytime. The presence of these mites on plants could somehow be related to the concurrent presence of the Gamasina on plants.

There is a large amount of information about the groups of mites effectively or potentially useful for biological pest control [15]. In a study conducted by Castilho et al. [5] in Norwegian strawberry fields, three species of some of the same genera as reported in our study (*Porrhostaspis* (Parasitidae), *Proprioseiopsis* (Phytoseiidae), and *Lasioseius* (Blattisociidae)) were collected. Among the Gamasina collected, some are members of the Phytoseiidae, a family that contains several species extensively used for biological pest control [37]. Of the four phytoseiid species found in this study, *N. cucumeris* has been widely commercialized for the control of the two-spotted spider mite and thrips on plants in different countries, including Norway [38]. Of the other families collected, Parasitidae is also commonly found in studies conducted worldwide, but these have not been reported as plant mites. Instead, they are soil inhabitants that feed mostly on nematodes and immature flies [39–41]. Based on the evaluation of the mites present in both types of straw, it was expected that parasitid species could be some of the predominant mites on strawberry leaves. This was not the case, most certainly because, despite their dominance in the straw, strawberry leaves and associated organisms in the climatic room were not attractive to them. Parasitid species have been evaluated for the biological control of pest organisms, especially of thrips [41], and the parasitid *P. longicornis* was previously reported from several coastal habitats in Norway [42]. Mites of the same genus have also been found in soil of strawberry fields in Brazil [20].

Ascidae, Blattisociidae, and Melicharidae, Gamasina groups found in this study, have not been used commercially, but some studies have demonstrated their potential as biological control agents, especially in humid habitats [43]. Mites of these groups seem to be generalists, being able to feed on fungi and to prey on mites and small insects. Melicharids have been reported to feed on immatures of small fruit flies (members of the Drosophilidae family), an insect group that contains an important direct strawberry pest, present in several countries, including Norway [43]. The melicharid *P. pygmaeus* was found in Norway in a coniferous forest [44]. In a study carried out in Brazil, this mite was found in large numbers in coffee husk used as mulching in strawberry fields [20]. It has been reported to consume eggs of *Drosophila* spp. [43] and two-spotted spider mite [45].

The Rhodacaridae have been inadequately studied for use as a biocontrol agent, but an important characteristic of species of this group is their small size, facilitating their movement below the soil surface, where they can encounter plant feeding nematodes, which they have been shown to be able to consume [40]. Eviphidids and veigaiids have also received little attention as biocontrol agents. This is apparently the first report of the eviphidid *A. halleri* in Norway, but this mite has already been reported in neighboring Sweden [46]. Eviphidids are commonly found in agricultural soils, phoretically on insects, especially beetles, and in manure. Interest in this group stems in part from the fact that it contains species known to prey on plant-feeding nematodes [47]. The veigaiid *V. nemorensis* was previously reported in Norway, both in natural and modified environments [40]. Mites of this family are common in litter, feeding on other mites and small arthropods [48].

Some of the Prostigmata collected (especially Anystidae and Cunaxidae) are also known to be predators [36]. However, they have not been developed as commercially available biocontrol agents, probably because of the difficulties in mass production.

4.2. Mites on Strawberry Plants

What would cause mites to move from the ground surface to plants and vice-versa? In general, the movement could occur to escape stress factors of relatively long duration. An example is excess humidity, as evidenced by the movement of Oribatida from the soil to the trees in the Amazon forest, in the rainy season of each year [49]. Other stressors may lead to the occurrence of diapause, common in phytoseiids [37,50] and tetranychids [51], stimulating the mites to move away from their usual habitat while active.

However, factors of shorter duration might also be involved and related to daily movements of mites. They could move from plants to soil during the day to escape low levels of relative humidity. Conversely, they could move back to plants at night, among other reasons, to search for food and/or places with lower competition (due to the diversity and abundance of predators, this is usually lower on plants than in the soil) and less intra-guild predation. Such movement could promote the biological control of potential pest organisms on plants.

The diversity of mites on plants was quite high in this study compared to another study conducted in Norwegian strawberry fields [5]. About eight species were found when a few single leaflets were visually examined; this number was roughly doubled when all plants were washed in alcohol, therefore, changing the plant volume and the method. However, with few exceptions, the number of specimens was low. Regarding the most common groups on plants (the blattisociid *Lasioseius* sp., the melicharid *Proctolaelaps* sp., and the anystids), the same response was observed in the evaluation of mites on leaflets and on the whole plants. However, there was a difference concerning the parasitid *P. consanguineus*, which was not collected from leaflets, but in relatively large numbers on whole plants. The main reason for this difference is probably the occurrence of this species in the crown of the strawberry plants, and not on leaves. In the present study, anystids were only found in the 'Mid-season' experiment, in which they were observed in significant numbers on plants with young cereal straw, and only at daytime. The anystid *Anystis agilis* (Banks) has been cited as a predator of the mites *Panonychus citri* (McGregor), two-spotted spider mite, and the thrips *Scirtothrips citri* Moultan [52]. These pests are found in the aerial part of the plants, although *S. citri* is also found in the soil or litter in the non-feeding resting stages. The anystids found in this study could have helped to reduce the number of two-spotted spider mite on the strawberry leaflets on plants with young cereal straw of the 'Mid-season' experiment, but reports of predation of anystids on two-spotted spider mite are not satisfactory in the literature, and this should be further investigated.

The phytoseiids *N. cucumeris*, *P. okanagensis*, and *T. masseei* have previously been collected on strawberry in Norway [5,53,54], but this is the first time *N. cucumeris* has been found in a Norwegian strawberry crop where it has never been released as a biocontrol agent of mites (including two-spotted spider mite [38]) or thrips. Only one specimen of this species has ever been recorded as naturally occurring in Norway (on *Corylus avellana* L. [55]). Thus, *N. cucumeris* is either naturally occurring in the area or it has followed the plant material. Regarding *P. okanagensis*, this species is definitely native in Norway. According to Meshkov [56], it does not feed on two-spotted spider mite to a significant extent.

Other studies have also reported the difference between the mite fauna on the leaves of certain plants in daytime and nighttime observations [20,25,57,58]. The movement of certain Gamasina biological control agents onto strawberry plants at night and the lower two-spotted spider mite numbers in straw treatments indicates that edaphic Gamasina prey on two-spotted spider mite and suggest that the use of straw with Gamasina may be an interesting conservational biocontrol strategy.

However, not all groups of Gamasina may be inclined to climb plants at night. The results of this study suggested that while some Gamasina apparently venture to visit the regions of more extensive surface in the plant, that is, the leaflets (Blattisociidae, Melicharidae), others visit only the parts closest to the soil, that is, the region of the crown of the plants, as observed for the parasitid *P. consanguineus*. The results also indicate that some Gamasina, such as Eviphididae, even when relatively abundant in the substrate, do not appear to climb the plants often. This suggests that their role as biological control agents on strawberry plants may not be significant.

5. Conclusions

Our study suggests that the use of cereal mulching by strawberry producers in Norway allows the maintenance of a larger number of predatory mites in the environment than would be possible without mulching. Further, it suggests that cereal straw maintained in the field for a longer time (2 years) hosts more predatory mites than cereal straw maintained for a shorter period (1 year). Organic soil mulch facilitates the maintenance of predators in the field, as extensively reported in the literature [10–15]. The difference between the abundance of Gamasina mites in the soil and in the straw was expected, as these mites are not only predators (usually unspecific), but also feed on other organisms found in the mulching, including fungi [44,59,60].

Although the main finding of the present study was the confirmation of nocturnal movement of the Gamasina from straw mulch to strawberry plants, the opposite behavior was also observed. Unlike Gamasina, anystids were observed on plants only during the day, suggesting an inverse movement of the Gamasina, perhaps helping in the reduction of intra-guild predation.

The results of this study may be useful in subsequent research to evaluate the performance of strawberry crops that incorporate the use of cereal straw used in previous years as mulching, favoring the maintenance of edaphic predatory mites in the field. In subsequent research, it seems recommendable to evaluate the potential predation and oviposition of Gamasina mites found in the aerial part of the strawberry plants when offered common strawberry pests. Moreover, in future studies, when sampling predatory mites on strawberry plants, the time of the day should be taken into account, and sampling should also be done at night.

Results could have been different, had we inoculated the plants with other organisms, onto which the edaphic mite might feed. Further, it might be worthwhile to conduct a similar study under field conditions, where environmental factors alter more slowly and in concert, which was not possible to incorporate in the present study.

Author Contributions: Conceptualization, F.d.C.N.E., N.T., I.K. and G.J.d.M.; Data curation, F.d.C.N.E., N.T., I.K., J.C.S., I.D.J. and G.J.d.M.; Formal analysis, F.d.C.N.E., J.C.S. and G.J.d.M.; Funding acquisition, N.T., I.K., I.D.J. and G.J.d.M.; Investigation, F.d.C.N.E., N.T., I.K. and G.J.d.M.; Methodology, F.d.C.N.E., N.T., I.K. and G.J.d.M.; Project administration, F.d.C.N.E., N.T., I.K. and G.J.d.M.; Resources, N.T., I.K., I.D.J. and G.J.d.M.; Supervision, N.T., I.K. and G.J.d.M.; Writing—original draft, F.d.C.N.E., N.T., I.K., I.D.J. and G.J.d.M. All authors have read and agreed to the published version of the manuscript.

Funding: This research was funded by CNPq (Conselho Nacional de Desenvolvimento Científico e Tecnológico) and The Research Council of Norway through the SMARTCROP project (Project Number 244526).

Acknowledgments: We are grateful to Marta Bosque Fajardo and Marit Helgheim at NIBIO for their practical assistance in conducting this work; to the José Bruno Malaquias at ESALQ-USP for help with statistics, and to

Marina Ferraz de Barbosa Camargo at ESALQ-USP for help in identifying Astigmatina mites. Last but not least we thank the Saxebøl family for answering lots of questions and letting us sample straw mulching from their strawberry field.

Conflicts of Interest: The authors declare no conflict of interest.

References

1. Chakraborty, D.; Nagarajan, S.; Aggarwal, P.; Gupta, V.K.; Tomar, R.K.; Garg, R.N.; Sahoo, R.N.; Sarkar, A.; Chopra, U.K.; Sarma, K.S.S.; et al. Effect of mulching on soil and plant water status: And the growth and yield of wheat (*Triticum aestivum* L.) in a semi-arid environment. *Agric. Water Manag.* **2008**, *95*, 1323–1334. [CrossRef]
2. Haslestad, J.; (Norsk Landbruksrådgivning, Ribadu, Hedmark, Norway). Personal communication, 2020.
3. Vi Spiser Mer Utenlandske Enn Norske Jordbær. Available online: https://www.ssb.no/jord-skog-jakt-og-fiskeri/artikler-og-publikasjoner/vi-spiser-mer-utenlandske-enn-norske-jordbaer (accessed on 22 January 2020).
4. Statistics Norway. Available online: https://www.ssb.no/en (accessed on 24 May 2020).
5. Castilho, R.C.; Duarte, V.S.; Moraes, G.J.; Westrum, K.; Trandem, N.; Rocha, L.C.D.; Delalibera, I., Jr.; Klingen, I. Two-spotted spider mite and its natural enemies on strawberry grown as protected and unprotected crops in Norway and Brazil. *Exp. Appl. Acarol.* **2015**, *66*, 509–528. [CrossRef] [PubMed]
6. Trandem, N. Greenhouse production of strawberries and blackberries in Norway–arthropod pests and biological control. In Proceedings of the IOBC/ WPRS Working Group "Integrated Plant Protection in Orchards" Subgroup "Soft Fruits", Dundee, Scotland, 18–21 September 2001; Gordon, S.C., Cross, J.V., Eds.; Scottish Crop Research Institute: Dundee, Scotland, 2003; Volume 26, pp. 45–50.
7. Klingen, I.; Westrum, K.; Meyling, N.V. Effect of Norwegian entomopathogenic fungal isolates against *Otiorhynchus sulcatus* larvae at low temperatures and persistence in strawberry rhizospheres. *Biol. Control.* **2015**, *81*, 1–7. [CrossRef]
8. Aasen, S.S.; Trandem, N. Strawberry blossom weevil *Anthonomus rubi* Herbst (Col.: Curculionidae): Relationships between bud damage, weevil density, insecticide use, and yield. *J. Pest. Sci.* **2006**, *79*, 169–174. [CrossRef]
9. Kader, M.A.; Senge, M.; Mojid, M.A.; Ito, K. Recent advances in mulching materials and methods for modifying soil environment. *Soil Till. Res.* **2017**, *168*, 155–166. [CrossRef]
10. Johnson, J.M.; Hough-Goldstein, J.A.; Vangessel, M.J. Effects of straw mulch on pest insects, predators, and weeds in watermelons and potatoes. *Environ. Entomol.* **2004**, *33*, 1632–1643. [CrossRef]
11. Larentzaki, E.; Plate, J.; Nault, B.; Shelton, A. Impact of straw mulch on populations of onion thrips (Thysanoptera: Thripidae) in onion. *J. Econ. Entomol.* **2008**, *101*, 1317–1324. [CrossRef]
12. Hoddle, M.S.; Morse, J.G.; Phillips, P.A.; Faber, B.A.; Jetter, K.M. Avocado thrips: A new challenge for growers. *Cali Agric.* **2002**, *56*, 103–107. [CrossRef]
13. Jensen, L.; Simko, B.; Shock, C.; Saunders, L. *Alternative Methods for Controlling Onion Thrips (Thrips tabaci) in Spanish Onions*; Malheur Experiment Station Annual Report 2001; Oregon State University Agricultural Experiment Station: Ontario, OR, Canada, 2002; Volume 1038, pp. 104–111.
14. Jamieson, L.E.; Stevens, P.S. The effect of mulching on adult emergence of Kelly's citrus thrips (*Pezothrips kellyanus*). *N. Z. Plant. Prot.* **2006**, *59*, 42–46.
15. Sánchez-Moreno, S.; Ferris, H. Suppressive service of the soil food web: Effects of environmental management. *Agric. Ecosyst. Environ.* **2007**, *119*, 75–87. [CrossRef]
16. Carrillo, D.; Moraes, G.J.; Peña, J.E. *Prospects for Biological Control. of Plant. Feeding Mites and Other Harmful Organisms*; Springer International Publishing: Cham, Switzerland, 2015; p. 337.
17. Walter, D.E.; Proctor, H.C. *Mites: Ecology, Evolution and Behaviour Life at a Microscale*, 2nd ed.; Springer Science and Business Media: Dordrecht, The Netherlands, 2013; p. 509.
18. Bottinelli, N.; Jouquet, P.; Capowiez, Y.; Podwojewski, P.; Grimaldi, M.; Peng, X. Why is the influence of soil macrofauna on soil structure only considered by soil ecologists? *Soil Till. Res.* **2015**, *146*, 118–124. [CrossRef]
19. Raven, P.H.; Evert, R.F.; Eichorn, S.E. *Biologia Vegetal*; Guanabara-Koogan: Rio de Janeiro, Brasil, 2001; p. 906.

20. Esteca, F.C.N.; Rodrigues, L.R.; Moraes, G.J.; Júnior, I.D.; Klingen, I. Mulching with coffee husk and pulp in strawberry affects edaphic predatory mite and spider mite densities. *Exp. Appl. Acarol.* **2018**, *76*, 161–183. [CrossRef] [PubMed]
21. Karg, W. Syno¨kologische Untersuchungen von Bodenmilben aus forstwirtschaftlich und landwirtschaftlich genutzten Böden. *Pedobiologia* **1967**, *7*, 198–214.
22. van de Bund, C.F. Gamasides as predators of phytophagous nematodes. In *Problems of Acarology, Symposium*; Bozek, J., Suski, Z.W., Jakubowska, J., Eds.; Publisher: Polska Akademia Nauk, Polska, 1970; p. 280.
23. Walter, D.E.; Behan-Pelletier, V. Mites in forest canopies: Filling the size distribution shortfall? *Annu. Rev. Entomol.* **1999**, *44*, 1–19. [CrossRef] [PubMed]
24. Onzo, A.; Hanna, R.; Zannou, I.; Sabelis, M.W.; Yaninek, J.S. Dynamics of refuge use: Diurnal, vertical migration by predatory and herbivorous mites within cassava plants. *Oikos* **2003**, *101*, 59–69. [CrossRef]
25. Parecis-Silva, P.V.; Nuvoloni, F.M.; Feres, R.J. Day vs. night: The importance of the circadian cycle over metacommunities and predator–prey densities. *Int. J. Acarol.* **2016**, *42*, 141–148. [CrossRef]
26. Fagan, L.L.; Didham, R.K.; Winchester, N.N.; Behan-Pelletier, V.; Clayton, M.; Lindquist, E.E.; Ring, R.A. An experimental assessment of biodiversity and species turnover in terrestrial versus canopy leaf litter. *Oecologia* **2006**, *147*, 335–347. [CrossRef]
27. Britto, E.P.J.; Gago, E.; Moraes, G.J. How promising is *Lasioseius floridensis* as a control agent of *Polyphagotarsonemus latus*? *Exp. Appl. Acarol.* **2012**, *56*, 221–231. [CrossRef]
28. Meier, U.; Graf, H.; Hess, M.; Kennel, W.; Klose, R.; Mappes, D.; Seipp, D.; Stauss, R.; Streif, J.; van den Boom, T. Phänologische Entwick-lungsstadien des Kernobstes (*Malusdomestica* Borkh. Und *Pyrus communis* L.), des Steinobstes (Prunus-Arten), der Johannisbeere (Ribes-Arten) und der Erdbeere (*Fragaria* x *ananassa* Duch.). *Nachrichtenbl. Deut. Pflanzenschutzd* **1994**, *46*, 141–153.
29. Oliveira, A.R.; Moraes, G.J.; Demétrio, C.G.E.; Nardo, E.A.E. Efeito do vírus de poliedrose nuclear de *Anticarsia gemmatalis* sobre Oribatida edáficos (Arachnida: Acari) em um campo de soja. *Embrapa Meio Ambiente* **2001**, *13*, 5–31.
30. Krantz, G.W.; Walter, D.E. *A Manual of Acarology*, 3rd ed.; Texas Tech University Press: Lubbock, TX, USA, 2009; p. 411.
31. R Development Core Team. *R: A Language and Environment for Statistical Computing, v.3.0.1*; R foundation for Statistical Computing: Vienna, Austria, 2013.
32. Pinzón, J.; Spence, J.R. Bark-dwelling spider assemblages (Araneae) in the boreal forest: Dominance, diversity, composition and life-histories. *J. Insect. Conserv.* **2010**, *14*, 439–458. [CrossRef]
33. Sjursen, H.; Michelsen, A.; Jonasson, S. Effects of long term soil warming and fertilization on microarthropod abundances in three sub-artic ecosystems. *Appl. Soil Ecol.* **2005**, *30*, 148–161. [CrossRef]
34. Blankinship, J.C.; Niklaus, P.A.; Hungate, B.A. A meta-analysis of responses of soil biota to global change. *Oecologia* **2011**, *165*, 553–565. [CrossRef]
35. Bedano, J.C.; Ruf, A. Sensitivity of different taxonomic levels of soil Gamasina to land use and anthropogenic disturbances. *Agric. Forest Entomol.* **2010**, *12*, 203–212. [CrossRef]
36. Hernandes, F.A.; Castro, T.M.M.; Venancio, R. Prostigmata (Acari: Trombidiformes) as biological control agents. In *Prospects for Biological Control of Plant Feeding Mites and Other Harmful Organisms*; Carillo, D., Moraes, G.J., Peña, J.E., Eds.; Springer International Publishing: Cham, Switzerland, 2015; pp. 151–184.
37. McMurtry, J.A.; Sourassou, N.F.; Demite, P.R. The Phytoseiidae (Acari: Mesostigmata) as biological control agents. In *Prospects for Biological Control of Plant Feeding Mites and Other Harmful Organisms*; Carillo, D., Moraes, G.J., Peña, J.E., Eds.; Springer International Publishing: Cham, Switzerland, 2015; pp. 133–149.
38. Knapp, M.; van Houten, Y.; van Baal, E.; Groot, T. Use of predatory mites in commercial biocontrol: Current status and future prospects. *Acarologia* **2018**, *58*, 72–82.
39. Szlendak, E.; Lewandowski, M. Development and reproductive capacity of the predatory mite *Parasitus consanguineus* (Acari: Parasitidae) reared on the larval stages of *Megaselia halterata* and *Lycoriella ingenua*. *Exp. Appl. Acarol.* **2009**, *47*, 285–292. [CrossRef]
40. Castilho, R.C.; Venancio, R.; Narita, J.P.Z. Mesostigmata as biological control agents, with emphasis on Rhodacaroidea and Parasitoidea. In *Prospects for Biological Control of Plant Feeding Mites and Other Harmful Organisms*; Carillo, D., Moraes, G.J., Peña, J.E., Eds.; Springer International Publishing: Cham, Switzerland, 2015; pp. 1–31.

41. Rueda-Ramírez, D.; Rios-Malaver, D.; Varela-Ramírez, A.; Moraes, G.J. Biology and predation capacity of *Parasitus bituberosus* (Acari: Mesostigmata: Parasitidae) on *Frankliniella occidentalis* (Thysanoptera: Thripidae), and free-living nematodes as its complementary prey. *Pest. Manag. Sci.* **2019**, *75*, 1819–1830. [CrossRef]
42. Bolger, T.; Devlin, M.; Seniczak, A. First records of ten species of Mesostigmata (Acari, Mesostigmata) added to the published Norwegian species list. *Nor. J. Entomol.* **2018**, *65*, 94–100.
43. Moraes, G.J.; Venancio, R.; dos Santos, V.L.V.; Paschoal, A.D. Potential of Ascidae, Blattisociidae and Melicharidae (Acari: Mesostigmata) as biological control agents of pest organisms. In *Prospects for Biological Control of Plant Feeding Mites and Other Harmful Organisms*; Carillo, D., Moraes, G.J., Peña, J.E., Eds.; Springer International Publishing: Cham, Switzerland, 2015; pp. 33–75.
44. Slomian, S.; Gulvik, M.E.; Madej, G.; Austad, I. Gamasina and Microgyniina (Acari, Gamasida) from soil and tree hollows at two traditional farms in Sogn og Fjordane, Norway. *Norw. J. Entomol.* **2005**, *52*, 39–48.
45. Lindquist, E.E.; Krantz, G.W.; Walter, D.E. Order Mesostigmata. In *A Manual of Acarology*, 3rd ed.; Krantz, G.W., Walter, D.E., Eds.; Texas Tech University Press: Lubbock, TX, USA, 2009; pp. 124–232.
46. Halliday, R.B. *Alliphis siculus* (Oudemans 1905) is not a synonym of *Alliphis halleri* (G. & R. Canestrini 1881) (Acari: Eviphididae). *Syst. Appl. Acarol.* **2008**, *13*, 51–64.
47. Karg, W.; Grosse, E. Raubmilben als Antagonisten von Nematoden. *Nachr. Pflanzenschutz DDR* **1983**, *37*, 208–212.
48. Mathys, G.; Tencalla, Y. Note préliminaire sur la biologie et la valeur prédatice de *Proctolaelaps hypudaei* Oudms (Acarien: Mesostigmata: Aceosejidae). *Stn. Fédérales Essais Agric.* **1959**, *600*, 645–654.
49. Moraes, J.; Franklin, E.; Morais, J.W.; Souza, J.L.P. Species of edaphic mites (Acari: Oribatida) and effects of topography, soil properties and litter gradients on their qualitative and quantitative composition in 64 km^2 of forest in Amazonia. *Exp. Appl. Acarol.* **2011**, *55*, 39–63. [CrossRef] [PubMed]
50. Overmeer, W.P.J. *Diapause. Spider Mites: Their Biology, Natural Enemies and Control*; Elsevier: Amsterdam, The Netherlands, 1985; Volume 2, pp. 95–102.
51. Veerman, A. *Diapause. Spider Mites: Their Biology, Natural Enemies and Control*; Elsevier: Amsterdam, The Netherlands, 1985; Volume 1, pp. 279–316.
52. Laing, J.E.; Knop, N.F. Potential use of predaceous mites other than Phytoseiidae for biological control of orchard pests. In *Biological Control of Pests by Mites, Special Publications*; Hoy, M.A., Cunningham, G.L., Knutson, L., Eds.; University of California: Berkeley, CA, USA, 1983; pp. 12–20.
53. Karg, W.; Edland, T. Neue Raubmilbenarten der Phytoseiidae Berlese, 1916. *Dtsch. Entomol. Z.* **1987**, *34*, 387–395. [CrossRef]
54. Trandem, N.; Klingen, I.; Haukeland, S.; Moraes, G.J. The occurrence of two pest mites and three groups of biocontrol agents in organic and conventional strawberry fields (abstract only). *IOBC Bull.* **2011**, *70*, 96.
55. Denmark, H.A.; Edland, T. The subfamily Amblyseiinae Muma (Acari: Phytoseiidae) in Norway. *Int. J. Acarol.* **2002**, *28*, 195–220. [CrossRef]
56. Meshkov, Y.I. Phytoseiid Mites (Parasitiformes, Phytoseiidae) on Main Berry Plantings. Ph.D. Thesis, Bolshie Vjazemi, Moscow, Russia, 1996.
57. Novotny, V.; Basset, Y.; Auga, J.; Boen, W.; Dal, C.; Drozd, P.; Kasbal, M.; Isua, B.; Kutil, R.; Manumbor, M. Predation risk for herbivorous insects on tropical vegetation: A search for enemy-free space and time. *Aust. J. Ecol.* **1999**, *24*, 477–483. [CrossRef]
58. Saigusa, M.; Oishi, K.; Ikumoto, A.; Iwasaki, H.; Terajima, M. Emergence patterns of small subtidal arthropods in relation to day/night, tidal, and surface/bottom factors: Investigations in the Boreal Sea, Japan (Akkeshi, Hokkaido). *J. Oceanogr.* **2000**, *56*, 295–310. [CrossRef]
59. Zaitsev, A.S.; Chauvat, M.; Pflug, A.; Wolters, V. Oribatid mite diversity and community dynamics in a spruce chronosequence. *Soil Biol. Biochem.* **2002**, *34*, 1919–1927. [CrossRef]
60. Walter, D.E.; Hunt, H.W.; Elliott, E.T. Guilds or functional groups? An analysis of predatory arthropods from a shortgrass steppe soil. *Pedobiologia* **1988**, *31*, 247–260.

diversity

MDPI

Article

Patterns of Distribution of Phoretic Deutonymphs of Uropodina on Longhorn Beetles in Białowieża Primeval Forest, Central Europe

Szymon Konwerski [1,*], Jerzy M. Gutowski [2] and Jerzy Błoszyk [1,3]

1 Natural History Collections, Faculty of Biology, Adam Mickiewicz University, 61-614 Poznań, Poland; bloszyk@gmail.com
2 Forest Research Institute, Department of Natural Forests, 17-230 Białowieża, Poland; j.gutowski@ibles.waw.pl
3 Department of General Zoology, Faculty of Biology, Adam Mickiewicz University, 61-614 Poznań, Poland
* Correspondence: szymkonw@amu.edu.pl

Received: 29 April 2020; Accepted: 10 June 2020; Published: 12 June 2020

check for updates

Abstract: We studied the distributional patterns of phoretic deutonymphs of the genera *Oodinychus* and *Trichouropoda* (Uropodina) on the longhorn beetles *Monochamus sartor urussovii*, *Plagionotus detritus*, *Tetropium castaneum*, and *Te. fuscum*, based on large samples of specimens (992 beetles and 25,587 mites) collected in the Białowieża Primeval Forest in Central Europe in the years 2008 and 2012–2016. All the studied phoretic phoront-host associations are characterized by different patterns of the attachment sites of mites on beetle's body. In the case of *O. ovalis* and *M. sartor urussovii* association, the deutonymphs were found mostly on the pronotum and dorsal surface of the elytra. This is the only instance in which phoronts were absent on the legs. Deutonymphs of *T. sociata* on *P. detritus* preferred the abdomnen (both tergites and ventrites) and the hindlegs. Only in this case the phoronts were attached inside the subelytral space. The highest number of deutonymphs of *T. shcherbakae* on both *Tetropium* species occured on the legs. In the case of *Te. castaneum*, similar proportions of mites were recorded on all pairs of legs, while the preferred location of mites phoretic on *Te. fuscum* were the forelegs. Both the preferences of phoretic deutonymphs for specific parts of the host's body and the participation of carriers transporting deutonymphs on particular parts of their bodies were very consistent.

Keywords: mites; Cerambycidae; phoresy; natural forest; Oodinychus; Trichouropoda; Monochamus; Plagionotus; Tetropium

1. Introduction

Mites, even those inhabiting well-studied systems are often overlooked or ignored, although they are strong interactors and major components of biological diversity—not passive inhabitants of ecosystems [1]. This statement is particularly relevant to Uropodina mites in natural forest ecosystems. In this group of mites, the unique morphological adaptations for phoretic purposes, namely the production of an anal pedicel [2,3], enabled them to use the saproxylic beetles as carriers (dispersants). This, in turn, allowed for the efficient use of unstable, temporary, cyclical, and patchy merocenoses in dying and dead trees, including beetle-generated microhabitats. Close and specific relationships characterize the pioneer species colonizing dying and dead trees in early stages of decay [4], among which the most important as dispersants of mites are Scolytinae and Cerambycidae [5]. While Uropodina and bark beetles relationships have been the subject of numerous studies, and their role is relatively well known, both in terms of phoresy and species communities in beetle galleries [6–18], little attention has been paid to uropodids and longhorn beetles, so the importance of their ecological relationships for the forest ecosystem is essentially unknown [19].

The phoretic relationships between uropodid mites and Cerambycidae in the natural forest ecosystem has been studied in recent years in Białowieża Primeval Forest, which is considered the best-preserved forest in the European lowlands. The studies have been focused on new associations of uropodids and longhorn beetle species [20], relationships between phoronts and their carriers [21,22], the contribution of beetle dispersants in increasing of alpha diversity of mites in dead wood merocenoses [23], as well as the stability and repeatability of phoretic relationships [24,25].

The aim of this research was to find the distributional patterns of phoretic deutonymphs for longhorn beetle species utilized as dispersants by uropodids in the natural forest ecosystem of Białowieża Primeval Forest in Central Europe, based on a large data set and complementing earlier studies, in order to clarify our understanding of relationships between the Uropodina mites and their saproxylic Cerambycidae carriers.

The Białowieża Primeval Forest is unique within Europe because of its natural structure [26–28], and plays a role of a living laboratory for ecological research [29]. It is also a European hotspot for Coleoptera [30–35]. Studies of primary forest provide reference data for ecological research in managed forests and improve the knowledge on conservation of saproxylic organisms [36].

2. Materials and Methods

The study concentrated on the Uropodina mites (Acari: Mesostigmata) phoretic on the most numerous (over 30 specimens with attached mites) longhorn beetle species: *Monochamus sartor urussovii*, *Plagionotus detritus*, *Tetropium castaneum*, and *Tetropium fuscum*.

The nomenclature of cerambycids follows Löbl and Smetana [37] and Plewa et al. [38], and the nomenclature of Uropodina follows Błoszyk [39]. The terminology of beetle morphological structures follows Lawrence et al. [40].

The research was conducted in the old growth (natural) Białowieża Primeval Forest, in natural stands rich in dead wood. The study sites were located in forest compartments 338A (N52°44′, E23°45′), near Teremiski, about 8 km west-north of the village of Białowieża, 448C (N52°42′, E23°46′) in the protected area "Professor Władysław Szafer's Nature Reserve", and 496C (N52°40′, E23°47′), about 4 km south west of the village of Białowieża. Detailed information about the study sites are in Błoszyk et al. [21,23] and Konwerski et al. [24,25].

Beetles were caught in 66 IPM-Intercept traps and 48 12-unit Lindgren multiple funnel traps, provided by the Canadian Forest Service. They were set in forest stands from May to July in 2008 and 2012–2016, and emptied every two weeks. Cerambycidae specimens were kept in 70% ethanol. In the laboratory they were individually (each specimen separately) examined under a stereomicroscope, to determine their species and to record the presence of mites. Beetles with attached mites (carriers) were selected for further study. Using a stereomicroscope (Olympus SZX12), the number of deutonymphs on beetles was counted and their location of attachment to beetles (hosts) was determined. The possible locations were the legs, head, scutellum, dorsal surface of the elytra, subelytra, membranous wings, pronotum, prosternum, mesoventrite, metaventrite, abdominal ventrites, and tergites.

All the uropodid mites found attached to longhorn beetles were identified using the comparative collection of J. Błoszyk, deposited in the Natural History Collections at the Faculty of Biology, Adam Mickiewicz University in Poznań (Poland), and the keys to mite species identification [41–44]. For biometrical analyses, selected mites were mounted on microscope slides using Hoyer's medium. The remaining mites were identified from temporary preparations after clearing in the lactic acid using an open slide technique and later placed in 75% ethanol.

The voucher specimens of longhorn beetles and mites have been deposited in the Natural History Collections at the Faculty of Biology, Adam Mickiewicz University in Poznań, Poland.

Distribution of phoretic deutonymphs on various parts of host's body was compared using multivariate exploratory techniques: cluster analysis (distance: Euclidean, method: ward). Frequency distribution tables were used for calculation of the phoretic load for each beetle species. Statistical analyses were performed using the software Statistica 10 [45].

3. Results

During the six years of the research project, 992 specimens of Cerambycidae with 25,587 transported deutonymphs of Uropodina were collected. On 54 specimens of *M. sartor urussovii*, there were 1088 phoretic deutonymphs belonging to a single species *Oodinychus ovalis*. On 494 specimens of *P. detritus*, there were 19,145 phoretic deutonymphs belonging to a single species *Trichouropoda sociata*. On 97 specimens of *Te. castaneum* and 347 specimens of *Te. fuscum*, there were 839 and 4515 phoretic deutonymphs, respectively, all belonging to *Trichouropoda shcherbakae*.

The mites were found on the following body parts of the beetles: all pairs of legs, head, scutellum, dorsal surface of the elytra, subelytra, membranous wings, pronotum, prosternum, mesoventrite, metaventrite, abdominal ventrites, and tergites (Table 1, Figures 1 and 2).

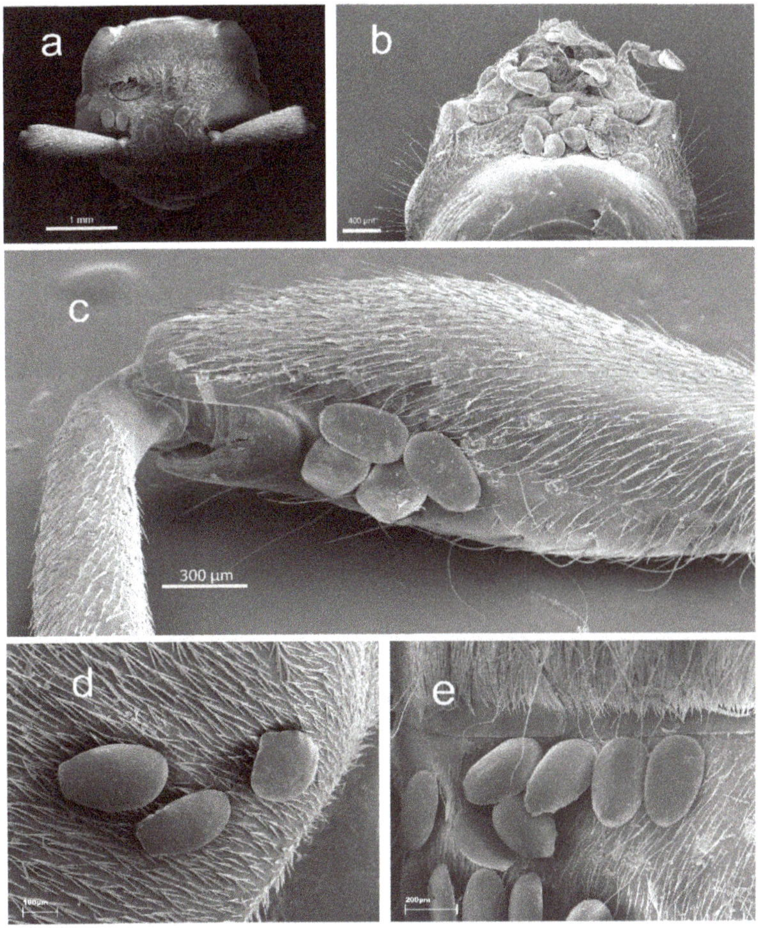

Figure 1. Deutonymphs of *Trichouropoda sociata* attached to *Plagionotus detritus*: (**a**) dorsal side of the head; (**b**) below the mouthparts; (**c**) foreleg; (**d**) dorsal surface of the elytra; (**e**) abdominal ventrites.

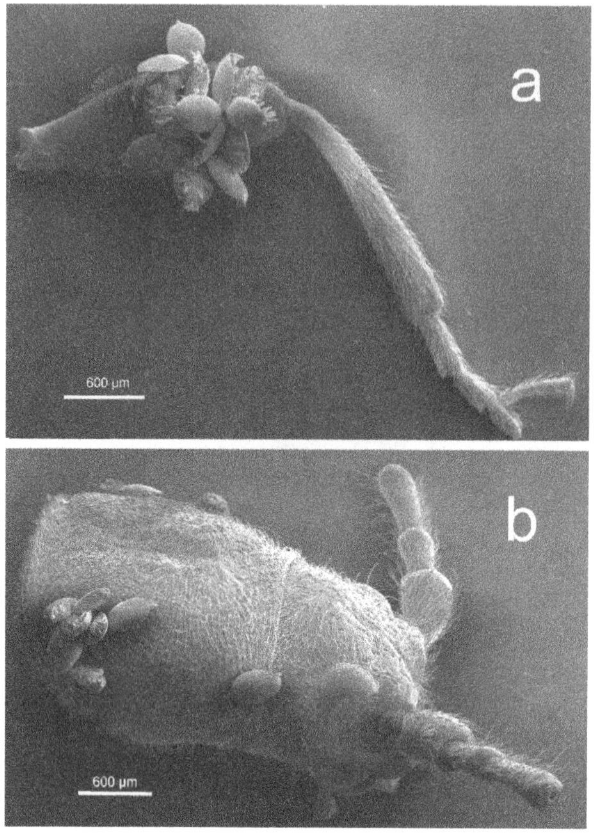

Figure 2. Deutonymphs of *Trichouropoda shcherbakae* attached to *Tetropium fuscum*: (**a**) foreleg; (**b**) pronotum.

Table 1. Number of deutonymphs attached to various parts of beetles' body: (**L1**) forelegs; (**L2**) midlegs; (**L3**) hindlegs; (**H**) head; (**Sc**) scutellum; (**Es**) surface of the elytra; (**Eu**) subelytra; (**W**) membranous wings; (**Pn**) pronotum; (**Ps**) prosternum; (**Me**) mesoventrite; (**Mt**) metaventrite; (**St**) abdominal ventrites; (**T**) abdominal tergites.

Beetle Species	L1	L2	L3	H	Sc	Es	Eu	W	Pn	Ps	Me	Mt	St	T
M. s. urussovii	0	0	0	0	0	141	0	0	930	0	13	0	4	0
P. detritus	37	688	1645	41	9	745	839	64	261	266	681	251	5989	7545
Te. Castaneum	214	265	231	5	0	21	0	0	40	13	12	16	21	1
Te. Fuscum	2262	1270	591	21	3	49	0	0	213	32	21	20	33	0

3.1. Distribution of Oodinychus ovalis on Monochamus sartor urussovii

The mites were found on dorsal surface of the elytra, pronotum, mesoventrite, and abdominal ventrites (Table 1). The number of deutonymphs attached to one beetle varied between one and 149.

The deutonymphs were found most frequently attached to the dorso-lateral side of the prothorax—the pronotum (85.4% of all carried deutonymphs, up to 141 individuals at one time). On the dorsal surface of the elytra, there were 12.9% of all deutonymphs (up to 122 individuals at one time), whereas on ventral side of the body the mites were attached only to the mesoventrite (1.2% of all carried deutonymphs, up to six individuals at one time) and abdominal ventrites (0.4%, four individuals—only once) (Figure 3).

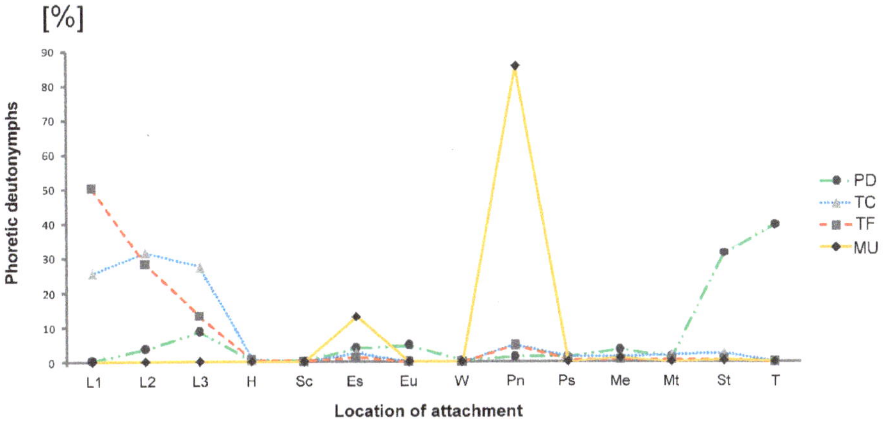

Figure 3. The proportion of phoretic deutonymphs on various parts of beetles' body: (**PD**) *Plagionotus detritus*; (**TC**) *Tetropium castaneum*; (**TF**) *Tetropium fuscum*; (**MU**) *Monochamus sartor urussovii*; (**L1**) forelegs; (**L2**) midlegs; (**L3**) hindlegs; (**H**) head; (**Sc**) scutellum; (**Es**) surface of the elytra; (**Eu**) subelytra; (**W**) membranous wings; (**Pn**) pronotum; (**Ps**) prosternum; (**Me**) mesoventrite; (**Mt**) metaventrite; (**St**) abdominal ventrites; (**T**) abdominal tergites. The colored lines serve as a guide to the eye.

The majority of collected beetles carried mites on the pronotum (92.6% of the carriers), whereas on the dorsal surface of the elytra the mites were found in 24.1% of the carriers. Only 5.5% of beetles carried deutonymphs on the mesoventrite, and 1.8% on the abdominal ventrites (Figure 4).

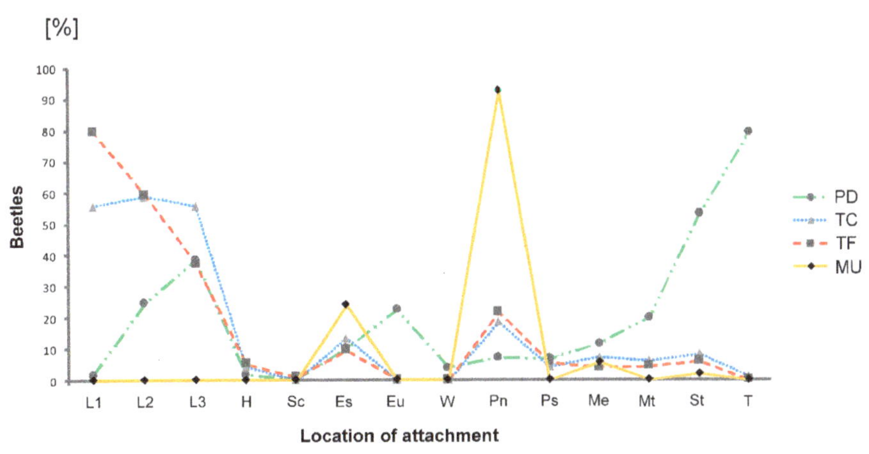

Figure 4. Frequency of individuals of beetles carrying the phoretic deutonymphs on various parts of their bodies: (**PD**) *Plagionotus detritus*; (**TC**) *Tetropium castaneum*; (**TF**) *Tetropium fuscum*; (**MU**) *Monochamus sartor urussovii*; (**L1**) forelegs; (**L2**) midlegs; (**L3**) hindlegs; (**H**) head; (**Sc**) scutellum; (**Es**) surface of the elytra; (**Eu**) subelytra; (**W**) membranous wings; (**Pn**) pronotum; (**Ps**) prosternum; (**Me**) mesoventrite; (**Mt**) metaventrite; (**St**) abdominal ventrites; (**T**) abdominal tergites. The colored lines serve as a guide to the eye.

The cluster analysis of distribution of deutonymphs on different parts of host's body shows that the pronotum and the surface of the elytra are similar (Figure 5A).

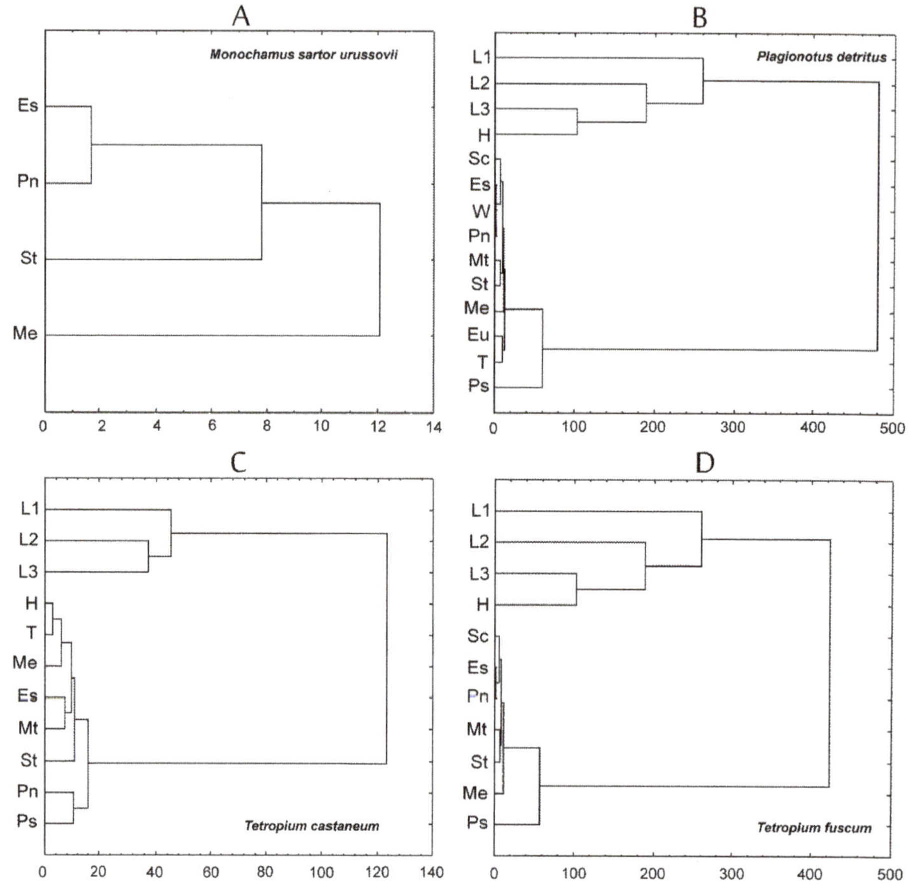

Figure 5. Cluster analysis of distribution of phoretic deutonymphs on various parts of beetles' body: (**A**) *Monochamus sartor urussovii;* (**B**) *Plagionotus detritus;* (**C**) *Tetropium castaneum,* (**D**) *Tetropium fuscum;* Distance: Euclidean, cluster method: ward; (**Es**) surface of the elytra; (**Eu**) subelytra; (**H**) head; (**L1**) forelegs; (**L2**) midlegs; (**L3**) hindlegs; (**Me**) mesoventrite; (**Mt**) metaventrite; (**W**) membranous wings; (**Pn**) pronotum; (**Ps**) prosternum; (**Sc**) scutellum; (**St**) abdominal ventrites; (**T**) abdominal tergites.

Analysis of the phoretic load shows that 81.5% of hosts carry 1–20 deutonymphs per beetle, whereas heavy phoretic load (over 100 deutonymphs per host) is observed in 7.4% of beetles.

3.2. Distribution of *Trichouropoda sociata* on *Plagionotus detritus*

The mites were found on all pairs of legs, head, scutellum, dorsal surface of the elytra, subelytra, membranous wings, pronotum, prosternum, mesoventrite, metaventrite, abdominal ventrites, and tergites (Table 1, Figure 1). The number of deutonymphs attached to one beetle varied between one and 502.

The deutonymphs were found most frequently attached to the abdominal tergites (39.6% of all carried deutonymphs) and ventrites (31.3%). These parts of the body were also characterized by the highest variation in the number of the attached mites (tergites 1–231 and ventrites 1–245 individuals). Most mites were attached to tergite III, followed by IV and II. Much fewer deutonymphs were carried on tergites I, VI, and V (Figure 6). In the case of the abdominal ventrites, the number decreased gradually from ventrite I to V (Figure 7).

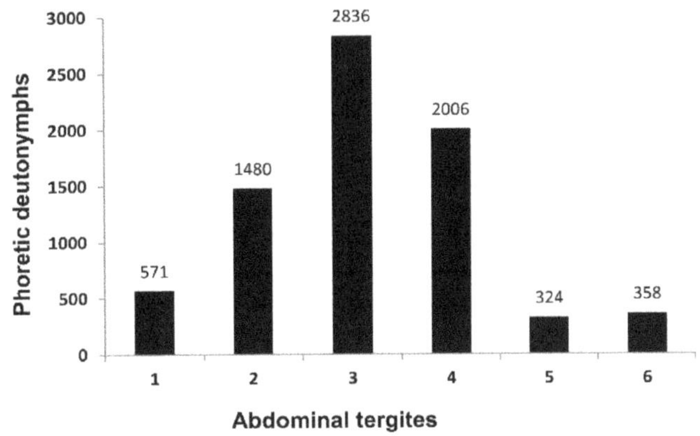

Figure 6. Number of phoretic deutonymphs of *Trichouropoda sociata* attached to the abdominal tergites of *Plagionotus detritus*.

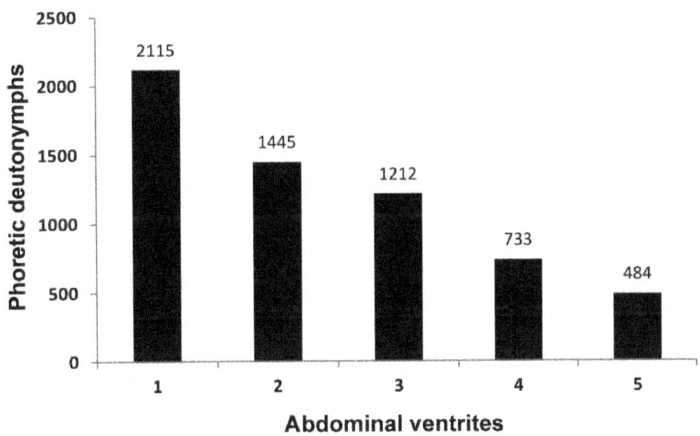

Figure 7. Number of phoretic deutonymphs of *Trichouropoda sociata* attached to the abdominal ventrites of *Plagionotus detritus*.

Mites attached to the legs, showed a strong preference for the hindlegs. On the forelegs there were 0.2% of all deutonymphs (up to 14 individuals at one time); midlegs: 3.6% (up to 36 individuals at one time); and hindlegs: 8.6% (up to 67 individuals at one time). The preferred locations of mites attached to the legs were the femora and metacoxae (Table 2, Figure 1c).

Table 2. Number of deutonymphs attached to various parts of beetles' legs: (**1-3**) pairs of legs; (**Co**) coxae; (**Tr**) trochanters; (**Fe**) femora; (**Ti**) tibiae; (**Ta**) tarsi.

Beetle Species	Co1	Tr1	Fe1	Ti1	Ta1	Co2	Tr2	Fe2	Ti2	Ta2	Co3	Tr3	Fe3	Ti3	Ta3
Plagionotus detritus	7	0	30	0	0	25	0	661	2	0	791	5	843	6	0
Tetropium castaneum	8	0	204	2	0	5	1	251	8	0	3	1	222	4	1
Tetropium fuscum	50	0	2209	3	0	4	0	1265	1	0	10	0	580	1	0

On the dorsal surface of the elytra, there were 3.9% of all deutonymphs (up to 122 individuals at one time), whereas on the subelytra there were 4.7% (up to 101 individuals at one time).

On the ventral side of the thorax, the most common location of attachment was the mesoventrite (3.6% of all deutonymphs, up to 75 individuals at one time). The prosternum and metaventrite had a similar percentage of attached deutonymphs, i.e., 1.4% (up to 34 individuals at one time) and 1.3% (up to 11 individuals at one time). On the dorso-lateral side of the thorax (the pronotum), there were 1.4% of all deutonymphs (up to 45 individuals at one time). On the head, scutellum, and membranous wings, there were fewer than 1% of all deutonymphs (Figure 3). Moreover, in the case of the wings, only one out of 64 individuals was attached to the anal pedicel.

The majority of collected beetles carried mites on the abdominal tergites (78.9% of the carriers) and ventrites (53% of the carriers). On the hindlegs, the mites were found in 38.5% of the carriers, whereas they were on the midlegs and forelegs in 24.7% and 1.6% of the carriers, respectively. In the case of the elytra, the mites occurred more often on the subelytra (22.5% of carriers) than on the dorsal surface (10.1% of carriers). In the case of the ventral side of the thorax, more beetles carried deutonymphs on the metaventrite (19.8%) than on the mesoventrite (11.5%) and the prosternum (6.7% of the carriers) (Figure 4).

The cluster analysis of distribution of deutonymphs on different parts of host's body shows that the legs, together with head, vary from all other parts. There is also a similarity between the subelytra and abdominal tergites as well as the metaventrite and abdominal ventrites (Figure 5B).

Analysis of the phoretic load shows that 76.1% of hosts carry 1–50 deutonymphs per beetle, whereas heavy phoretic load (over 200 deutonymphs per host) is observed in 1.4% of beetles.

3.3. Distribution of *Trichouropoda shcherbakae* on *Tetropium castaneum*

The mites were found on all pairs of legs, head, dorsal surface of the elytra, pronotum, prosternum, mesoventrite, metaventrite, abdominal ventrites, and tergites (Table 1). The number of deutonymphs attached to one beetle varied between one and 59.

The deutonymphs were found most frequently attached to the legs: on the forelegs there were 25.5% of all deutonymphs (up to 20 individuals at one time); midlegs: 31.6% (up to 18 individuals at one time); and hindlegs: 27.5% (up to 36 individuals at one time). The preferred locations of mites attached to the legs were the femora (Table 2). On the dorso-lateral side of the thorax (the pronotum), there were 4.8% of all deutonymphs (up to six individuals at one time). On the dorsal surface of the elytra and the abdominal ventrites, there were equally 2.5% of attached mites (up to three and five individuals at one time, respectively). On the prosternum, mesoventrite, and metaventrite, there were fewer than 2% of all deutonymphs, whereas on the head and abdominal tergites there were fewer than 1% (Figure 3).

Most beetles carried mites on all pairs of legs: on the midlegs: 58.8% of the carriers; on the forelegs and hindlegs: 55.7%. On the pronotum, the mites were found in 18.6% of the carriers. On the dorsal surface of the elytra deutonymphs occurred in 13.4% of beetles. All other locations were found in fewer than 10% of the carriers (Figure 4).

The cluster analysis of distribution of deutonymphs on different parts of host's body shows that the legs vary from all other parts, and the forelegs differ from the midlegs and hindlegs. There is also a similarity between the pronotum and prosternum, as well as between the surface of the elytra and metaventrite (Figure 5C).

Analysis of the phoretic load shows that 86.7% of hosts carry 1–20 deutonymphs per beetle, whereas heavy phoretic load (over 40 deutonymphs per host) is observed in 6.1% of beetles.

3.4. Distribution of *Trichouropoda shcherbakae* on *Tetropium fuscum*

The mites were found on all pairs of legs, head, scutellum, dorsal surface of the elytra, pronotum, prosternum, mesoventrite, metaventrite, and abdominal ventrites (Table 1, Figure 2). The number of deutonymphs attached to one beetle varied between one and 126.

The deutonymphs were found most frequently attached to the legs, with a strong preference for the first pair: on the forelegs, there were 50.1% of all deutonymphs (up to 61 individuals at one time); midlegs: 28.1% (up to 49 individuals at one time); and hindlegs: 13.1% (up to 25 individuals at one time). The preferred locations of mites attached to the legs were the femora (Table 2, Figure 2a).

On the dorso-lateral side of the thorax (the pronotum), there were 4.7% of all deutonymphs (up to 15 individuals at one time). On the dorsal surface of the elytra there were only 1.1% of attached mites (up to four individuals at one time). On the head, scutellum, prosternum, mesoventrite, metaventrite, and abdominal ventrites, there were fewer than 1% of all deutonymphs (Figure 3).

Most beetles carried mites on the forelegs (79.8% of the carriers), whereas on the midlegs and hindlegs the mites were found in 59.4% and 37.2%, respectively. Deutonymphs on the pronotum occurred in 21.6% of beetles. All other locations were found in fewer than 10% of the carriers (Figure 4).

The cluster analysis of distribution of deutonymphs on different parts of host's body shows that the legs together with head vary from all other parts and the forelegs vary from other legs. There is also a similarity between the surface of the elytra and pronotum, as well as between the metaventrite and abdominal ventrites (Figure 5D).

Analysis of the phoretic load shows that 81.8% of hosts carry 1–20 deutonymphs per beetle, whereas heavy phoretic load (over 40 deutonymphs per host) is observed in 8.1% of beetles.

4. Discussion

The results of our studies indicates that every analyzed carrier species has different pattern of distribution of phoretic deutonymphs of Uropodina attached to its body. This also applies to morphologically similar species of the genus *Tetropium*, utilized as dispersants of the single Uropodina mite species *T. shcherbakae*.

In the case of *O. ovalis* and *M. sartor urussovii* association, the deutonymphs were found mostly on the pronotum and dorsal surface of the elytra. This is the only instance in which phoronts were absent on the legs (Table 1). Interestingly, also in the case of two North American representatives of the genus *Monochamus* no uropodid mites were found on the legs, despite heavy phoretic load [46] and large analyzed samples [47].

Deutonymphs of *T. sociata* on *P. detritus* preferred the abdomen (both tergites and ventrites) and the hindlegs. Only, in this case, the phoronts were attached inside the subelytral space (abdominal tergites covered by elytra, subelytra, and membranous wings) (Table 1).

The highest number of deutonymphs of *T. shcherbakae* on both *Tetropium* species occurred on the legs. In the case of *Te. castaneum*, similar proportions of mites were recorded on all pairs of legs, while the preferred location of mites phoretic on *Te. fuscum* were the forelegs (Table 1). Konwerski et al. [24] proposed that the different preferences for spatial attachment of *T. shcherbakae*, depending on the species of carriers, may be due to different preferences of the two *Tetropium* species for host trees and microhabitats. This, in turn, could influence the mites' climbing behavior in various merocenoses.

Analysis of the presence of *T. sociata* on the abdominal tergites of *P. detritus* revealed the preferences for the middle part (tergites III, IV, and II) (Figure 6). We may suspect that mites use the relatively large subelytral space to avoid falling off from the carrier during the folding and unfolding of the membranous wings. The two last visible tergites (V and VI) are avoided, probably because of relatively high mobility of the end of abdomen, and the possibility of detaching of the mites by the ridges of the elytra.

The number of deutonymphs of *T. sociata* on the abdominal ventrites of *P. detritus* decreases gradually from ventrite I to V (Figure 7). The highest number of mites on the ventrite I is highly likely to be connected with the large number of phoronts attached to the metacoxal plates and metafemora.

The legs were the place of attachment for deutonymphs in the case of three out of four analyzed phoront-host associations, and the femora were the preferred location. The metacoxa of *P. detritus* also were utilized by a large number of mites. The trochanters, tibiae, and tarsi were avoided in all the analyzed species (Table 2). While it is easy to understand that the number of mites attached to the trochanters is limited by its size, the tibiae are similar in length to femora (Figure 2a). However, the tibiae are much thinner and more movable, so the risk of detachment during walking or cleaning is higher. High mobility and the segmented structure of tarsi can cause detachment of deutonymphs, thus, these parts of legs are avoided by the phoronts.

Our observations show that the deutonymphs attach to surfaces of various types: smooth (e.g., subelytra in *P. detritus*), with setae of low density (e.g., femora in *Tetropium* and *P. detritus*), and setae of high density (e.g., abdominal ventrites and surface of the elytra in *P. detritus*). This conforms to the results obtained by Bajerlein et al. [48], who analyzed carrier surface morphology and topography in the case of coprophilous beetles and deutonymphs of *Uropoda orbicularis*.

No phoronts were found on the antennae and protarsi of the carriers. It appears that avoidance of these body parts may be caused by the behavior of the host, that is, the ability to clean the protarsi with the mouthparts and the antennae (the most important sensory organ) with both, the mouthparts and protarsi. Also avoiding of the forelegs in *P. detritus* may stem from the cleaning behavior of the host [25].

Although the low number of mites attached to the beetle's head is a rule, in the case of fairly large carriers, there is a lot of space for small deutonymphs to attach in places where they would not cause any hindrance, e.g., on the ventral side of the head—below the mouthparts [25]. We found the presence of mites in this location in *P. detritus* (Figure 1b). It is noteworthy that this attachment site of Uropodina has been found also in the case of Passalidae, which are large saproxylic beetles [49,50].

In our research the majority of deutonymphs were found attached to the heavy sclerotized regions of carrier's body (e.g., elytra, pronotum, ventrites, legs). Only, in the case of *T. sociata* and *P. detritus*, the mites also preferred less sclerotized regions, namely abdominal tergites. In the abovementioned association one phoront was found attached to the membranous wing as well. Attaching to the heavy sclerotized parts of beetles' body should be considered a rule in phoretic relationships between Uropodina and saproxylic Coleoptera, although attachment to the flexible membrane at the joint between the head and prosternum in the case of *Lucanus cervus* and the intersegmental membrane under elytra of *Rhynchophorus ferrugineus* were also reported [51,52]. *Oodinychus ovalis*, which, in our research, was utilizing *M. sartor urussovii*, has been also recorded from Chilopoda [53]. Interestingly, most deutonymphs found on centipedes were attached to the membranous and flexible cuticle of the pleurites [54,55]. This, in turn, means that the same species of Uropodina has the ability to utilize various types of surfaces for attachment, and their preferences differ, depending on the carrier.

The presence of unattached deutonymphs on the membranous wings of *P. detritus* can be explained by the "scooping" mechanism proposed by Konwerski et al. [25]. As a result of the folding and unfolding of the membranous wings, some mites previously attached to the subelytra and abdominal tergites may be detached, especially when there are many of them.

The cluster analysis of distribution of phoretic deutonymphs on various parts of beetles' body shows some similarities and differences (Figure 5). However, its interpretation is only of a speculative nature, as no observations of mite attachment to the carrier were made. The legs differ from other body parts in all the cases when phoronts are attached to them (*T. sociata–P. detritus*, *T. shcherbakae–Te. castaneum*, and *Te. fuscum*). This may be caused by the specific structure of these attachment sites, which are very long and narrow. The similarity between the pronotum and elytral surface (*O. ovalis–M. sartor urussovii*), as well as the metaventrite and abdominal ventrites (*T. sociata–P. detritus*, *T. shcherbakae–Te. fuscum*) can be explained by the fact that these adjacent parts create wide dorsal and ventral surfaces, suitable for mite movement. Climbing or descending of deutonymphs on lateral surfaces may explain the similarities between the pronotum and prosternum, as well as the metaventrite and elytral surface (*T. shcherbakae–Te. castaneum*). The subelytra and abdominal tergites (*T. sociata–P. detritus*) together create specific closed subelytral space.

Small phoretic load should be considered a rule in all of the analyzed phoront–host associations. The data presented in this study show that the highest percentage of beetles (76.1–86.7%) transported relatively low number of deutonymphs per carrier, while a heavy phoretic load was observed in fairy low percentage of hosts (1.4–8.1%). Similar observations have been made in the case of Scolytinae [13,56–58].

It can be noticed that, in the case of all studied phoront–host associations, the strength of the relationship was also expressed by the fact that both the preferences of phoretic deutonymphs for specific parts of the host's body, and the participation of carriers transporting deutonymphs on particular parts of their bodies were very consistent (Figures 3 and 4).

5. Conclusions

Comparing the distribution of phoretic deutonymphs of Uropodina on the four longhorn beetle species, namely *M. sartor urussovii*, *P. detritus*, *Te. castaneum*, and *Te. fuscum*, it can be concluded that all the studied phoretic associations are characterized by different patterns of the attachment sites of mites on host's body.

Both the preferences of phoretic deutonymphs for specific parts of the host's body and the participation of carriers transporting deutonymphs on particular parts of their bodies were very consistent in the analyzed mite-longhorn beetle associations.

This is by far the most thorough analysis of patterns of phoretic Uropodina on Cerambycidae, based on the large data set collected in a natural forest over a long period of time. The study confirms the high stability and strength of the phoretic relationships between uropodid mites and their longhorn beetle dispersants.

Author Contributions: Conceptualization, S.K. and J.B.; methodology, S.K., J.M.G. and J.B.; software, S.K., J.M.G. and J.B.; validation, S.K., J.M.G. and J.B.; formal analysis, S.K., J.M.G. and J.B.; investigation, S.K., J.M.G. and J.B.; resources, S.K., J.M.G. and J.B.; data curation, S.K. and J.B.; writing—original draft preparation, S.K. and J.B.; writing—review and editing, S.K., J.M.G. and J.B.; visualization, S.K. and J.B.; supervision, S.K.; Project administration, J.M.G.; Funding acquisition, J.M.G.; All authors have read and agreed to the published version of the manuscript.

Funding: This work was partly funded by the project 680602 implemented under the agreement between the Canadian Forest Service in Canada and the Forest Research Institute in Poland.

Acknowledgments: We thank Jon Sweeney, the coordinator of entomological research, for his permission to use the samples in the acarological survey. We thank Krzysztof Sućko for his assistance with the trapping of beetles. We thank Zbigniew Adamski for taking the SEM images and Jerzy Kozak for valuable help in statistical analyses.

Conflicts of Interest: The authors declare no conflict of interest.

References

1. Walter, D.E.; Proctor, H.C. *Mites: Ecology, Evolution & Behaviour. Life at a Microscale*, 2nd ed.; Springer: Berlin/Heidelberg, Germany, 2013; pp. 1–494.
2. Bajerlein, D.; Witaliński, W. Anatomy and fine structure of pedicellar glands in phoretic deutonymphs of uropodid mites (Acari: Mesostigmata). *Arthropod Struct. Dev.* **2012**, *41*, 245–257. [CrossRef] [PubMed]
3. Bajerlein, D.; Witaliński, W.; Adamski, Z. Morphological diversity of pedicels in phoretic deutonymphs of Uropodina mites (Acari: Mesostigmata). *Arthropod Struct. Dev.* **2013**, *42*, 185–196. [CrossRef] [PubMed]
4. Athias-Binche, F. Dispersal in varying environments: The case of phoretic uropodid mites. *Can. J. Zool.* **1993**, *71*, 1793–1798. [CrossRef]
5. Hunter, P.E.; Rosario, R.M.T. Associations of Mesostigmata with other arthropods. *Annu. Rev. Entomol.* **1988**, *33*, 393–417. [CrossRef]
6. Vitzthum, H. Acari als Commensalen von Ipiden. (Der Acarologischen Beobachtungen 11. Reihe.). *Zool. Jahrb. Abt. Syst. Geogr. Biol. Tiere* **1926**, *52*, 407–503.
7. Gwiazdowicz, D.J. Mesostigmatid mites (Acari) associated with Scolytidae in Poland. In *Selected Problems of Acarological Research in Forests*, 1st ed.; Gwiazdowicz, D.J., Ed.; Wydawnictwo Uniwersytetu Przyrodniczego w Poznaniu: Poznań, Poland, 2008; pp. 59–95.
8. Moser, J.C.; Konrad, H.; Blomquist, S.R.; Kirisits, T. Do mites phoretic on elm bark beetles contribute to the transmission of Dutch elm disease? *Naturwissenschaften* **2009**, *97*, 219–227. [CrossRef]
9. Gwiazdowicz, D.J.; Kamczyc, J.; Teodorowicz, E.; Błoszyk, J. Mite communities (Acari, Mesostigmata) associated with Ips typographus (Coleoptera, Scolytidae) in managed and natural Norway spruce stands in Central Europe. *Open Life Sci.* **2012**, *7*, 910–916. [CrossRef]
10. Gwiazdowicz, D.J.; Błoszyk, J.; Gdula, A.K. Alpha diversity of mesostigmatid mites associated with the bark beetle Ips typographus (L.) in Poland. *Insect Conserv. Divers.* **2015**, *8*, 448–455. [CrossRef]

11. Hofstetter, R.W.; Moser, J.C.; Blomquist, S.R. Mites associated with bark beetles and their hyperphoretic ophiostomatoid fungi. In *The Ophiostomatoid Fungi: Expanding Frontiers*, 1st ed.; Seifert, K.A., de Beer, Z.W., Wingfield, M.J., Eds.; CBS-KNAW Biodiversity Centre: Utrecht, The Netherlands, 2014; pp. 165–176.
12. Chaires-Grijalva, M.P.; Estrada-Venegas, E.G.; Equihua-Martínez, A.; Moser, J.C.; Blomquist, S.R. Trophic habits of mesostigmatid mites associated with bark beetles in Mexico. *J. Acarol. Soc. Jpn.* **2016**, *25*, 161–167. [CrossRef]
13. Khaustov, A.A.; Trach, V.; Bobylev, A.N. Mites (Acari) phoretic on six-toothed spruce bark beetle, Pityogenes chalcographus Linnaeus (Coleoptera: Curculionidae: Scolytinae), in Western Siberia, Russia. *Acarina* **2016**, *24*, 137–151. [CrossRef]
14. Zach, P.; Kršiak, B.; Kulfan, J.; Parák, M.; Kontschán, J. Mites Trichouropoda and Uroobovella spp. (Uropodoidea) phoretic on bark beetles (Scolytinae): A comparison from a declining mountain spruce forest in Central Europe. *Int. J. Acarol.* **2016**, *42*, 212–217. [CrossRef]
15. Khaustov, A.A.; Klimov, P.B.; Trach, V.; Bobylev, A.N.; Salavatulin, V.M.; Khaustov, V.A.; Tolstikov, A.V. Review of mites (Acari) associated with the European spruce bark beetle, Ips typographus (Coleoptera: Curculionidae: Scolytinae) in Asian Russia. *Acarina* **2018**, *26*, 3–79. [CrossRef]
16. Paraschiv, M.; Martínez-Ruiz, C.; Fernández, M.M. Dynamic associations between Ips sexdentatus (Coleoptera: Scolytinae) and its phoretic mites in a Pinus pinaster forest in northwest Spain. *Exp. Appl. Acarol.* **2018**, *75*, 369–381. [CrossRef] [PubMed]
17. Kontschán, J.; Szőcs, G.; Kiss, B.; Khaustov, A.A. Bark beetle associated trematurid mites (Acari: Uropodina: Trematuridae) from Asian Russia with description of a new species. *Syst. Appl. Acarol.* **2019**, *24*, 1592–1603. [CrossRef]
18. Vissa, S.; Hofstetter, R.W.; Bonifácio, L.; Khaustov, A.; Knee, W.; Uhey, D.A. Phoretic mite communities associated with bark beetles in the maritime and stone pine forests of Setúbal, Portugal. *Exp. Appl. Acarol.* **2019**, *77*, 117–131. [CrossRef]
19. Hofstetter, R.W.; Vissa, S.; Moser, J.; Bonifácio, L. Interactions among mites, host trees, fungi and beetles. In *Interactions behind the Pine Wilt Disease (FCT NEMATRANSFER Project—PTDC/AGR-FOR/4391/2014), Proceedings of the Interactions behind the Pine Wilt Disease, Oeiras, Portugal, 21 June 2018*; Bonifácio, L., Henriques, J., Eds.; Instituto Nacional de Investigação Agrária e Veterinária: Oeiras, Portugal, 2018; pp. 31–37.
20. Gwiazdowicz, D.J.; Gutowski, J. Records of phoretic mesostigmatid mites (Acari: Mesostigmata) on beetles (Coleoptera: Carabidae, Cerambycidae, Elateridae, Erotylidae, Scolytinae) in the Białowieża Primeval Forest. *Pol. J. Èntomol.* **2012**, *81*, 305–310. [CrossRef]
21. Błoszyk, J.; Gutowski, J.; Gwiazdowicz, D.J.; Mądra-Bielewicz, A. Phoresy of Trichouropoda shcherbakae Hirschmann, 1972 (Acari: Mesostigmata) on beetles of the genus Tetropium Kirby, 1837 (Coleoptera: Cerambycidae) in Białowieża Forest, Poland. *Open Life Sci.* **2013**, *8*, 986–992. [CrossRef]
22. Gwiazdowicz, D.J.; Gutowski, J.M.; Kamczyc, J.; Teodorowicz, E. Phoretic relationships between Plagionotus detritus (Coleoptera: Cerambycidae) and Trichouropoda sociata (Acari: Mesostigmata). *Èntomol. Fenn.* **2013**, *24*, 59–64. [CrossRef]
23. Błoszyk, J.; Gutowski, J.M.; Gwiazdowicz, D.J.; Mądra-Bielewicz, A.; Konwerski, S.; Książkiewicz, Z. Monochamus sartor (Coleoptera: Cerambycidae) contributes to alpha diversity of Uropodina mites (Acari: Mesostigmata) in first stage of wood decay in Białowieża Primeval Forest. *Int. J. Acarol.* **2016**, *42*, 218–223. [CrossRef]
24. Konwerski, S.; Gutowski, J.M.; Książkiewicz-Parulska, Z.; Błoszyk, J. Repeatability of the phoretic relationships between mites of the genus Trichouropoda Berlese (Parasitiformes: Uropodina) and longhorn beetles of the genus Tetropium Kirby (Coleoptera: Cerambycidae) in Białowieża Primeval Forest, Central Europe. *Int. J. Acarol.* **2017**, *43*, 612–621. [CrossRef]
25. Konwerski, S.; Gutowski, J.M.; Błoszyk, J. Analysis of the phoretic relationships between mites of the genus Trichouropoda Berlese (Parasitiformes: Uropodina) and the longhorn beetle Plagionotus detritus (Linnaeus) (Coleoptera: Cerambycidae) based on multiannual observations in Białowieża Primeval Forest, Central Europe. *Int. J. Acarol.* **2018**, *45*, 29–40. [CrossRef]
26. Faliński, J.B. *Vegetation Dynamics in Temperate Lowland Primeval Forest: Ecological Studies in Białowieża Forest*, 1st ed.; Dr W. Junk Publishers: Dordrecht, The Netherlands, 1986; pp. 1–537.

27. Bobiec, A.; Van Der Burgt, H.; Meijer, K.; Zuyderduyn, C.; Haga, J.; Vlaanderen, B. Rich deciduous forests in Białowieża as a dynamic mosaic of developmental phases: Premises for nature conservation and restoration management. *For. Ecol. Manag.* **2000**, *130*, 159–175. [CrossRef]
28. Kujawa, A.; Orczewska, A.; Falkowski, M.; Blicharska, M.; Bohdan, A.; Buchholz, L.; Chylarecki, P.; Gutowski, J.M.; Latałowa, M.; Mysłajek, R.W.; et al. The Białowieża Forest—A UNESCO Natural Heritage Site—Protection priorities. *For. Res. Pap.* **2016**, *77*, 302–323. [CrossRef]
29. Jaroszewicz, B.; Cholewińska, O.; Gutowski, J.M.; Samojlik, T.; Zimny, M.; Latałowa, M. Białowieża Forest —A relic of the high naturalness of European forests. *Forests* **2019**, *10*, 849. [CrossRef]
30. Eckelt, A.; Müller, J.; Bense, U.; Brustel, H.; Bußler, H.; Chittaro, Y.; Cizek, L.; Frei, A.; Holzer, E.; Kadej, M.; et al. "Primeval forest relict beetles" of Central Europe: A set of 168 umbrella species for the protection of primeval forest remnants. *J. Insect Conserv.* **2017**, *22*, 15–28. [CrossRef]
31. Gutowski, J.M.; Królik, R.; Ługowoj, J.; Sućko, K.; Sweeney, J. New data on the occurrence of buprestid beetles (Coleoptera: Buprestidae) in the Białowieża Primeval Forest. *For. Res. Pap.* **2019**, *80*, 167–176. [CrossRef]
32. Gutowski, J.M.; Kurzawa, J. New data on the distribution, biology and morphology of Asemum tenuicorne Kraatz, 1879 (Coleoptera: Cerambycidae), with new records from Poland. *Èntomol. Fenn.* **2019**, *30*, 57–71. [CrossRef]
33. Plewa, R.; Jaworski, T.; Tarwacki, G.; Sućko, K.; Konwerski, S.; Królik, R.; Lasoń, A.; Melke, A.; Przewoźny, M.; Ruta, R.; et al. Beetles (Coleoptera) new for the fauna of the Białowieża Forest including a species new for Poland. *Èntomol. Fenn.* **2019**, *30*, 114–125. [CrossRef]
34. Gutowski, J.M.; Sućko, K.; Borowski, J.; Kubisz, D.; Mazur, M.A.; Melke, A.; Mokrzycki, T.; Plewa, R.; Żmihorski, M. Post-fire beetle succession in a biodiversity hotspot: Białowieża Primeval Forest. *For. Ecol. Manag.* **2020**, *461*, 117893. [CrossRef]
35. Plewa, R.; Jaworski, T.; Tarwacki, G.; Sućko, K.; Konwerski, S.; Królik, R.; Lasoń, A.; Melke, A.; Przewoźny, M.; Ruta, R.; et al. New records of beetle species (Coleoptera) from the Polish part of Białowieża Forest with special emphasis on the genus Episernus, C.G. Thomson, 1863 (Ptinidae) in Central Europe. *Pol. J. Èntomol.* **2020**, *1*, 26–42. [CrossRef]
36. Lachat, T.; Müller, J. Importance of Primary Forests for the Conservation of Saproxylic Insects. In *Parasitic Crustacea*; Springer Science and Business Media LLC.: Berlin/Heidelberg, Germany, 2018; pp. 581–605.
37. Löbl, I.; Smetana, A. *Catalogue of Palaearctic Coleoptera, Vol. 6—Chrysomeloidea*; Apollo Books: Stenstrup, Denmark, 2010; pp. 1–924.
38. Plewa, R.; Sikora, K.; Gutowski, J.M.; Jaworski, T.; Tarwacki, G.; Tkaczyk, M.; Rossa, R.; Hilszczański, J.; Magoga, G.; Kajtoch, Ł. Morphology, genetics and Wolbachia endosymbionts support distinctiveness of Monochamus sartor sartor and M. s. urussovii (Coleoptera: Cerambycidae). *Arthropod Syst. Phylo.* **2018**, *76*, 123–135.
39. Błoszyk, J. Uropodina. In *Fauna of Poland. Characteristics and Checklist of Species*, 1st ed.; Bogdanowicz, W., Chudzicka, E., Pilipiuk, I., Skibińska, E., Eds.; Muzeum i Instytut Zoologii PAN: Warszawa, Poland, 2008; Volume 3, pp. 61–64, 76–78.
40. Lawrence, J.F.; Ślipiński, A.; Seago, A.E.; Thayer, M.K.; Newton, A.F.; Marvaldi, A.E. Phylogeny of the Coleoptera Based on Morphological Characters of Adults and Larvae. In *Annales zoologici*; Museum and Institute of Zoology, Polish Academy of Sciences: Warszawa, Poland, 2011; Volume 61, pp. 1–217. [CrossRef]
41. Hirschmann, W.; Wiśniewski, J. Gangsystematik der Parasitiformes Teil 507. Weltweite Revision der Ganggatung Trichouropoda Berlese 1916—Geschichte, Bestimmungstabellen, Artenverzeichnisse (Trichouropodini, Uropodinae). *Acarologia* **1989**, *36*, 1–65.
42. Karg, W. *Acari (Acarina), Milben Unterordnung Parasitiformes (Anactinochaeta). Uropodina Kramer, Schildkrötenmilben*; VEB Gustav Fischer Verlag: Jena, Germany, 1989; Volume 67, pp. 1–203.
43. Kinn, D.; Linit, M. *A Key to Phoretic Mites Commonly Found on Long-Horned Beetles Emerging from Southern Pines*; Southern Forest Experiment Station: Asheville, NC, USA, 1989; Volume 357, pp. 1–8. [CrossRef]
44. Mašán, P. Mites of the kohort Uropodina (Acarina, Mesostigmata) in Slovakia. *Annot. Zool. Bot.* **2001**, *223*, 1–320.
45. StatSoft Inc. *Statistica (Data Analysis Software System)*, version 10; StatSoft Polska Sp. z o.o.: Kraków, Poland, 2011.
46. Judd, W.W. Nymphal Mites of the Genus Uropoda Attached to a Cerambycid Beetle. *Ann. Èntomol. Soc. Am.* **1950**, *43*, 608–609. [CrossRef]

47. Knee, W.; Hartzenberg, T.; Forbes, M.R.; Beaulieu, F. The natural history of mites (Acari: Mesostigmata) associated with the white-spotted sawyer beetle (Coleoptera: Cerambycidae): Diversity, phenology, host attachment, and sex bias. *Can. Èntomol.* **2012**, *144*, 711–719. [CrossRef]
48. Bajerlein, D.; Adamski, Z.; Kacalak, W.; Tandecka, K.; Wiesner, M.; Jurga, S. To attach or not to attach? The effect of carrier surface morphology and topography on attachment of phoretic deutonymphs of Uropoda orbicularis (Acari). *Naturwissenschaften* **2016**, *103*, 61. [CrossRef]
49. Hunter, P.E. Mites associated with New World passalid beetles (Coleoptera: Passalidae). *Acta Zool. Mex.* **1993**, *58*, 1–37. [CrossRef]
50. Barrios-Torres, P.L.; Villegas-Guzmán, G.A. Ácaros asociados a dos species de Odontotaenius (Coleopetra: Passalidae) en México. *Rev. Biol. Trop.* **2015**, *63*, 659–671. [CrossRef]
51. Hawes, C. Hitching a lift: Stag beetles as carriers of mites. *White Admiral* **2010**, *77*, 16–17.
52. Mazza, G.; Cini, A.; Cervo, R.; Longo, S. Just phoresy? Reduced lifespan in red palm weevils Rhynchophorus ferrugineus (Coleoptera: Curculionidae) infested by the mite Centrouropoda almerodai (Uroactiniinae: Uropodina). *Ital. J. Zool.* **2011**, *78*, 101–105. [CrossRef]
53. Bloszyk, J.; Klimczak, J.; Lesniewska, M. Phoretic relationships between Uropodina (Acari: Mesostigmata) and centipedes (Chilopoda) as an example of evolutionary adaptation of mites to temporary microhabitats. *Eur. J. Èntomol.* **2006**, *103*, 699–707. [CrossRef]
54. Napierała, A.; Książkiewicz, Z.; Lesniewska, M.; Gwiazdowicz, D.J.; Mądra-Bielewicz, A.; Błoszyk, J. Phoretic relationships between uropodid mites (Acari: Mesostigmata) and centipedes (Chilopoda) in urban agglomeration areas. *Int. J. Acarol.* **2015**, *41*, 250–258. [CrossRef]
55. Leśniewska, M.; (Adam Mickiewicz University, Poznań, Poland). Personal communication, 2020.
56. Moser, J.C. Phoretic carrying capacity of flying southern pine beetles (Coleoptera: Scolytidae). *Can. Èntomol.* **1976**, *108*, 807–808. [CrossRef]
57. Moser, J.C. Surveying mites (Acarina) phoretic on the southern pine beetle (Coleoptera: Scolytidae) with sticky traps. *Can. Èntomol.* **1976**, *108*, 809–813. [CrossRef]
58. Knee, W.; Forbes, M.R.; Beaulieu, F. Diversity and Host Use of Mites (Acari: Mesostigmata, Oribatida) Phoretic on Bark Beetles (Coleoptera: Scolytinae): Global Generalists, Local Specialists? *Ann. Èntomol. Soc. Am.* **2013**, *106*, 339–350. [CrossRef]

diversity

MDPI

Communication

Revisiting the Evolution of Arboreal Life in Oribatid Mites

Sylvia Schäffer *, Stephan Koblmüller and Günther Krisper

Institute of Biology, University of Graz, Universitätsplatz 2, 8010 Graz, Austria; stephan.koblmueller@uni-graz.at (S.K.); guenther.krisper@uni-graz.at (G.K.)

* Correspondence: sylvia.schaeffer@uni-graz.at

Received: 8 May 2020; Accepted: 22 June 2020; Published: 22 June 2020

check for updates

Abstract: Though mostly soil dwelling, oribatid mites are found in all kind of habitats, with several species exclusively living on trees. Using previously published DNA sequences and eco-morphological data available from the literature, we inferred the number of transitions between soil dwelling to a truly arboreal lifestyle in oribatid mites and the shape evolution of a particular morphological structure of a sense organ (bothridial seta (= sensillus) of a trichobothrium), the shape of which was previously reported to be associated with an arboreal lifestyle. Our data suggest that a truly arboreal lifestyle evolved several times independently in oribatid mites, but much less often than previously proposed in the past. Even though all truly arboreal species indeed seem to possess a capitate sensillus, this character is not exclusive for arboreal taxa. Nonetheless, since all truly arboreal species do have a capitate sensillus, this might be considered an important (pre-)adaptation to a life on trees. We further provide guidelines on how the term "arboreal" should be applied in future mite research and emphasize the importance of exact microhabitat characterization, as this will greatly facilitate comparisons across studies.

Keywords: lifestyle; tree related microhabitats; 18S rRNA; phylogeny; ancestral state reconstruction; sensillus

---

1. Introduction

Forest ecosystems harbor a strikingly large number of mite species, with particularly high biodiversity in forest soil. This tremendous species richness and adaptations to micro-environments within this ecosystem have led researchers to emphasize the potential of mites as environmental indicators [1–4].

For most tree-associated organisms, a tree is not a single, homogeneous environment. It comprises different sub-areas or regions. For example, the tree trunk itself can be divided into a basal, median, and upper part. The branches, twigs, and leaves of the canopy represent other distinct regions. The ecologically most important structures, however, are the tree-related microhabitats, which are natural, tree-borne features. They provide specific conditions and substrates, like different food resources or structures for breeding and shelter. Since different types of microhabitat support different organismic communities [5–7], they have gained considerable attention in forest management and conservation, especially in the last two decades. While nests of birds and mammals serve as home for nidicolous species, the boreholes and galleries of xylobiotic insects are important habitats and/or food sources for special communities consisting of housemates, commensals, and parasites of these xylobionts. Similar to knot-holes or holes in the trunk, which contain decaying and decomposed organic material, so-called suspended soils developed on large trees with broad branch forks contain their own, often specialized fauna.

As already shown by Wunderle [8], trees provide a habitat for several primarily soil-dwelling microarthropods, especially for oribatid mite species.

It has been known for a long time that Oribatida also occur on trees [9], and researchers soon recognized that species composition differs between soil and tree canopy [10,11]. Since then, many eco-faunistic studies targeted the diversity of oribatids on trees [12–15]. However, many previous studies provided often confusing, disparate information on individual species or species communities inhabiting trees, which indicates that abundance, distribution and diversity of tree-living species vary considerably even under presumably similar conditions. These differences in community composition can be mainly attributed to (i) habitat types (e.g., rain forest, "perpetual mist" forests, deciduous and mixed forests; [16]), (ii) tree species (e.g., [8]), or (iii) zones and age of a tree, which determines tree diameter and bark thickness and has significant impact on the number of available microhabitats [17]. Nonetheless, the diversity of tree-living oribatid mites is undoubtedly much higher than known so far. Several recent studies showed the level of cryptic diversity is high in (tree-living) oribatids, even in supposedly well known, common, and easily identifiably taxa [18–20], providing an important contribution to our knowledge of the biodiversity in forest ecosystems.

While many eco-faunistic studies dealt with tree-living oribatid communities, only one study aimed at a comprehensive analysis of arboreal lifestyle evolution in oribatids, focusing on the transition from soil dwelling to arboreal life and potentially associated phenotypic adaptations [21]. Specifically, the authors suggested that arboreal life evolved independently at least 15 times and that tree-living is more common in evolutionary younger taxa with strong sclerotization, sexual reproduction, and capitate sensilli [21].

Microhabitats are known to harbor a disproportionately high number of (likely unknown) species, not only among oribatid mites. Therefore, it is important to recognize and conserve these microhabitats [6,7]. So far, several studies have demonstrated the important role of tree-associated microhabitats for the taxonomic diversity in oribatid mites [22–27]. In all these studies, the authors use the term "arboreal" (= corticolous or arboricole), which means growing on or living in trees, to indicate that a specimen/species was collected on a tree. However, the exact use of this term differs among studies. Some use it to refer to strictly tree-living species, others apply it to more generalist species found both in soil and on trees [10,28,29]. Moreover, it is still questionable whether the species found on trees are tree or rather microhabitat specific (see also [30]). These uncertainties complicate generalizations regarding the evolution of a tree-living lifestyle in oribatid mites.

Against this background and utilizing a comprehensive phylogenetic tree based on previously published 18S rDNA sequences, the present study aims to (i) revisit the origin of a tree-living lifestyle in oribatids, (ii) test the hypothesis (according to [11]) that a capitate sensillus is a special morphological feature of arboreal life, and (iii) discuss and redefine the term "arboreal" in oribatid mites.

2. Material and Methods

2.1. Data Acquisition

We chose the 18S rRNA (18S) gene for our meta-analysis because of (i) the largest amount of sequence data available for a particular locus, covering all major groups of the oribatid mite tree of life, and (ii) its low substitution rate; the 18S gene is generally more suitable for resolving phylogenetic relationships at higher taxonomic levels. Only sequences with a minimum length of 1600 bp were obtained from GenBank. The final dataset included sequences of 165 oribatid species (see Table S1). Based on Klimov et al. [31], we used three species of Paleosomata, *Ctenacarus araneola, Palaeacarus hystricinus,* and *Stomacarus ligamentifer* as outgroup.

Prior to phylogenetic analysis, sequences were aligned using the MAFFT v.7 web version [32]. Poorly aligned regions were removed from the alignment using the program trimAl v1.2 [33] applying the heuristic "automated1" method (final length of dataset = 1619 bp).

Character coding of traits for ancestral state reconstruction (ASR) of lifestyle and sensillus type was primarily based on Weigmann [34] as well as on original species descriptions. Additional literature was only used in case of uncertainties or missing information (see Table S2). Ancestral character state reconstruction for lifestyle was coded as follows: soil (0), tree- and soil-living (1), exclusively tree-living (= arboreal, 2), littoral (3), mangrove (4), and limnic (5). The coding of the sensillus types followed roughly the classification scheme of Aoki [11]: no sensillus present (0), capitate and strongly clavate (1), clavate (2), slightly clavate (3), with rami (4), rod-like thin (5), and not specified (6, for details see below) (Figure 1).

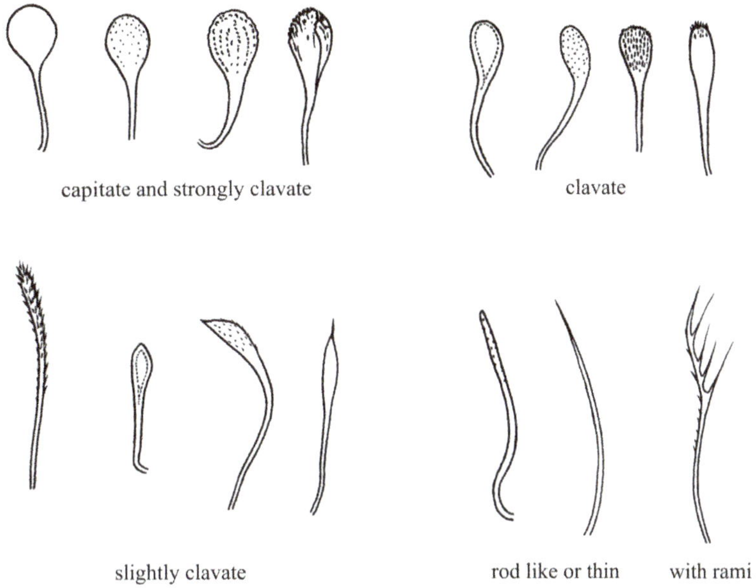

Figure 1. Drawing of sensillus types used for character coding in the present study.

2.2. Analysis

The best-fitting model of molecular evolution was selected based on the Akaike Information Criterion (AIC) in the "Smart Model Selection" tool (SMS; [35]) implemented in the PhyML 3.0 online execution ([36]; http://www.atgc-montpellier.fr/phyml/). Phylogenetic inference was based on Maximum Likelihood (ML), implemented in PhyML and Bayesian inference (BI), implemented in MrBayes 3.2.7 [37], applying the GTR+I+G model selected by SMS. ML analyses were run under default parameter settings and nodal support was assessed by means of bootstrapping (1000 replicates). For BI analyses, posterior probabilities were obtained from a Metropolis-coupled Markov chain Monte Carlo simulation (2 independent runs; 4 chains with 25 million generations each; trees sampled every 1000 generations). Run convergence and stationarity of parameters were checked in Tracer v1.7 ([38]; available at http://beast.community/tracer) and by means of the standard deviation of split frequencies (<0.01) in MrBayes. The first 12,500 (25%) trees were discarded as burn-in prior to constructing an "allcompat" consensus tree from the remaining 37,502 trees.

We traced the evolution of two characters (lifestyle, sensillus shape) over the molecular phylogeny using maximum parsimony (MP) and maximum likelihood (ML)-based reconstructions conducted in Mesquite v.3.51 [39].

We applied the "trace character over trees" option to account for topological uncertainty. The Markov k-state 1 (Mk1) parameter model was employed for the analysis of both characters, with equal probability for any particular character state. Both reconstructions were integrated over

20,001 randomly selected post burn-in trees of the BI analysis and the ancestral states were summarized on the BI consensus tree.

Due to intra-generic polymorphism of sensillus shape in the genera "*Liodes*" and *Eremaeozetes* ([34] and Schatz pers. communication), we refrained from assigning a particular morphology to *Liodes* sp. (The denomination *Liodes* sp. is not a valid name for a genus of the oribatid mite family Neoliodidae [40]. It is unclear whether this record in GenBank from 1997 refers to a species of *Neoliodes* or to another genus) and *Eremaeozetes* sp. The higher-level classification of oribatids follows Schatz et al. [41].

3. Results

As both the ML and BI tree inference resulted in very similar tree topologies, only the Bayesian consensus tree is shown (Figure 2). Compared to the traditional classification, our phylogenetic tree revealed several discrepancies, for example the infraorders Parhyposomata and Mixonomata, as well as the superfamilies Plateremaeoidea, Cepheoidea, Cymbaeremaeoidea, Licneremaeoidea, Gustavioidea and Ameronothroidea appear as para- or polyphyletic, but often with only low statistical support, indicating a fair amount of topological uncertainty (which was accounted for in our ASR analyses).

ASR of lifestyle and sensillus based on ML and MP reconstructions revealed very similar results, except for some nodes that were reconstructed with greater uncertainty (equivocal) in the ML approach (Figure 3, Figure S1). Our results clearly indicate that the preconditions for colonizing tree trunks and canopies were already present in the ancestors of the higher Oribatida; e.g., *Camisia* species can be found both in soil and tree samples (Figure 3a). Nevertheless, to live nearly exclusively on trees seems to be the exception (see also [29]). However, our analysis suggests that arboreal life evolved seven times independently within the investigated taxa. Furthermore, our analysis supports a multiple independent evolution of the capitate sensillus (Figure 3b) and indicates that capitate sensilli are not restricted to arboreal species.

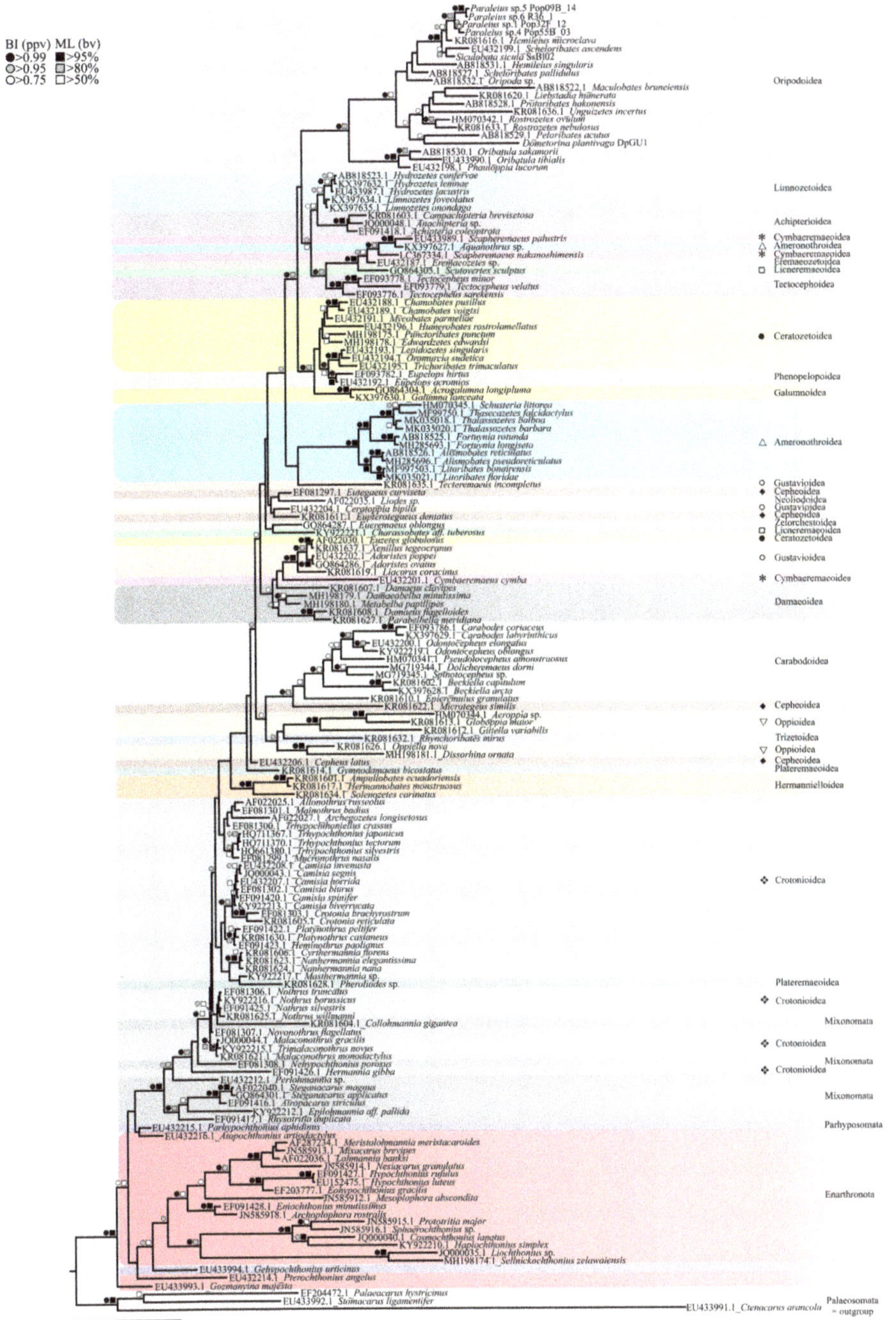

Figure 2. Bayesian "allcompat" consensus tree of 165 representatives covering all major groups of the oribatid mite tree of life. The tree is based on sequences of the 18S rRNA gene. Symbols at nodes represent posterior probability values (ppv) for Bayesian inference (BI) and bootstrap values (bv) for Maximum Likelihood (ML). Symbols on the right side of the tree indicate para- or polyphyletic superfamilies and mark their affiliation.

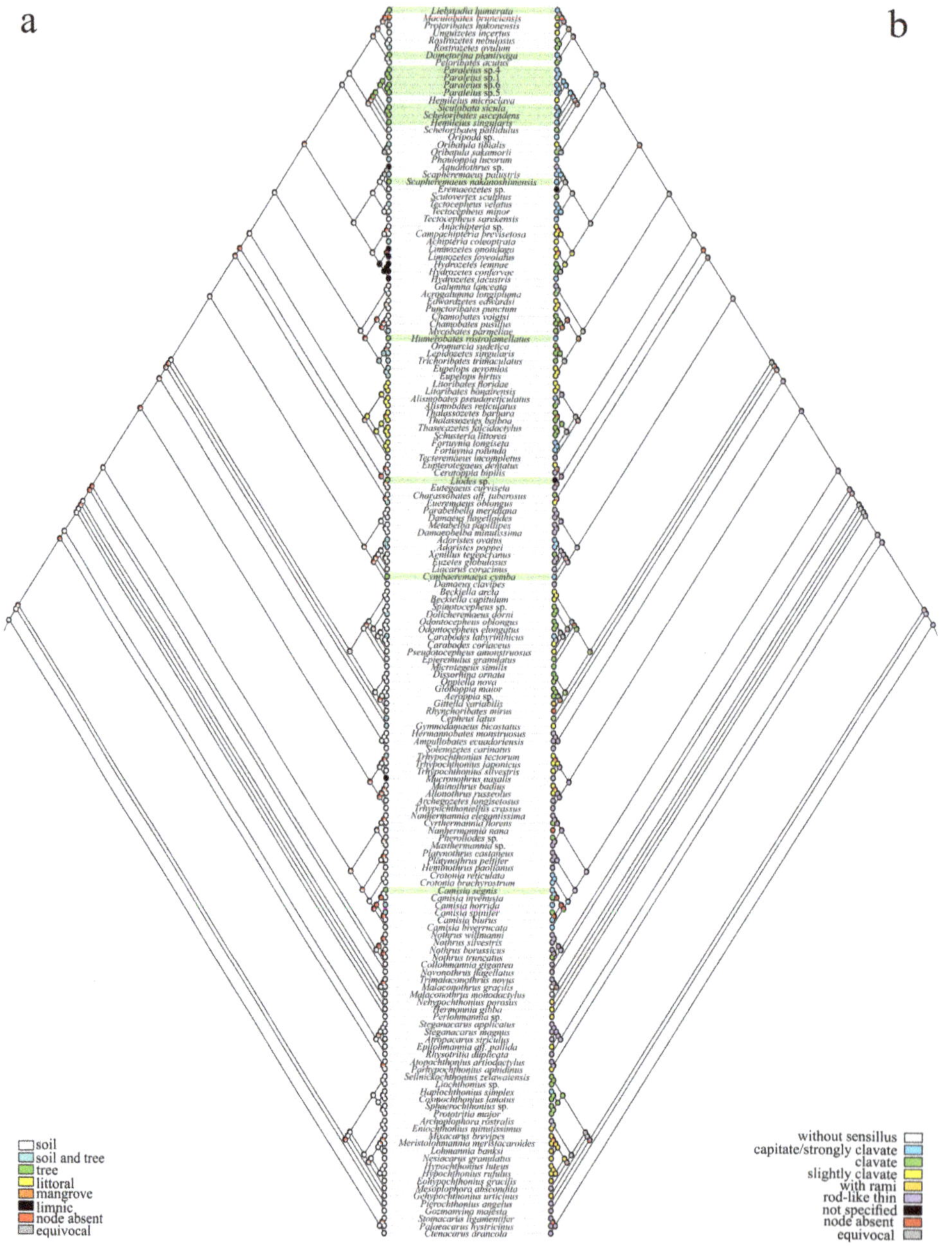

Figure 3. Ancestral state reconstructions based on likelihood of (**a**) lifestyle and (**b**) sensillus type found in oribatid mites. Tree-living taxa are highlighted in green.

4. Discussion

4.1. Revisiting the Tree-Living Lifestyle in Oribatids

Ancestral state reconstruction can give important information on the history of traits that do not fossilize, such as particular ecologies or behaviors [42]. However, consequently, inconsistent interpretation of morphological features can lead to erroneous results and conclusions. Notwithstanding some discrepancies between our 18S phylogeny and classical taxonomy (partially apparent also in [43,44]), which is due to a lack of resolution at some inner nodes and a partly inappropriate classification scheme [45], our ancestral state reconstructions, taking topological uncertainty into account, unambiguously indicate seven transitions between soil- and tree-dwelling lifestyle in the present oribatid dataset (Figure 3a). These are much fewer transitions than inferred by a former study that proposed that life on trees evolved at least 15 times independently in oribatid mites [21], even though our study includes many more taxa. The different findings can be explained by different character coding of the lifestyle type. Whereas [21] differentiated mite species into either soil- or tree-living organisms, the majority of these taxa can be found both in soil and on trees. We used, apart from the standard classification book for oribatids [34], additional literature in cases of uncertainties (see Table S2). Moreover, we employed the character trait "tree-living" in a stricter sense, i.e., only species exclusively found on trees were assigned this character trait. As in [21], however, we can conclude that life on tree already evolved within the Crotonioidea. In this context it is also interesting to note that representatives of Enarthronota and Mixonomata, *Liochthonius* sp., and *Epilohmannia* sp. have been collected from suspended soil samples in North America [15,46], and *Liochthonius pseudolaticeps* has been found on bark of trees in Spain [47]. This suggests that the preconditions for colonizing tree trunks and canopies were already present in the early-derived oribatids, indicating a repeated independent evolution of tree life.

Many taxa included in our study are known to occur both in soil and on trees, e.g., [15]; only 14 of the 165 included species are almost exclusively tree-living. The occurrence of soil-dwelling oribatids on trees might be explained for example by so called suspended soils on trees, which provide ecological conditions similar to the forest floor [15]. Wallwork [48] and Gjelstrup [49] mentioned that mosses and lichens on rocks and trees represent such suspended soils. Thus, different kinds of suspended soils apparently facilitated the expansion of suitable habitats, especially for hemi- and epedaphic species. Most of these suspended soil dwellers are also able to survive and reproduce on the ground; therefore, many species can be found on trees as well as in soil samples. However, soil-dwelling species on trees could also be the result of introduction by other organisms. In general, colonizing a tree takes place mainly via three mechanisms: (1) the stem and suspended soils from soil and litter, (2) the upper stem and canopy by wind, and (3) the upper stem and canopy by phoresy. It has been shown that especially the transfer by birds [50,51] and by wind [52] might play an important role for the exchange of species from soil to tree.

Of course, there are many more strictly tree-living species or species occurring both in soil and on trees for which no genetic data are available so far [11,53–55]. Including an increasing number of these species in future studies will be of great importance for providing a more accurate picture of the evolution of arboreal life in oribatid mites.

4.2. Capitate Sensillus—A Special Morphological Feature of Tree-Dwelling Oribatid Mites?

Trichobothria of oribatids are complex sensory organs, highly mechanosensitive sensilli, which are of both taxonomical and functional relevance [56,57]. They are regarded as vibration receptors reacting to substrate vibrations or air currents [57]. These organs are constructed in a way that the movement of the seta (sensillus) is allowed only in two directions [57]. The shape of the sensillus has been proposed to correlate with the preferred habitat type. Aoki [11] already stated that the sensilli of arboreal taxa tend to be short capitate to strongly clavate, suggesting a function as gravity receptors. The directionality of the movement of the sensillus could be suitable to perceive the direction of gravitation, especially if the distal end of the sensillus is capitate and therefore relatively heavy. In this case, the animals would be able to "know" whether they are crawling up or down. Experimental data to verify this assumption and the hypothesized correlation of sensillus type with (micro-)habitat [57], however, are still lacking.

In general, the sensilli of oribatid mites show a broad variety in shape from filiform to club-like and globose. When the club is slender the sensillus is termed to be clavate, when the distal part broadens quickly and shows an approximately rounded or spherical club it represents a so-called capitate sensillus (see Figure 1). Clavate (large and often ornamented) sensilli were proposed to predominate in "wandering species" between soil and trees, whereas they should be capitate in arboreal species [11]. Whether an associated shortening of the stalk of the sensillus, to reduce the irritation by air currents [58]—in extreme cases the club disappears in the bothridium (e.g., in *Camisia abdosensilla* [59])—represents a further adaptation to the life on trees needs to be tested in future studies. Our study supports a multiple evolution of the capitate sensillus (Figure 3b) and indicates that capitate sensilli are found in, but are not restricted to, arboreal species, a hypothesis already discussed previously [14]. While all herein investigated arboreal species possess capitate sensilli, except for *Liodes* sp. (but see Material and Methods), the same sensillus type can also be found in soil dwellers, limnic and littoral species. Thus, even if particular sensillus types do not appear to be, associated with particular habitat types (Figure 3), the fact that all truly arboreal species seem to possess a capitate sensillus suggests that this sensillus type should indeed be considered an important (pre-)adaptation to a life on trees.

4.3. Discussing the Term "Arboreal" in Oribatid Mites

Oribatid mites can be frequently found on trees. When comparing the results of eco-faunistic studies, however, it becomes apparent that abundance, distribution and diversity of tree-living species varies considerably. Moreover, very often it is still unclear if the species found on trees are truly tree dwellers or rather wandering species in search for food, mating partner, or shelter. It is therefore necessary to clarify which species should be designated as arboreal taxa. In other words, when is a species truly arboreal?

Arboreal species *sensu stricto* are herein defined as those taxa that live on the bark of stems and branches without a layer of lichens or mosses, in the canopy, or on twigs and leaves. Furthermore, these species should undergo their whole life cycle on the tree. Among the many oribatid species (~10,000 described species [60]), only very few can be classified as arboreal *s.s.* For example, *Adhaesozetes polyphyllos* [53] as well as *Phylleremus leei* and *Ph. hunti* [61], all of which bear a short and capitate sensillus, might be considered as strictly arboreal. Unfortunately, no genetic data are available for these Australian species. All other tree-living species that also utilize different microhabitats on trees should be referred to as arboreal *sensu lato* and they should be further defined by the microhabitat they inhabit. These species are usually not found in litter and soil on the ground, except for, in rare cases, single specimens probably fallen from the tree. Some of the taxa included in our study, such as *Cymbaeremaeus cymba*, *Camisia segnis*, and *Liebstadia humerata*, can be found on stems or twigs without epiphytic layer as well as in lichens. Therefore, these species should not be defined as arboreal *s.s.* according to the above given definition. Taxa that are mainly associated with mosses or lichens on trees should be referred to as by their substrate, thus muscicolous or lichenicolous, as e.g., *Dometorina plantivaga*, which mainly occurs on lichens. All *Paraleius* species studied so far are associated with bark beetles and can be found in their galleries, where the mites live and reproduce. According to their microhabitat, *Paraleius* spp. should thus be defined as gallery-living species. Information on ecology and (micro-)habitat is available for some species, allowing for a microhabitat-based classification. This, however, is not the case for many other oribatids, which highlights the need for more detailed ecological investigations. The application of our proposed classification system has of course the consequence that only few taxa will be referred to as arboreal *sensu stricto*.

In the case of temperate and Mediterranean European forests, we suggest the application of Larrieu et al.'s [7] scheme, categorizing the tree related microhabitats into 15 main groups according to 12 substrates and four microclimatic conditions. As conditions differ among ecosystems (e.g., rainforest vs. temperate forest); however, one must also differentiate between the ecosystems. Due to different environmental conditions like moisture, temperature, or radiation, it is likely that the number of existing microhabitats differs significantly. We expect a considerably higher number of microhabitats in rainforests, even considering the suspended soils, the variety of epiphytes or the growth of the trees. Previous studies have already shown that species found on trees are rather microhabitat specific than tree specific [24,27,30,62], pointing to the need of standardized classification schemes of microhabitats that will allow for a more straight forward and, importantly, comparable classification of tree-living taxa. This in turn will facilitate a more accurate comparison between microhabitats of different tree species, geographic areas, etc. Moreover, as mentioned at the beginning, microhabitats are important substrates contributing to the internal heterogeneity of forests [7]. Their detection and investigation will improve our knowledge about oribatid biodiversity in the future.

Reliable interpretation of evolutionary trajectories crucially depends on exact and detailed knowledge of the biological characteristics and the taxonomic classification/phylogenetic relationships of the studied taxa. Unfortunately, in many instances important information is lacking. These difficulties are exemplified by *Micreremus brevipes*, a well-known "arboreal" species, which has been reported also from Svalbard (Norway) in a permafrost and tundra habitat with litter of mosses, lichens, and vascular plants near the seashore, but definitively no trees [63]. One must assume that such different habitats show morphologically very similar, but (genetically) different species. This phenomenon of cryptic diversity might in part explain the presumed occurrence of one and the same "euryoecious species" on the ground and on trees.

Compared to their soil-living relatives, tree-living microarthropods often have to cope with markedly different climatic conditions, though this difference appears to be only minor in some ecosystems. Salavatulin [27] hypothesized that only a few oribatid mite species adapted to a life on trees. Reasons therefore would be an increased vulnerability of microhabitats to wind or solar radiation, but also a more frequent change of dryness and moisture [27]. While this holds true for habitats with extreme conditions, like forest-tundra [27] or exposed tree stands in alpine regions, drought and drought resistance should not play an important role as limiting factor in tropical and temperate rainforests. Because of the high phylogenetic age of oribatid mites, drought resistance might have been evolved in pioneer species, which colonized raw soils with changing humid conditions in ancient times. If this physiological feature is important for arboreal life, then the ability to survive dry periods should be an ancient trait. Furthermore, the food resources on virgin soils are, similar to those on the bark of trees, bacteria, algae, and lichens. The ability to feed on these food resources and the drought resistance enabled species not only to colonize litter and upper layers of soil but also to colonize trees and their different microhabitats (also see [64]).

5. Conclusions

To conclude, our analysis revealed that a truly arboreal lifestyle *sensu lato* is not particularly common, evolved several times independently in oribatid mites, but much less often than formerly proposed. We propose that only taxa that live on the bark of stems and branches without a layer of lichens or mosses, in the canopy or on twigs and leaves and undergo their whole life cycle on the tree should be regarded as "arboreal *s.s.*" A refined microhabitat-based classification scheme will be particularly important for future comparative ecological and evolutionary (meta-)analyses, as the different tree-associated microhabitats certainly require different phenotypic adaptations. Nonetheless, the presence of a capitate sensillus appears to be a morphological (pre)-adaptation to life of trees (arboreal *s.l.*), potentially functioning as gravity receptor. Yet, it is unlikely that this is the sole adaptation to an arboreal lifestyle in oribatid mites and potential specific morphological adaptations for the life on trees should be investigated in more detail, e.g., claws and adhesive organs on legs as well as the ultrastructure and sensitivity of trichobothria.

The biodiversity of tree-living oribatid mites is undoubtedly much higher than known so far. Especially the canopies of old trees in temperate deciduous forests are not investigated thoroughly. It is well known that large old trees as well as tree-born structures represent ecologically important sites [65,66]. So-called small natural features contribute to biodiversity in a disproportionate manner regarding their small size [67]. Therefore, a uniform classification scheme of microhabitats will be of crucial importance in mite research, as these small natural features become more and more important for preserving forest biodiversity.

Supplementary Materials: The following are available online at http://www.mdpi.com/1424-2818/12/6/255/s1, Table S1. Specimens, GenBank accession numbers and coding of morphological characters for the analysed samples. Table S2. Literature used for the character coding of morphological characters. Figure S1. Ancestral state reconstructions based on parsimony of (a) lifestyle and (b) sensillus type found in oribatid mites.

Author Contributions: Conceptualization, S.S., S.K., and G.K.; data curation, G.K.; formal analysis, S.S; funding acquisition, S.S.; writing—original draft, S.S., S.K., and G.K. All authors have read and agreed to the published version of the manuscript.

Funding: This research was funded by the Austrian Science Fund (FWF, project number P27843-B25).

Acknowledgments: We would like to thank three anonymous reviewers for their positive and helpful comments on a previous version of this manuscript. We further acknowledge the financial support for Open Access Funding by the University of Graz.

Conflicts of Interest: The authors declare no conflict of interest.

References

1. Ruf, A. A maturity index for predatory soil mites (Mesostigmata: Gamasina) as an indicator of environmental impacts of pollution on forest soils. *Appl. Soil Ecol.* **1998**, *9*, 447–452. [CrossRef]
2. Van Straalen, N.M. Evaluation of bioindicator systems derived from soil arthropod communities. *Appl. Soil Ecol.* **1998**, *9*, 429–437. [CrossRef]
3. Behan-Pelletier, V.M. Oribatid mite biodiversity in agroecosystems: Role for bioindication. *Agric. Ecosyst. Environ.* **1999**, *74*, 411–423. [CrossRef]
4. Gergócs, V.; Hufnagel, L. Application of oribatid mites as indicators. *Appl. Ecol. Environ. Res.* **2009**, *7*, 79–98. [CrossRef]
5. Kraus, D.; Bütler, R.; Krumm, F.; Lachat, T.; Larrieu, L.; Mergner, U.; Paillet, Y.; Schuck, A.; Winter, S. *Catalogue of Tree Microhabitats: Reference Field List*; European Forest Institute: Joensuu, Finland, 2016.
6. Paillet, Y.; Archaux, F.; Boulanger, V.; Debaive, N.; Fuhr, M.; Gilg, O.; Gosselin, F.; Guilbert, E. Snags and large trees drive higher tree microhabitat densities in strict forest reserves. *For. Ecol. Manag.* **2017**, *389*, 176–186. [CrossRef]
7. Larrieu, L.; Paillet, Y.; Winter, S.; Bütler, R.; Kraus, D.; Krumm, F.; Lachat, T.; Micheli, A.K.; Regnery, B.; Vandekerkhove, K. Tree related microhabitats in temperate and Mediterranean European forests: A hierarchical typology for inventory standardization. *Ecol. Indic.* **2018**, *84*, 194–207. [CrossRef]
8. Wunderle, I. Die Oribatiden-Gemeinschaften (Acari) der verschiedenen Habitate eines Buchenwaldes. *Carolinea* **1992**, *50*, 79–144.
9. Travé, J. Écologie et biologie des Oribates (Acariens) saxicoles et arboricoles. *Vie et Milieu Suppl.* **1963**, *14*, 267.
10. Niedbala, W. Tree moss-mite fauna in the outskirts of Poznan. In *Oribatei and Their Role in the Process of the Soil Formation*; Lithuanian Academy of Sciences: Vilnius, Lithuania, 1970; pp. 103–112.
11. Aoki, J.I. Soil mites (oribatids) climbing trees. In *Proceedings of the 3rd International Congress of Acarology, Prague 1971*; Daniel, M., Rosický, B., Eds.; Springer: Dordrecht, The Netherlands, 1973; pp. 59–65.
12. Behan-Pelletier, V.M.; Winchester, N.N. Arboreal oribatid mite diversity: Colonizing the canopy. *Appl. Soil Ecol.* **1998**, *9*, 45–54. [CrossRef]
13. Winchester, N.N.; Behan-Pelletier, V.; Ring, R.A. Arboreal specificiy, diversity and abundance of canopy-dwelling oribatid mites (Acari: Oribatida). *Pedobiologia* **1999**, *43*, 391–400.
14. Behan-Pelletier, V.M.; Walter, D.E. Biodiversity of oribatid mites (Acari: Oribatida) in tree canopies and litter. In *Invertebrates as Webmasters in Ecosystems*; Coleman, D.C., Hendrix, P.F., Eds.; CABI Publishing: Wallingford, UK, 2000; pp. 187–202.
15. Lindo, Z.; Winchester, N.N. Oribatid mite communities and foliar litter decomposition in canopy suspended soils and forest floor habitats of western redcedar forests, Vancouver Island, Canada. *Soil Biol. Biochem.* **2007**, *39*, 2957–2966. [CrossRef]
16. Arroyo, J.; Kenny, J.; Bolger, T. Variation between mite communities in Irish forest types—Importance of bark and moss cover in canopy. *Pedobiologia* **2013**, *56*, 241–250. [CrossRef]

17. Michel, A.K.; Winter, S.; Linde, A. The effect of tree dimension on the diversity of bark microhabitat structures and bark use in Douglas-fir (*Pseudotsuga menziesii var. menziesii*). *Can. J. For. Res.* **2011**, *41*, 300–308. [CrossRef]
18. Heethoff, M.; Domes, K.; Laumann, M.; Maraun, M.; Norton, R.A.; Scheu, S. High genetic divergences indicate ancient separation of parthenogenetic lineages of the oribatid mite *Platynothrus peltifer* (Acari, Oribatida). *J. Evol. Biol.* **2007**, *20*, 392–402. [CrossRef]
19. Pfingstl, T.; Baumann, J.; Lienhard, A. The Caribbean enigma: The presence of unusual cryptic diversity in intertidal mites (Arachnida, Acari, Oribatida). *Org. Divers. Evol.* **2019**, *19*, 609–623. [CrossRef]
20. Schäffer, S.; Kerschbaumer, M.; Koblmüller, S. Multiple new species: Cryptic diversity in the widespread mite species *Cymbaeremaeus cymba* (Oribatida, Cymbaeremaeidae). *Mol. Phylogenet. Evol.* **2019**, *135*, 185–192. [CrossRef]
21. Maraun, M.; Erdmann, G.; Schulz, G.; Norton, R.A.; Scheu, S.; Domes, K. Multiple convergent evolution of arboreal life in oribatid mites indicates the primacy of ecology. *Proc. R. Soc. B* **2009**, *276*, 3219–3227. [CrossRef] [PubMed]
22. Seyd, E.L.; Seaward, M.R.D. The association of oribatid mites and lichens. *Zool. J. Linn. Soc.* **1984**, *80*, 369–420. [CrossRef]
23. André, H.M. Associations between corticolous microarthropod communities and epiphytic cover on bark. *Holarct. Ecol.* **1985**, *8*, 113–119. [CrossRef]
24. Nicolai, V. The bark of trees: Thermal properties, microclimate and fauna. *Oecologia* **1986**, *69*, 148–160. [CrossRef]
25. Root, H.T.; McGee, G.G.; Norton, R.A. Arboreal mite communities on epiphytic lichens of the Adirondack Mountains of New York. *Northeast. Nat.* **2007**, *14*, 425–438. [CrossRef]
26. Taylor, A.R.; Ranius, T. Tree hollows harbour a specialised oribatid mite fauna. *J. Insect Conserv.* **2014**, *18*, 39–55. [CrossRef]
27. Salavatulin, V. Microhabitat distribution of arboreal oribatid mites (Oribatida), associated with the Siberian pine (*Pinus sibirica*) of Western Siberia. *Exp. Appl. Acarol.* **2019**, *78*, 469–483. [CrossRef] [PubMed]
28. Vanek, J. Arboreal oribatid mites (Acari, Oribatei) in young stands of *Abies alba*. *Sbornik Vysoke Skoly Zemedelske a Lesnicke Fakulty v Praha* **1972**, *5*, 291–300.
29. Krivolutsky, D.A. Arboricular (Tree-dwelling) oribatid mites as bioindicators of environment quality. *Dokl. Biol. Sci.* **2004**, *399*, 470–473. [CrossRef] [PubMed]
30. Behan-Pelletier, V.M.; John, M.G.S.; Winchester, N. Canopy Oribatida: Tree specific or microhabitat specific? *Eur. J. Soil Biol.* **2008**, *44*, 220–224. [CrossRef]
31. Klimov, P.B.; O'Connor, B.M.; Chetverikov, P.E.; Bolton, S.J.; Pepato, A.R.; Mortazavi, A.; Tolstikov, A.V.; Bauchan, G.R.; Ochoa, R. Comprehensive phylogeny of acariform mites (Acariformes) provides insights on the origin of the four-legged mites (Eriophyoidea), a long branch. *Mol. Phylogenet. Evol.* **2018**, *119*, 105–117. [CrossRef]
32. Katoh, K.; Rozewicki, J.; Yamada, K.D. MAFFT online service: Multiple sequence alignment, interactive sequence choice and visualization. *Brief. Bioinform.* **2019**, *20*, 1160–1166. [CrossRef]
33. Capella-Gutiérrez, S.; Silla-Martínez, J.M.; Gabaldón, T. TrimAl: A tool for automated alignment trimming in large-scale phylogenetic analyses. *Bioinformatics* **2009**, *25*, 1972–1973. [CrossRef]
34. Weigmann, G. *Hornmilben (Oribatida)*; Goecke and Evers: Keltern, Germany, 2006.
35. Lefort, V.; Longueville, J.-E.; Gascuel, O. SMS: Smart Model Selection in PhyML. *MBE* **2017**, *34*, 2422–2424. [CrossRef]
36. Guindon, S.; Dufayard, J.-F.; Lefort, V.; Anisimova, M.; Hordijk, W.; Gascuel, O. New algorithms and methods to estimate Maximum-Likelihood phylogenies: Assessing the performance of PhyML 3.0. *Syst. Biol.* **2010**, *59*, 307–321. [CrossRef] [PubMed]
37. Ronquist, F.; Teslenko, M.; van der Mark, P.; Ayres, D.L.; Darling, A.; Höhna, S.; Larget, B.; Liu, L.; Suchard, M.A.; Huelsenbeck, J.P. MrBayes 3.2: Efficient Bayesian phylogenetic inference and model choice across a large model space. *Syst. Biol.* **2012**, *61*, 539–542. [CrossRef] [PubMed]
38. Rambaut, A.; Drummond, A.J.; Xie, D.; Baele, G.; Suchard, M.A. Posterior summarization in Bayesian phylogenetics using Tracer1.7. *Syst. Biol.* **2018**, *67*, 901–904. [CrossRef] [PubMed]
39. Maddison, W.P.; Maddison, D.R. Mesquite: A Modular System for Evolutionary Analysis. Version 3.51. Available online: http://www.mesquiteproject.org (accessed on 28 June 2018).

40. Halliday, R.B.; Majka, C.G. Clarification of the status of the genus names *Leiodes* Latreille 1796 (Coleoptera), *Liodes* von Heyden 1826 (Acari) and *Neoliodes* Berlese 1888 (Acari). *Zootaxa* **2010**, *2600*, 61–65. [CrossRef]
41. Schatz, H.; Behan-Pelletier, V.M.; O'Connor, B.M.; Norton, R.A. Suborder Oribatida van der Hammen, 1968. In *Animal Biodiversity: An Outline of Higher-Level Classification and Survey of Taxonomic Richness*; Zhang, Z.Q., Ed.; Zootaxa 3148; Magnolia Press: Auckland, New Zealand, 2011; pp. 141–147.
42. Pagel, M. Inferring the historical patterns of biological evolution. *Nature* **1999**, *401*, 877–884. [CrossRef]
43. Krause, A.; Pachl, P.; Schulz, G.; Lehmitz, R.; Seniczak, A.; Schaefer, I.; Scheu, S.; Maraun, M. Convergent evolution of aquatic life by sexual and parthenogenetic oribatid mites. *Exp. Appl. Acarol.* **2016**, *70*, 439–453. [CrossRef]
44. Pachl, P.; Lindl, A.C.; Krause, A.; Scheu, S.; Schaefer, I.; Maraun, M. The tropics as ancient cradle of oribatid mite diversity. *Acarologia* **2017**, *57*, 309–322. [CrossRef]
45. Schäffer, S.; Koblmüller, S.; Pfingstl, T.; Sturmbauer, C.; Krisper, G. Ancestral state reconstruction reveals multiple independent evolution of diagnostic morphological characters in the "Higher Oribatida" (Acari), conflicting with current classification schemes. *BMC Evol. Biol.* **2010**, *10*, 246. [CrossRef]
46. Lindo, Z.; Winchester, N.N. A comparison of microarthropod assemblages with emphasis on oribatid mites in canopy suspended soils and forest floors associated with ancient western redcedar trees. *Pedobiologia* **2006**, *50*, 31–41. [CrossRef]
47. Subias, L.S. Taxonomía y Ecología de los Oribátidos Saxícolas y Arborícolas de la Sierra de Guadarrama (Acarida, Oribatida). Ph.D. Thesis, Universidad Complutense de Madrid, Madrid, Spain, 1977; p. 375.
48. Wallwork, J.A. *The Distribution and Diversity of Soil Fauna*; Academic Press: London, UK; New York, NY, USA; San Francisco, CA, USA, 1976.
49. Gjelstrup, P. Epiphytic cryptostigmatid mites on some beech- and birch-trees in Denmark. *Pedobiologia* **1979**, *19*, 1–8.
50. Krivolutsky, D.A.; Lebedeva, N.V. Oribatid mites (Oribatei, Acariformes) in bird feathers: Non-passerines. *Acta Zool. Litu.* **2004**, *14*, 26–47. [CrossRef]
51. Krivolutsky, D.A.; Lebedeva, N.V. Oribatid mites (Oribatei, Acariformes) in bird feathers: Passeriformes. *Acta Zool. Litu.* **2004**, *14*, 19–38. [CrossRef]
52. Lehmitz, R.; Russell, D.; Hohberg, K.; Christian, A.; Xylander, W.E. Wind dispersal of oribatid mites as a mode of migration. *Pedobiologia* **2011**, *54*, 201–207. [CrossRef]
53. Walter, D.E.; Behan-Pelletier, V.M. Systematics and ecology of *Adhaesozetes polyphyllos* sp.nov. (Acari: Oribatida: Licneremaeoidea), a leaf-inhabiting mite from Australian rainforests. *Can. J. Zool.* **1993**, *71*, 1024–1040. [CrossRef]
54. Lindo, Z.; Clayton, M.; Behan-Pelletier, V.M. Systematics and ecology of the genus *Dendrozetes* (Acari: Oribatida: Peloppiidae) from arboreal habitats in Western North America. *Zootaxa* **2010**, *2403*, 10–22. [CrossRef]
55. Bayartogtokh, B.; Itioka, T.; Kitora, H.; Meleng, P.; Shimano, S. New findings of poronotic oribatid mites (Acari: Oribatida) from the high canopy of a Bornean tropical rain forest. *Int. J. Acarol.* **2020**, *46*, 73–82. [CrossRef]
56. Van der Hammen, L. *Glossary of Acarological Terminology. Vol. I. General Terminology*; Dr. W. Junk Publishers: The Hague, The Netherlands, 1980; p. 244.
57. Alberti, G.; Moreno, A.I.; Kratzmann, M. The fine structure of trichobothria in moss mites with special emphasis on *Acrogalumna lonigpluma* (Berlese, 1904) (Oribatida, Acari, Arachnida). *Acta Zool.* **1994**, *75*, 57–74. [CrossRef]
58. Norton, R.A.; Palacios-Vargas, J.G. Nueva *Belba* (Oribatei: Damaeidae) de musgos epifitos de Mexico. *Folia Entomol. Mex.* **1982**, *52*, 61–73.
59. Olszanowski, Z.; Clayton, M.R.; Humble, L.M. New species of the genus *Camisia* (Acari: Oribatida): An arboreal mite with enclosed sensilli. *Can. Entomol.* **2002**, *134*, 707–721. [CrossRef]
60. Schatz, H.; Behan-Pelletier, V. Global diversity of oribatids (Oribatida: Acari: Arachnida). *Hydrobiologia* **2008**, *595*, 323–328. [CrossRef]
61. Behan-Pelletier, V.M.; Walter, D.E. *Phylleremus* n. gen., from leaves of deciduous trees in eastern Australia (Oribatida: Licneremaeoidea). *Zootaxa* **2007**, *1386*, 1–17. [CrossRef]
62. Walter, D.E.; O'Dowd, D.J. Beneath biodiversity: Factors influencing the diversity and abundance of canopy mites. *Selbyana* **1995**, *16*, 12–20.

63. Niedbala, W. Oribatei (Acari) of Spitzbergen. *Bull. Acad. Polon. Sci. Série Sci. Biologique* **1971**, *19*, 737–742.
64. Erdmann, G.; Otte, V.; Langel, R.; Scheu, S.; Maraun, M. The trophic structure of bark-living oribatid mite communities analysed with stable isotopes (15N, 13C) indicates strong niche differentiation. *Exp. Appl. Acarol.* **2007**, *41*, 1–10. [CrossRef]
65. Bauer, D.M.; Bell, K.P.; Nelson, E.J.; Calhoun, A.J. Managing small natural features: A synthesis of economic issues and emergent opportunities. *Biol. Conserv.* **2017**, *211*, 80–87. [CrossRef]
66. Lindenmayer, D.B. Conserving large old trees as small natural features. *Biol. Conserv.* **2017**, *211*, 51–59. [CrossRef]
67. Hunter, M.L. Conserving small natural features with large ecological roles: An introduction and definition. *Biol. Conserv.* **2017**, *211*, 1–2. [CrossRef]

diversity

MDPI

Article

Endemism of Uropodina Mites: Spurious or Real?

Jerzy Błoszyk [1,2] and Agnieszka Napierała [1,*]

1 Department of General Zoology, Faculty of Biology, Adam Mickiewicz University, ul. Uniwersytetu Poznańskiego 6, 61-614 Poznań, Poland; bloszyk@gmail.com
2 Natural History Collections, Faculty of Biology, Adam Mickiewicz University, ul. Uniwersytetu Poznańskiego 6, 61-614 Poznań, Poland
* Correspondence: agan@amu.edu.pl

Received: 23 June 2020; Accepted: 14 July 2020; Published: 18 July 2020

Abstract: Analyzing the data from the existing literature about geographic distribution of mites from the suborder Uropodina (Acari: Mesostigmata), one can get the impression that this group of mites is characterized by an unusual extent of endemism on a global scale. This observation encouraged the authors of this study to ascertain whether endemism in Uropodina mites is a real feature of this group or whether it stems from the current state of affairs in this field of research. The study is based on evidence from the literature on the topic and data obtained from long-term research conducted on extensive materials from all over the globe (over 40,000 samples). The discussion presented in the article is supported by many examples, showing that both hypotheses can in fact be proved right. The major point of reference in this study is the fairly well-known fauna of Uropodina in Europe, whereas South America is the testing area for the two hypotheses.

Keywords: Acari; biogeography; geographical variability; Mesostigmata; neotropical region; palearctic species; range of occurrence; *Rotundabaloghia*; species identification; *Uropoda* (*Phaulodinychus*) *penicillata*

1. Introduction

Mites from the suborder Uropodina (Acari: Mesostigmata) are quite characteristic and easy to distinguish from other species of mites. Due to their large body size and high form diversity, these arachnids have been the object of research interests of many acarologists for a long time. The first descriptions of species were written over 240 years ago [1], and most of them were described at least 100 years ago.

As for now, over 2000 species from different geographical locations of the whole globe have already been classified as Uropodina and described in various studies [2]. However, many of the described species were found only in one location. The fact is that despite considerable dispersion abilities of these mites (phoresy), most Uropodina species have a restricted range of occurrence [2–5]. Most research in this field conducted so far has focused on the fauna of Europe, especially on Central Europe, where in some countries the distribution of particular species has been thoroughly described in numerous studies. Among such countries, there are Poland and Slovakia, where the ranges of Uropodina species were analyzed by means of the Universal Transverse Mercator (UTM) grid [3,5–8] or a different grid system (e.g., the grid system of *Databank of the Fauna of Slovakia*) [4]. Furthermore, the eastern states of Australia have also been recently thoroughly examined in this respect [9–12]. Nevertheless, little is still known about the exact distribution of Uropodina mites in other parts of the globe, where researchers usually describe only random occurrences of the species they describe, which in turn may suggest immense endemism in this group of organisms.

The authors of the current article present their own observations concerning the zoogeography of this group of mites on the basis of long-term research and materials obtained from all over the globe. The major aim of this study is therefore to verify the hypothesis that the endemism of species from suborder Uropodina implied by the data available in the existing literature on the topic is in fact spurious.

2. Materials and Methods

The material used in this study comprises over 40,000 samples (obtained mainly from sieved litter and soil, as well as from unsieved material from different types of unstable microhabitats such as dead wood, nests of birds and small mammals, anthills, etc.) collected in different parts of the globe by different collectors, in the period 1941–2019. All samples were sorted out, and the extracted mites were identified by the first author or coworkers. The obtained samples contained over 300,000 specimens of mites from the suborder Uropodina. The core of the material analyzed for the purpose of this study contained samples from Europe (mainly from Poland—30,000 samples), and also from Spain (30 samples), France (50 samples), Belgium (40 samples), the Netherlands (100 samples), Germany (50 samples), Denmark (100 samples), Sweden (30 samples), Norway (500 samples), the Czech Republic (150 samples), Slovakia (170 samples), and the Ukraine (30 samples). The extensive materials of exotic species from all continents, which in the past were used mainly to describe new species, in this study were analyzed to ascertain the geographical distribution of particular species. These materials allowed us to put forward the hypothesis about the "spurious endemism" of Uropodina mites all over the globe: in Australia and Tasmania (700 samples), New Guinea (50 samples), New Zealand (30 samples), South America (300 samples), Africa (100 samples), Madagascar (30 smples), India (80 samples), Nepal (50 samples); Indonesia (20 samples), Thailand (10 samples), Vietnam (20 samples), Mongolia (10 samples), Russia (Syberia—50 samples), Afganistan (20 samples), and Pakistan (5 samples).

The whole material used in this study has been deposited in The Invertebrate Databank, (Natural History Collections, Faculty of Biology AMU, Poznań, Poland). The obtained metadata are also available through the on-line platform developed within the digital project AMUNATCOLL: AMU Nature Collections on the Internet: Digitization and the biodiversity data resources of the Faculty of Biology at Adam Mickiewicz University in Poznań are available at https://rhus-76.man.poznan.pl:3000/.

3. Results and Discussion

3.1. Geographical Ranges of Uropodina

Interestingly, there are virtually no cosmopolitan species among Uropodina mites. Their range of occurrence is usually confined to one geographical region, and therefore, chiefly to one continent. One of the few exemptions in this respect are the palearctic species, which occur in Europe and Asia. The examples of such species are: *Oodinychus ovalis* (C.L.Koch, 1839), *Trichouropoda patavina* (G. Canestrini, 1885), *T. polytricha* (Vitzthum, 1923), *Uroobovella marginata* (C. L. Koch, 1829), *Apionoseius infirmus* Berlese, 1887 (in Europe and Kazakhstan); *Dinychus perforatus* Kramer, 1882, *Oplitis paradoxa* Berlese, 1919, *Trachytes aegrota* (C.L.Koch, 1841), *T. patavina*, *T. tuberosa* (Hirschmann et Zirngiebl-Nicol, 1961), *A. infirmus*, *Uroseius acuminatus* (C.L. Koch, 1847) (in Europe and Mongolia); *U. hunzikeri* Schweizer, 1922 (in Europe and Tajikistan); and *Leiodinychus orbicularis* (C. L. Koch, 1839) (in Europe and India). However, due to the fact that ranges of Uropodina mites have been determined to a different extent in each country on all continents, the available data are still fragmentary. Moreover, one cannot be certain whether the ranges of some Uropodina species, which are much broader than one might expect, are in fact results of incorrect identification of those species. For example, a quite common European species, *T. aegrota*, has also been found in the USA [2]. However, the lack of any comparative material in this case (museum collections are extremely helpful in such cases) prevents us from checking whether indeed it is the same species. This species could have been transported from Europe with ballast soil of sailing ships during the colonization period, when many cargo ships started regular trips from Europe to North America. This means that *T. aegrota* is probably an extraneous species in this geographical region, and it is also possible that the found specimens were incorrectly identified. In 1996, Błoszyk and Szymkowiak [13] described a species in Utah (*T. kaliszewskii* from Provo) which is very similar to *T. aegrota*. It is very likely that the species has a broader range of occurrence and was mistakenly identified as the European species *T. aegrota*. A similar situation has been observed when specimens of extraneous species were found in Polish seaports with warehouses containing wood from exotic trees [14–16].

In the area of Poland, some Uropodina species have their northern or western boundaries of occurrence range. Among the species with such ranges there are *T. irenae* (Pecina, 1970) and *T. minima* (Trägårdh, 1910) [17,18]. *Trachytes irenae* occurs only in the southern parts of Poland (Figure 1), and in Europe the range of occurrence of this species is also narrow (Figure 1A). A similar situation can be also observed in the case of *T. minima* (Figure 2). Moreover, *T. montana* (Willmann, 1953) [19] is a mountainous species, which in Europe usually occurs at the altitude over 1000 m a.s.l., and in Poland it also occurs only in the mountains in the southern parts of the country (Figure 3).

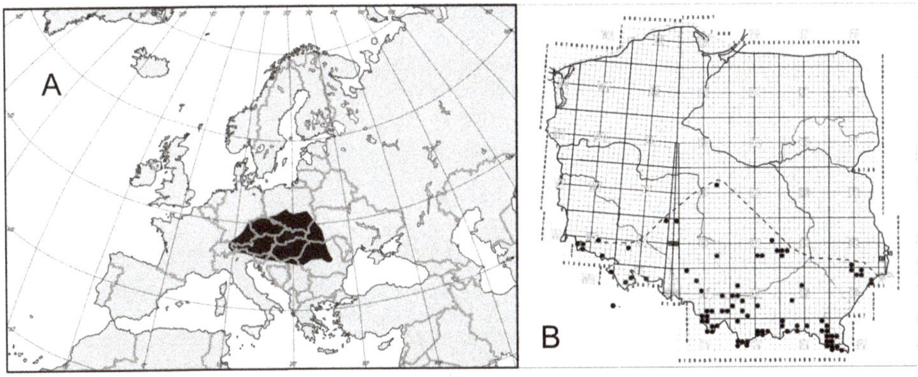

Figure 1. Range of *Trachytes irenae*: (**A**) in Europe, (**B**) in Poland; black dots—recorded occurrences of the species, dotted line—northern range boundary.

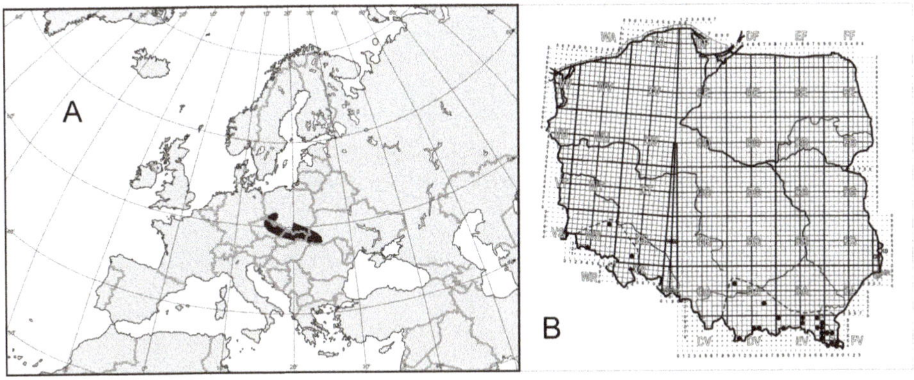

Figure 2. Range of *Trachytes minima*: (**A**) in Europe, (**B**) in Poland; black dots—recorded occurrences of the species, dotted line—northern range boundary.

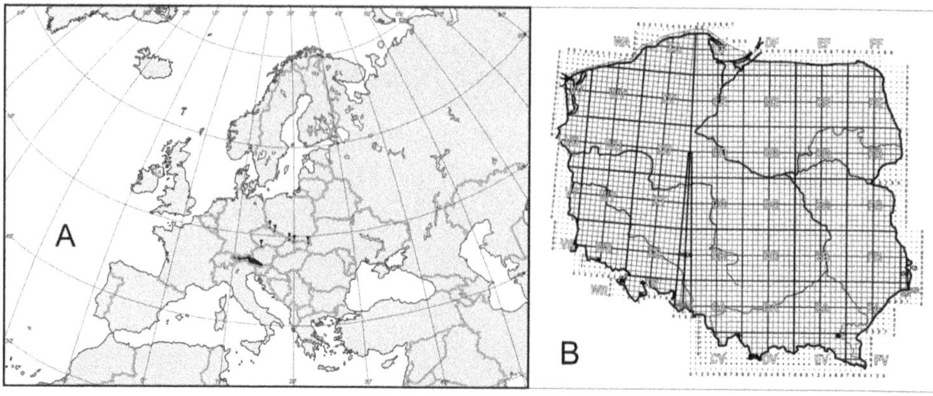

Figure 3. Range of *Trachytes montana*: (**A**) in Europe, (**B**) in Poland; black—recorded occurrences of the species.

The disjunctive range of occurrence is observed in the case of *Neodiscopoma splendida* Vitzthum, 1941. Southern and northern boundaries of range of occurrence of this species are separated by an area with no specimens of this species in the central part of Poland (Figure 4).

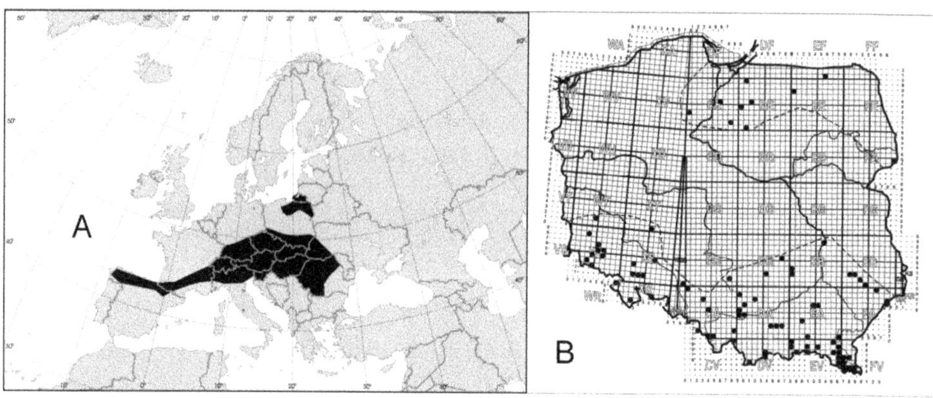

Figure 4. Range of *Neodiscopoma splendida*: (**A**) in Europe, (**B**) in Poland; black dots—recorded occurrences of the species, dotted line—occurrence range of southern and northern populations.

Among the species from the eastern Carpathians which occur in Poland, there are *Trachytes splendida* (Huţu, 1973), *Urodiaspis stammeri* (Hirschmann and Zirgiebl-Nicol, 1969) (Figure 5), and *Cilliba selnicki* (Hirschmann and Zirgiebl-Nicol, 1969) [3,4].

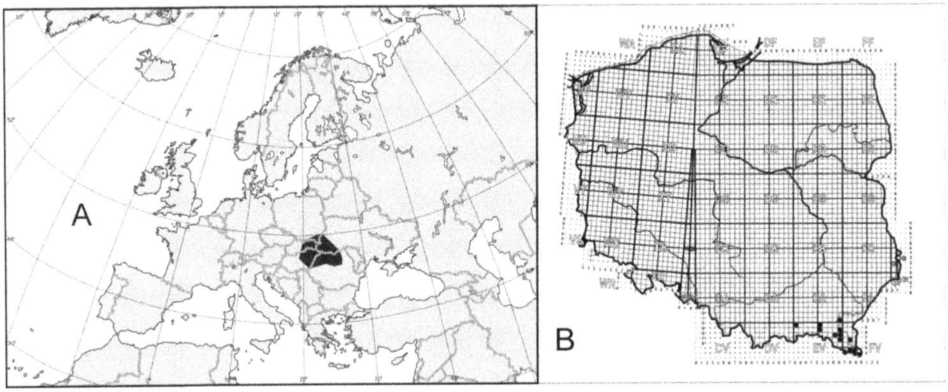

Figure 5. Range of *Urodiaspis stammeri*: (**A**) in Europe, (**B**) in Poland; black dots—recorded occurences of the species.

The so-called "Atlantic" species which occur in the area of Poland are such species as *Cilliba erlangensis* (Hirschmann, Zirngiebl-Nicol, 1969), *Olodiscus misella* (Berlese, 1916), and *Polyaspinus cylindricus* (Berlese, 1916), which have their eastern range boundaries in this country (Figure 6). The ranges of occurrence of the abovementioned species are similar to that of the European beech—*Fagus sylvatica* L.

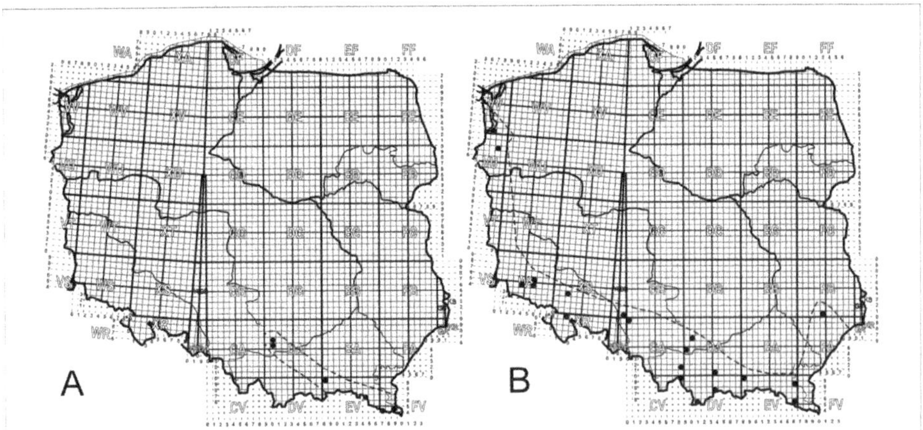

Figure 6. Species with their eastern range boundaries in Poland: (**A**) *Olodiscus misella*, (**B**) *Cilliba erlangensis*; black dots—recorded occurrences of the species.

The examples of range boundaries of selected Uropodina species in Poland and Europe adduced the above result from the geological history of this region (especially the Peistocene glaciations), and subsequently the habitat requirements of the species (including, for example, the range of tree species and type of the forest preferred by the mite species as habitats), as well as the dispersion abilities and reproduction strategy of the species (parthenogenesis vs. sexual reproduction). For this reason, it is very important to consider every discovery of a new species in a given area in a broader context, preferably in relation to the biology of the species and its associations with the environment. Recording occurrences of already described species in new locations is also very important, as this will allow us to determine more precisely their exact geographical distribution in the future.

In this study, we also analyzed the range of occurrence of some Uropodina species from Australia. The results of the analyses show that particular genera within Uropodina form in each of the discerned types of rain forests found in Australia vicariants, which means that each type of forest has its own specific Uropodina species. A good example can be species from the genus *Acroseius* occurring in rainforests on the eastern coast. Every type of rainforest on this continent has its own specific species of Uropodina from this genus. Moreover, the ranges of these species overlap with the biogeographical boundaries of Australia (Figure 7) [10,11].

Figure 7. Distribution of species of *Acroseius*: Aw—*A. weiri*, At—*A. tuberculatus*, Awo—*A. womersleyi* against biogeographical boundaries of Australia (according to Błoszyk et al. 2013 [11], slightly changed).

For example, *Acroseius weiri* occurs only in an isolated area of a tropical rainforest in Eungella National Park, separated from more southerly areas of forest by the dry St. Lawrence Gap, which is also recognized as a biogeographic barrier for reptiles [20], glow-worms [21], and land snails [22]. The southern range boundary of *A. tuberculatus* (Clyde Mountain) is close to the northern range of *A. womersleyi* (Brown Mountain). The two species are separated there by the valley of the Tuross River, which does not appear to offer a significant biogeographic barrier. However, the rainforest in this area is confined to a series of very small scattered patches among the Eucalyptus forest, and there is a large gap in the distribution of these patches inland from Narooma [23], which coincides with the gap between *A. tuberculatus* and *A. womersleyi*. All the records of *A. tuberculatus* for which we have adequate data are from rainforest litter, so this gap in the distribution of rainforest may be a barrier to its dispersal [11].

This issue looks a bit different in the case of genera. The ranges of the genera interchange along the parallels of altitude, and the equator is the boundary. Some of the genera occur only in one of the two hemispheres. For example, the most typical genera in the northern hemisphere are *Trachytes*, *Cilliba*, *Urodiaspis*, which do not occur in the southern hemisphere. Species from such genera as *Rotundabaloghia*, *Castriidinychus*, *Platysetosus*, *Acroseius*, and *Capricornelia* do not occur on the northern hemisphere. The other genera can have different proportions of species occurring on the northern and southern hemisphere (Figure 8).

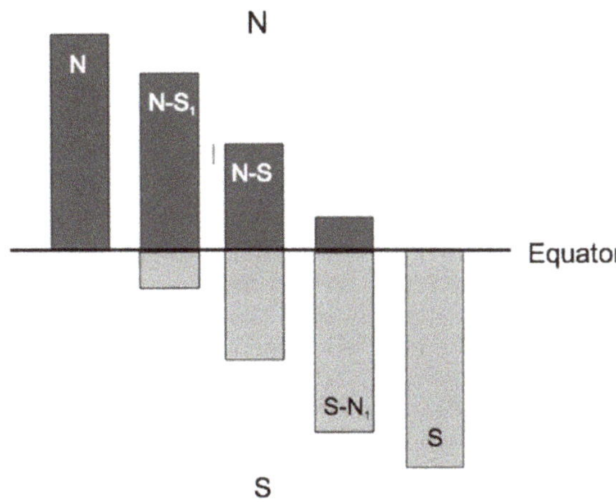

Figure 8. Occurrence of species from particular genera on both hemispheres: N—only in northern hemisphere, N-S_1—mainly in northern hemisphere, N-S—with the same rate in both hemispheres, S-N_1—with prevailance in southern hemisphere, S—only in southern hemisphere.

3.2. Spurious or Real Endemism

As has been said, geographical ranges of occurrence of most European species of Uropodina have already been described in detail [2–5]. In the case of the other continents, there are in fact no detailed accounts with information about the exact geographical distribution of Uropodina species. Most of the species from other continents were found only in one location and as single specimens. This can be said especially about exotic materials from South America, Australia and Oceania, Africa, and south-eastern Asia. A bit more information is now available about the ranges of Uropodina mites in Australia. More precise studies based on a large number of samples from different locations show the actual ranges of some new species discovered recently [10–12,24]. Thus, the question is whether sporadic occurrences of some species from this group on other continents reflect unique endemism of these mites or whether they stem from the extent to which the geographical distribution of these mites has been already surveyed on every continent, or whether it is a combination of these two factors.

In the first period of research into this group of organisms during the period between 1800 and 1930, the number of described new taxa from all continents was no higher than 100 species in the next decade, whereas between 1970 and 1990 several new species were described every year. The researchers from previous generations, such as Berlese, Oudemans and Sellnick, who only had materials with specimens of Uropodina from places where these mites had not been recorded earlier, did not give any synonyms due to the low number of species described in the literature. The taxa they described at that time (despite the poor quality of these descriptions and very often incorrect systematic status diagnoses) were indeed new species.

The descriptions of new Uropodina species in Europe made during that period (until the Second World War) in many cases have caused a lot of ambiguity. Due to the large number of acarologists interested in this group of mites and the lack of precision in the first descriptions of the oldest species, many species were often described several times under different names, which is evident in the high number of synonyms. Furthermore, contemporary acarologists also often cannot avoid forging new synonyms. This situation stems mainly from the fact that the quality of new descriptions is still often poor and that researchers ignore the evidence preserved in type specimens, but base their descriptions on earlier accounts available in the literature. Unfortunately, it may seem that some acarologists are now focused only on describing new species and they are not interested at all in conducting any research into the biology, ecology, and zoogeography of Uropodina mites. The terse descriptions of new species, which frequently provide no basic information about the biology of the species such as variability (geographical and ecological) and range of occurrence, do not allow researchers to classify the species correctly. Besides this, many acarologists interpret all types of variability in the same way, which has an identical impact on speciation, and, as a result, they often end up with another synonym for the species. It should be borne in mind that the verification and re-classification of such data post factum is extremely hard and laborious. In many cases, any verification of the description is very hard or simply impossible due to the lack of access to the type specimen used by the researcher. Finally, new species used to be (and also these days often are) described on the basis of just one specimen, developmental (juvenile) stages, or specimens of only one sex. In this situation, synonyms have already become a commonplace in the literature on Uropodina mites.

Thus, thorough verification of synonyms is in fact one of the major taxonomic problems in this group of mites. The name of a species can indicate whether there is any necessity to check the diagnosis and designation of the species. Hirschmann and his co-workers described a lot of new Uropodina species between 1961 and 1993. Unfortunately, in their descriptions of new species, they did not take into account the geographical and ecological variability of the taxa, regarding all kinds of morphological variability of the specimens as indication of speciation. Moreover, as many of the species that they described turned out to be similar to other species that had been earlier described, they often used "*similis*" in the name or other terms suggesting similarity to a different species (Table 1).

Most of the 252 species enumerated in Table 1 were recorded only in one location and the number of found specimens was usually very low. It is also noteworthy that almost all the species described as new to science with names suggesting similarity to other taxa, which had been described earlier, usually come from one geographical location (often from the same country). This fact allows us to raise the question whether in both cases it is the same species or not. In such a situation, the overall biodiversity of Uropodina mites would be obviously lower on a given continent or country, but the ranges of occurrence obtained from the verified data on particular taxa would probably give more reliable results, showing their actual geographical distribution.

Table 1. Morphologically similar species with names suggesting their close genetic relationship listed in the catalogue by Wiśniewski and Hirschmann (1993) [1].

Proper Species	Similar Species
Baloghibrasiluropoda foveolata Hirschmann, 1973—Brazil	*Baloghibrasiluropoda foveolatasimilis* Hirschmann, 1973—Brazil; *Baloghibrasiluropoda foveatoides* Hirschmann, 1973—Brazil
Baloghkaszabia baloghi Hirschmann, 1973—Brazil	*Baloghkaszabia baloghisimilis* Hirschmann, 1973—Brazil
Castriidinychus castri Hirschmann, 1973—Chile	*Castriidinychus castrisimilis* Hirschmann, 1973—Chile
Castriidinychus dentatus (Hirschmann, 1972)—Chile	*Castriidinychus similidentatus* Hirschmann, 1973—Chile *Castriidinychus dentatoides* Hirschmann, 1973—Chile
Castriidinychus eupunctatus Hirschmann, 1972—Chile	*Castriidinychus eupunctatosimilis* Hirschmann, 1972—Chile
Clausiadinychus cristatus Sellnick, 1930—Martinique	*Clausiadinychus similicristatus* Hirschmann, 1973—Brazil
Cyllibula (Baloghcyllibula) alta (Sellnick, 1973)—Trinidad	*Cyllibula (Baloghcyllibula) altasimilis* Hirschmann, 1977)—Bolivia
Deraiophorus hirschmanni Hiramatsu, 1977—Japan	*Deraiophorus hirschmannisimilis* Hiramatsu, 1977—Japan
Deraiophorus piriformis Hirschmann, 1973—?	*Deraiophorus piriformoides* Hirschmann and Hiramatsu, 1990—Philippines
Deraiophorus dicornutus Hirschmann, 1973—Bolivia	*Deraiophorus dicornutosimilis* Hirschmann, 1973—Bolivia
Deraiophorus hexacornutus Hirschmann, 1973—New Guinea	*Deraiophorus hexacornutosimilis* Hirschmann, 1973—New Guinea
Deraiophorus penicillatus Hirschmann, 1973—Cejlon	*Deraiophorus penicillatasimilis* Hirschmann, 1973—Ceylon
Deraiophorus loksai Hirschmann, 1973—Brazil, Paragwaj	*Deraiophorus loksaisimilis* Hirschmann, 1973—Paraguay
Deraiophorus kaszabi Hirschmann, 1973—Brazil, Chile	*Deraiophorus kaszabisimilis* Hirschmann, 1973—Brazil
Deraiophorus stammeri Hirschmann and Zirngiebl-Nicol, 1969—Brazil	*Deraiophorus stammerisimilis* Hirschmann, 1973—Brazil
Dinychus carinatus Berlese, 1903—Italy, France, Poland	*Dinychus bincheaecarinatus* Hirschmann, Wągrowska-Adamczyk and Zirngiebl-Nicol, 1984—Germany, France, Slovakia *Dinychus carinatus var. magnus* (Athias-Binche, 1980)—France
Dicourella baloghi Hirschmann and Zirngiebl-Nicol, 1969—Hungary, Romania, Poland	*Dicourella baloghisimilis* Wiśniewski, 1984—Poland
??	*Discourella crucisimilis* Hirschmann, 1972—Paraguay, Brazil
Discourella caputmedusae (Berlese and Leonardi, 1901)—Chile	*Discourella caputmedusaesimilis* Hirschmann, 1972—Chile
Discourella modesta (Leonardi, 1899)—Europe	*Discourella modestasimilis* (Hiramatsu and Hirschmann, 1979)—Canada
Discourella rotunda Hirschmann, 1972—Brazil	*Discourella rotunda* Hirschmann, 1973—Brazil *Discourella rotundiformis* Hirschmann, 1973—Brazil

Table 1. *Cont.*

Proper Species	Similar Species
Hutufeideria hirschmanni Hiramatsu, 1978—New Guinea	*Hutufeideria hirschmannisimilis* Hiramatsu, 1980—New Guinea
Hutufeideria feideri Hirschmann and Hiramatsu, 1977—New Guinea	*Hutufeideria feiderisimilis* Hiramatsu, 1981—New Guinea
Jerzywisniewska depilata (Trouessart, 1902)—Brazil	*Jerzywisniewska depilatasimilis* (Wiśniewski, 1902)—Brazil
Kaszabjbaloghia kaszabi Hirschmann, 1873—Ecuador	*Kaszabjbaloghia kaszabisimilis* Hirschmann, 1873—Peru
Nenteria oudemansi Hirschmann and Ziringiebl—Nicol, 1969—Netherlands	*Nenteria oudemansiformis* Hirschmann, 1985—?
Nenteria pilosella (Berlese, 1903)—Italy	*Nenteria pilosellapides* Hirschmann and Hiramatsu, 1978—Italy
Nenteria ritzemai (Oudemans, 1903)—Netherlands, Belgia, Niemcy	*Nenteria ritzemaisimilis* Hirschmann and Hiramatsu, 1978—Japan
Nenteria tropica (Oudemans, 1905)—Togo	*Nenteria tropicasimilis* Wiśniewski and Hirschamnn, 1985—Guinea
Oplitis baloghi Zirngiebl–Nicol and Hirschmann, 1973—Paraguay	*Oplitis baloghisimilis* Zirngiebl–Nicol and Hirschmann, 1973—Chile
Oplitis bispirata (Sellnick, 1954)—Brazil	*Oplitis similibispirata* Zirngiebl–Nicol, 1973—Bolivia
Oplitis castrii Zirngiebl—Nicol and Hirschmann, 1973—Chile	*Oplitis castriisimilis* Zirngiebl–Nicol and Hirschmann, 1973—Brazil
Oplitis mahunkai Zirngiebl—Nicol and Hirschmann, 1973—Brazil	*Oplitis mahunkaisimilis* Zirngiebl–Nicol and Hirschmann, 1973—Brazil
Oplitis minutissima (Berlese, 1903)—Netherlands, Italy, United Kingdom, Ukraine	*Oplitis japanominutissima* Hiramatsu, 1979—Japan *Oplitis similiminutissima* Hiramatsu, 1979—Japan
Oplitis dimidiata Hirschmann, 1991—Belize	*Oplitis dimidiatasimilis* Hirschmann and Wiśniewski, 1991—Cuba
Oplitis kaszabii Zirngiebl—Nicol and Hirschmann, 1973—Peru	*Oplitis kaszabiisimilis* Zirngiebl—Nicol and Hirschmann, 1973—Brazil
Oplitis pecki Hirschmann, 1991—Galapagos Islands	*Oplitis peckisimilis* Hirschmann, 1991—Galapagos Islands
Rotundabaloghia baloghi Hirschmann, 1975—New Guinea	*Rotundabaloghia baloghisimilis* Hirschmann, 1975—New Guinea *Rotundabaloghia baloghioides* Hirschmann, 1975—New Guinea
Rotundabaloghia incisa Hirschmann, 1992—Peru	*Rotundabaloghia incisasimilis* Hirschmann, 1992—Peru
Rotundabaloghia leteciae Hirschmann, 1992—Colombia	*Rotundabaloghia leteciaesimilis* Hirschmann, 1992—Colombia
Rotundabaloghia soliformis Hirschmann, 1992—Ecuador	*Rotundabaloghia soliformoides* Hirschmann, 1992—Ecuador
Rotundabaloghia haradai Hiramatsu, 1983—Borneo	*Rotundabaloghia haradaisimilis* Hiramatsu and Hirschmann, 1992—Philippines
Rotundabaloghia campanella Hirschmann, 1992—Cameroon	*Rotundabaloghia campanellasimilis* Hirschmann, 1992—Cameroon
Rotundabaloghia kaszabi Hirschmann, 1975—New Guinea	*Rotundabaloghia kaszabi* Hirschmann, 1975—New Guinea
Tetrasejaspis baloghi Hirschmann, 1973—Brazil	*Tetrasejaspis baloghisimilis* Hirschmann, 1973—Brazil

Table 1. *Cont.*

Proper Species	Similar Species
Trachyuropoda arculata Hirschmann, 1975—Brazil	*Trachyuropoda similiarculata* Hirschmann, 1975—Brazil
Trachyuropoda baloghi Hirschmann, 1975—Chile	*Trachyuropoda baloghisimilis* Hirschmann, 1975—Chile
Trachyuropoda coccinea (Michael, 1891)—Europe	*Trachyuropoda similicoccinea* Hiramatsu, 1979—Japan
Trachyuropoda longicornuta Hirschmann, 1976—Spain	*Trachyuropoda longicornutasimilis* Hirschmann, 1976—??
Trachyuropoda formicaria (Lubbock, 1881)—Europe	*Trachyuropoda formicariasimilis* Hirschmann, 1975—Russia
Trachyuropoda mesofovea Hirschmann, 1976—Paragway	*Trachyuropoda mesofoveasimilis* Hirschmann, 1976—??
Trachyuropoda schusteri Hirschmann, 1976—Brazil	*Trachyuropoda schusterisimilis* Hirschmann, 1976—??
Trachyuropoda multituberosa (Willmann, 1951)—Austria	*Trachyuropoda multituberculata* Hirschmann, 1976—Spain, Austria *Trachyuropoda tuberosa* Hirschmann, 1976—Spain
Trachyuropoda dicarinata Hirschmann, 1976—Peru	*Trachyuropoda dicarinatasimilis* Hirschmann, 1976—Peru
Trichouropoda polytricha (Vitzthum, 1923)—Central Europe, Turkiestan	*Trichouropoda polytrichasimilis* Hirschmann, 1972—Portugal
Trichouropoda lindquisti Hirschmann, 1978—Canada	*Trichouropoda lindquistisimilis* Hirschmann, 1978—Canada
Trichouropoda afossalis Hirschmann, 1978—Ghana	*Trichouropoda afossalisimilis* Hirschmann and Wiśniewski, 1987—Indie
Trichouropoda denticulata Hirschmann, 1978—USA	*Trichouropoda denticulatasimilis* Wiśniewski and Hirschmann, 1987—Honduras
Trichouropoda ditricha Hirschmann and Wiśniewski, 1987—Ghana	*Trichouropoda ditrichasimilis* Hirschmann and Wiśniewski, 1987—Cameroon
Trichouropoda bipilis (Vitzthum, 1921)—Austria	*Trichouropoda similibipilis* Hirschmann, 1972—USA
Trichouropoda obscura (C. L. Koch, 1836)—Austria	*Trichouropoda obscurasimilis* Hirschmann and Zirngiebl–Nicol, 1961—Central Europe
Trichouropoda punctata Hirschmann and Zirngiebl–Nicol, 1961—Spain, Norway	*Trichouropoda punctatasimilis* Wiśniewski and Hirschmann, 1986—Poland
Trichouropoda javensis (Oudemans, 1901)—Java	*Trichouropoda similijavensis* Hiramatsu and Hirschmann, 1979—Philippines *Trichouropoda philippino javensis* Hirschmann and Wiśniewski, 1988—Philippines
Trichouropoda adifaxa (Vitzthum, 1921)—Spanish Guinea	*Trichouropoda adifaxasimilis* Hirschmann and Wiśniewski, 1986—Spanish Guinea
Trichouropoda endroedyi Hirschmann and Wiśniewski, 1986—Ghana	*Trichouropoda endroedyioides* Hirschmann and Wiśniewski, 1986—Ghana
Trichouropoda ruehmi Hirschmann, 1972—Brazil	*Trichouropoda ruehmisimilis* Wiśniewski and Hirschmann, 1987—Brazil
Trichouropoda serrata Hirschmann and Zirngiebl–Nicol, 1961—Germany	*Trichouropoda serratasimilis* Hirschmann and Zirngiebl–Nicol, 1961—Canada
Trichouropoda stammeri Hirschmann and Zirngiebl–Nicol, 1969—Hungary	*Trichouropoda stammerisimilis* Hirschmann, 1978—Poland
Trichouropoda sturmii Hirschmann and Wiśniewski, 1987—Colombia	*Trichouropoda sturmiisimilis* Hirschmann and Wiśniewski, 1987—Colombia

Table 1. *Cont.*

Proper Species	Similar Species
Trichouropoda sturmii Hirschmann and Zirngiebl–Nicol, 1961 Central and Western Europe, Mongolia, Siberia	*Trichouropoda sturmiisimilis* Hirschmann and Wiśniewski, 1987—Germany
Trichouropoda sellnicki Hirschmann and Zirngiebl–Nicol, 1969—Bahamas	*Trichouropoda sellnickioides* Wiśniewski and Hirschmann, 1988—Cuba
Trichouropoda solaris Hirschmann, 1972—Brazil	*Trichouropoda solarissima* Hirschmann, 1978—Bolivia
Trichouropoda vanna (Lombardini, 1928)—Brazil	*Trichouropoda vannaoides* Hirschmann, 1978—Bolivia
Trichouropoda quadritricha Hirschmann and Wiśniewski, 1988—Ghana	*Trichouropoda quadritrichasimilis* Hirschmann and Wiśniewski, 1988—Ghana
Trigonuropoda crucistructura Hirschmann, 1975—Sri Lanka	*Trigonuropoda crucistructuraoides* Hirschmann, 1975—Sri Lanka
Trigonuropoda schizostructura Hirschmann, 1975—Sri Lanka	*Trigonuropoda schizostructurasimilis* Hirschmann, 1975—Sri Lanka
Trigonuropoda tuberosa Hirschmann, 1975—New Guinea	*Trigonuropoda tuberosasimilis* Hirschmann, 1975—New Guinea
Trigonuropoda monofoveolata Hirschmann, 1975—New Guinea	*Trigonuropoda monofoveolatasimilis* Hirschmann, 1975—New Guinea
Trigonuropoda sanguinea Hirschmann and Hiramatsu, 1977—Japan	*Trigonuropoda sanguineasimilis* Hirschmann and Hiramatsu, 1977—Taiwan
Trigonuropoda tuberculata Hirschmann, 1975—New Guinea	*Trigonuropoda tuberculatasimilis* Hiramatsu, 1979—Japan
Trigonuropoda terra–reginae Domrow, 1957—Australia	*Trigonuropoda terra–reginaesimilis* Hirschmann, 1975—Australia
Trigonuropoda trioculata Hirschmann, 1975—Australia	*Trigonuropoda trioculatasimilis* Hirschmann, 1975—Australia
Trigonuropoda trichobaloghia Hirschmann, 1975—Peru	*Trigonuropoda trichobaloghiasimilis* Hirschmann, 1975—New Guinea
Trigonuropoda pontina Hirschmann, 1975—Ceylon	*Trigonuropoda trichopontina* Hirschmann, 1975—Sri Lanka
Uroactinia hippocrepea (Berlese, 1918)—Thaiti	*Uroactinia hippocrepoides* (Vitzthum, 1935)—Marquesas Islands
Urodiaspis religiosa Hiramatsu, 1979—Japan	*Urodiaspis similireligiosa* Hiramatsu, 1979—Japan
Uroobovella costai Hirschmann and Zirngiebl–Nicol, 1972—Israel, Romania, Greenland	*Uroobovella costaisimilis* Wiśniewski, 1980—New Guinea
Uroobovella faceta Hiramatsu and Hirschmann, 1978—Ecuador	*Uroobovella facetaoides* Hiramatsu and Hirschmann, 1978—New Guinea
Uroobovella pectinata (Hirschmann, 1973)—New Guinea	*Uroobovella pectinatasimilis* Hiramatsu, 1980—Indonezja
Uroobovella fimicola (Berlese, 1903)—Europe	*Uroobovella fimicolasimilis* Hirschmann and Zirngiebl–Nicol, 1972—Chile
Uroobovella flagelliger (Berlese, 1910)—Italy, Sweden, Germany, Switzerland, Europe	*Uroobovella flagelligerformis* Hirschmann, 1979—Canada
Uroobovella foveolata Hirschmann and Zirngiebl–Nicol, 1972—France	*Uroobovella foveolatasimilis* Hiramatsu, 1980—Japan
Uroobovella ipidis (Vitzthum, 1923)—Europe, Lebanon	*Uroobovella ipidisimilis* Hirschmann and Zirngiebl–Nicol, 1962—Germany, Poland

Table 1. *Cont.*

Proper Species	Similar Species
Uroobovella orri Hirschmann, 1972—USA	*Uroobovella orrisimilis* Hirschmann and Zirngiebl–Nicol, 1975—Australia
Uroobovella obovata (Canestrini and Berlese, 1884)—Europe	*Uroobovella similiobovata* Hirschmann and Zirngiebl–Nicol, 1962—Germany, Rumunia
Uroobovella ovalis Hirschmann and Zirngiebl–Nicol, 1962—Germany	*Uroobovella similiovalis* Hirschmann and Zirngiebl–Nicol, 1979—Germany
Uroobovella pauxila Hiramatsu, 1981—New Guinea	*Uroobovella pauxiloides* Hirschmann, 1981—Vietnam
Uroobovella takakii Hiramatsu, 1980—Japan	*Uroobovella similitakakensis* Hirschmann, 1981—Vietnam
Uroobovella portalis Hirschmann, 1973—Brazil	*Uroobovella portalisimilis* Hirschmann, 1981—Brazil
Uroobovella nova (Oudemans, 1902—Europe	*Uroobovella novasimilis* Hiramatsu, 1979—Japan
Uroobovella Zaireensis Hirschmann, 1981—Zaire	*Uroobovella similiZaireensis* Hirschmann, 1981—USA
Uroobovella vitzthumi Hirschmann and Zirngiebl–*Nicol, 1962—??*	*Uroobovella vitzthumisimilis* Hirschmann, 1973—Paraguay
Uropoda (Metadinychus) serrata Hirschmann, 1972—Paraguay	*Uropoda (Metadinychus) serratasimilis* Hiramatsu and Hirschmann, 1972—Bolivia
Uropoda (Metadinychus) argasiformis (Berlese, 1916)—Brazil, Bolivia	*Uropoda (Metadinychus) similiargasiformis* Hirschmann, 1981—Zaire
Cilliba cassidea (Hermann, 1804)—Europe	*Cilliba cassideasimilis* Błoszyk, Stachowiak and Halliday, 2006—Central Europe
Uropoda (Phaulodinychus) amani Hirschmann, 1973—East Africa	*Uropoda (Phaulodinychus) amanisimilis* Wiśniewski, 1980—Bismarck Archipelago
Uropoda (Phaulodinychus) difoveolata Hirschmann and Zirngiebl–Nicol, 1969—Brazil	*Uropoda (Phaulodinychus) difoveolatasimilis* Hirschmann, 1972—Brazil
Uropoda (Phaulodinychus) quadridentata Hirschmann, 1973—Ecuador	*Uropoda (Phaulodinychus) quadridentatasimilis* Hiramatsu, 1980—Brazil
Uropoda (Phaulodinychus) hiramatsui Hirschmann, 1976—New Guinea	*Uropoda (Phaulodinychus) hiramatsuiformis* Hirschmann, 1976—New Guinea *Uropoda (Phaulodinychus) hiramatsuioides* Hirschmann, 1976—New Guinea *Uropoda (Phaulodinychus) hiramatsuisimilis* Hirschmann, 1976—New Guinea
Uropoda (*Phaulodinychus*) *hamulifera* Michael, 1894—Europe	*Uropoda* (*Phaulodinychus*) *similihamulifera* Hiramatsu, 1894—Japan
Uropoda (*Phaulodinychus*) *stolida* Hiramatsu and Hirschmann, 1978—Peru	*Uropoda* (*Phaulodinychus*) *stolidasimilis* Hiramatsu and Hirschmann, 1979—Peru
Uropoda (*Uropoda*) *orbicularis* (Muller, 1776)—Europe	*Uropoda* (*Uropoda*) *japanoorbicularis* Hiramatsu, 1979—Japan *Uropoda* (*Uropoda*) *onishiiorbicularis* Hiramatsu, 1980—Japan *Uropoda* (*Uropoda*) *similiorbicularis* Hiramatsu, 1980—Japan
Uropoda (*Phaulodinychus*) *penicillata* Hirschmann and Zirngiebl–Nicol, 1969—Panama	*Uropoda* (*Phaulodinychus*) *penicillatasimilis* Hirschmann, 1972—Brazil
Uropoda (*Phaulodinychus*) *procera* Hiramatsu and Hirschmann, 1979—Meksyk	*Uropoda* (*Phaulodinychus*) *procerasimilis* Hiramatsu and Hirschmann, 1979—Mexico
Uropoda (*Phaulodinychus*) *regia* (Vitzthum, 1921)—Bolivia	*Uropoda* (*Phaulodinychus*) *regiasimilis* Hirschmann, 1972—Brazil
Uropoda (*Phaulodinychus*) *Braziliensis* (Sellnick, 1962)—Brazil	*Uropoda* (*Phaulodinychus*) *similiBraziliensis* Hirschmann, 1972– Paragway
Uropoda (*Phaulodinychus*) *insulana* Hiramatsu, 1979—Japan	*Uropoda* (*Phaulodinychus*) *insulanasimilis* Hiramatsu, 1981—Japan

Table 1. *Cont.*

Proper Species	Similar Species
Uropoda (*Phaulodinychus*) *porticensis* (Berlese, 1903)—Italy	*Uropoda* (*Phaulodinychus*) *porticensoides* Hirschmann, 1993—Switzerland
Uropoda (*Phaulodinychus*) *morikawai* Hiramatsu, 1978—Japan	*Uropoda* (*Phaulodinychus*) *similimorikawai* Hiramatsu, 1979—Japan
Uropoda (*Phaulodinychus*) *tropicana* Hiramatsu, 1978—New Guinea	*Uropoda* (*Phaulodinychus*) *tropicanasimilis* Hiramatsu, 1981—New Guinea
Uropoda (*Phaulodinychus*) *ungulata* Hirschmann and Hiramatsu, 1977—Ecuador	*Uropoda* (*Phaulodinychus*) *ungulatasimilis* Hirschmann and Hiramatsu, 1979—Ecuador

[1] The names given here are from the catalogue.

3.2.1. Example I: Mites from the Genus Rotundabalogia from South America

To analyze the extent of endemism among Uropodina mites on the basis of the evidence in the literature on the topic, we analyzed mites from the genus *Rotundabalogia*. In 1992, Hirschmann [25] described 65 species from this genus from a few countries in South America. The specimens were extracted from samples collected at 18 sites located in Colombia, Ecuador, Venezuela, Brazil, Peru, and Bolivia (Figure 9).

Figure 9. Locations of the species from the genus *Rotundabaloghia* described by Hirschmann in 1992 [25] from the area of South America.

The most striking thing here is the number of found species in the locations marked on the map—the number fluctuated between one and nine (Table 2). For example, in Europe it is very hard to find in one sample more than two species of the same genus (Błoszyk—unpublished data). Finding between seven and nine species from the genus *Rotundabaloghia* in a few sites (1, 13, 14, 15) can indicate a very fast pace of speciation in a given area or incorrect classification at the level of genus. However, the lack of any descriptions of the microhabitats in which the samples were collected does not allow us to verify the latter presumption. On the other hand, it is also possible that rainforests offer extremely favorable environmental conditions, enhancing the fast speciation of these mites, which can easily adapt to different types of microhabitats by modification of their body size and morphology.

Table 2. List of species from the genus *Rotundabaloghia* described by Hirschmann in 1992 [25] from the area of South America (numbers of sites are the same as on the map). Species in bold were found in more than one location.

No	Country	Location	Species
1	Colombia	Monserrate	*Rotundabalogia lamellosa* Hirschmann, 1992; ***Rotundabaloghia tetraclavata*** **Hirschmann, 1992;** *Rotundabaloghia chisacaensis* Hirschmann, 1992; *Rotundabaloghia monserratensis*Hirschmann, 1992; *Rotundabaloghia monterredondoensis* Hirschmann, 1992; *Rotundabaloghia silvacola* Hirschmann, 1992; *Rotundabaloghia humicola* Hirschmann, 1992; *Rotundabaloghia bosquensis* Hirschmann, 1992; *Rotundabaloghia fincae* Hirschmann, 1992;
2	Colombia	Alto Belem	*Rotundabaloghia forcipata* Hirschmann, 1992; ***Rotundabaloghia hexaspinosa*** **Hirschmann, 1992;** *Rotundabaloghia belemensis* Hirschmann, 1992; *Rotundabaloghia altoensis* Hirschmann, 1992;
3	Colombia	El Guerrero	*Rotundabaloghia pituitosa* Hirschmann, 1992; ***Rotundabaloghia sexspinosa*** **Hirschmann, 1992;** *Rotundabaloghia diclavata* Hirschmann, 1992; *Rotundabaloghia flava* Hirschmann, 1992; *Rotundabaloghia guerreroensis* Hirschmann, 1992;
4	Colombia	Paramo	***Rotundabaloghia hexaspinosa*** **Hirschmann, 1992;**
5	Colombia	La Rusia	***Rotundabaloghia hexaspinosa*** **Hirschmann, 1992;**
6	Colombia	Letecia	***Rotundabaloghia hexaspinosa*** **Hirschmann, 1992;** *Rotundabaloghia amazonasae* Hirschmann, 1992; *Rotundabaloghia leteciae* Hirschmann, 1992; *Rotundabaloghia leteciasimilis* Hirschmann, 1992;
7	Colombia	Chingaza	***Rotundabaloghia sexspinosa*** **Hirschmann, 1992;** *Rotundabaloghia octospinosa* Hirschmann, 1992; ***Rotundabaloghia tetraclavata*** **Hirschmann, 1992;** *Rotundabaloghia chingazensis* Hirschmann, 1992;
8	Colombia	La Calera; Bogota; Chisaca	***Rotundabaloghia tetraclavata*** **Hirschmann, 1992;** *Rotundabaloghia diclavata* Hirschmann, 1992; *Rotundabaloghia chisacaensis* Hirschmann, 1992; *Rotundabaloghia pajonalis* Hirschmann, 1992;
9	Colombia	La Tagua	*Rotundabaloghia tague* Hirschmann, 1992;
10	Colombia	Huila Resina	*Rotundabaloghia hullae* Hirschmann, 1992; *Rotundabaloghia resinae* Hirschmann, 1992;
11	Brazil	Manaus	*Rotundabaloghia tetraunguiseta* Hirschmann, 1992; *Rotundabaloghia manausensis* Hirschmann, 1992;
12	Brazil	Santos	*Rotundabaloghia hexaunguiseta* Hirschmann, 1992;
13	Peru	Lima-Pucallpa	*Rotundabaloghia limae* Hirschmann, 1992; *Rotundabaloghia duodecimsetae* Hirschmann, 1992; *Rotundabaloghia limae* Hirschmann, 1992; *Rotundabaloghia magnioperculi* Hirschmann, 1992; *Rotundabaloghia ucayali* Hirschmann, 1992; *Rotundabaloghia vonalis* Hirschmann, 1992; *Rotundabaloghia peruensis* Hirschmann, 1992;
14	Ecuador	Quito-Bacza	*Rotundabaloghia ovaligynella* Hirschmann, 1992; *Rotundabaloghia soliformoides* Hirschmann, 1992; *Rotundabaloghia quitoaensis* Hirschmann, 1992; *Rotundabaloghia magna* Hirschmann, 1992; *Rotundabaloghia soliformis* Hirschmann, 1992; *Rotundabaloghia linguaeformis* Hirschmann, 1992; *Rotundabaloghia baczaensis* Hirschmann, 1992; *Rotundabaloghia maculosa* Hirschmann, 1992; *Rotundabaloghia equadorensis* Hirschmann, 1992;

Table 2. *Cont.*

No	Country	Location	Species
15	Peru	Iquitos	*Rotundabaloghia incisasimilis* Hirschmann, 1992; *Rotundabaloghia incisa* Hirschmann, 1992; *Rotundabaloghia iquitosensis* Hirschmann, 1992; *Rotundabaloghia iquitosensoides* Hirschmann, 1992; *Rotundabaloghia pucallpae* Hirschmann, 1992; *Rotundabaloghia maranonensis* Hirschmann, 1992; *Rotundabaloghia moyobambe* Hirschmann, 1992; *Rotundabaloghia cajamarcae* Hirschmann, 1992;
16	Peru	Machu Picchu	*Rotundabaloghia picchuensis* Hirschmann, 1992;
17	Venezuela	?	*Rotundabaloghia venezuelae* Hirschmann, 1992;
18	Bolivia	?	*Rotundabaloghia boliviensis* Hirschmann, 1992;

Out of 65 species described as being new to science, Hirschmann (1992) found only three of them in more than one location (Table 2). The most striking thing is that the species which were found in several different sites occurred in Colombia, which is the country that Hirschmann examined most thoroughly. This means that the extent of endemism in this group decreases proportionally to the extent to which a given country has been examined.

3.2.2. Example II: Uropoda (Phaulodinychus) penicillata Hirschmann et Zirgiebl-Nicol, 1969 from Ecuador

In order to prove empirically our hypothesis, we also used extensive material collected in the area of Ecuador focusing on one species—*Uropoda* (*Phaulodinychus*) *penicillata* Hirschmann and Zirgiebl-Nicol, 1969 (Figure 10).

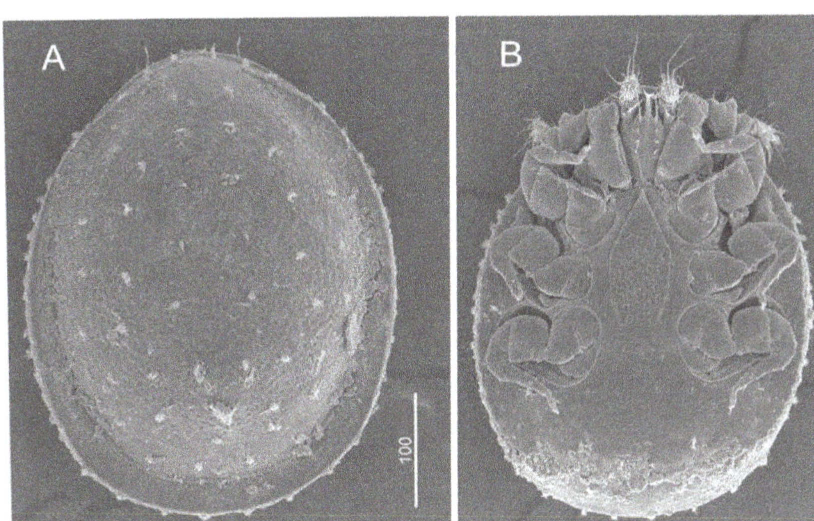

Figure 10. *Penicillaturopoda pennicillata*, female from Ecuador: (**A**) dorsal side, (**B**) ventral side.

The name of this species was used for the first time in the literature in 1964 [26], but with no detailed description and information about the place of occurrence (only a drawing of the genital shield and the ratio of its length to width were given by the author), and thus, it should be regarded as a nomenum nudum. The second time this name was used in Acarologie Folge 8, Teil 9 one year later

(1965) [27]. This time, a figure showing the hypostome and the dorsal side was given, but still with no information about the exact location in which the species was found. A more detailed description of the species with a location (Panama) was given in Acarologie Folge 12 (1969) (Hirschmann, Zirngiebl-Nicol 1969) [28], with figures showing the ventral side, chelicerae, and epistome. However, the description contained an error in the reference to the figure presented in Folge 7. Hirschmann (1972a,b) [29,30] established the *Uropoda* (*Phaulodinychus*) *penicillata* species group. In 1972, Hirschamnn described a morphologically identical species from Brazil (Figure 11). Due to the fact that the locations of the two species are 5000 km away from each other, he thought that they were two separate species. However, he noticed the striking similarity of the described taxon to *U.* (*P.*) *penicillata*, classifying it as a new species and he gave it a new name *Uropoda* (*Phaulodinychus*) *penicillatasimilis*. Then he changed the taxonomic status of the *U.* (*P.*) *penicillata* species group to the genus level, as *Penicillaturopoda* [31]. Finally, the *penicillata* species group [2] included six species (including Berlese species *complicate*).

Figure 11. Distribution of recorded sites with *Penicillaturopoda pennicillata*; 1—specimens described by Hirschmann and Zirgiebl-Nicol in 1969 as *Uropoda* (*Phaulodinychus*) *penicillata*; 2—specimens described by Hirschamnn in 1972 as *Uropoda* (*Phaulodinychus*) *penicillatasimilis*; •—unmarked sites on which soil samples were collected in Ecuador in 2014. Gray colour is used to mark areas with potential occurrence of species.

In 2014, the Natural History Collections (Faculty of Biology at AMU, Poznań, Poland) acquired a large collection of over 70 soil samples from Ecuador. The material, which was collected in different locations, contained numerous specimens of *U.* (*P.*) *penicillata* (see Appendix A).

Having analyzed the variability of the morphological characteristics used by Hirschmann (1972a,b) [29,30] and Hirschmanna and Hiramatsu (1990) [32] as the basis for discerning new species, we claim that both *U.* (*P.*) *pennicillata* from Panama and *U.* (*P.*) *pennicillatasimilis* from Brazil, as well as the specimens from Ecuador that we analyzed, should be in fact classified as one species. In our opinion, the observed differences (e.g., body size of the specimens, chaetotaxy, and surface sculpture) meet the criteria of population variability, and they stem from the very broad range of occurrence of this species. In Ecuador, this species occurred in 18 out of 78 analyzed samples (frequency 24%). The species was recorded mainly in synantropic habitats, with interference of humans, such as greenbelts near busy roads, city parks and agriculture fields. Thus, the species is presumably very common in the tropical zone in two subregions of the Neotropical region (Mexican and Brazilian) (Figure 11).

The examples given above are only a few of many other instances of the so-called "spurious endemism" which can be observed among Uropodina species. This phenomenon can be explained by the fact that the distribution of most species from this group presented in the literature on some continents has been recorded so far only in a few places. As the examples given above show, this situation stems mainly from the current state of knowledge in this field of research and erroneous designation of some species, and not from the actual range of occurrence of the described species.

The results of research into the geographical distribution of particular species of Uropodina mites in Europe, especially in Poland, show, however, that these mites in many cases apparently have a narrow range of occurrence (Figures 2–6). This means that they are more prone to endemism than, for example, Oribatida [33–35]. Other groups of mites which are similar in this respect to Uropodina are in Poland species from the family Labidostommidae (Prostigmata) and Zerconidae (Mesostigmata) [36,37]. Species from both these groups have their range boundaries in Poland, which often overlap with those of Uropodina mites. However, it is noteworthy that the area of Europe has been more thoroughly analyzed in this respect than other continents, and therefore, the ranges given in the literature reflect the actual geographical distribution of the species than only occasional occurrences of a large number of species, which can be observed in tropical regions.

One should also bare in mind that rain forests on all continents due to their unique ecological diversity of niches are extremely favorable ecosystems for fast speciation of many groups of organisms, including Uropodina mites, and therefore, they are also places with the highest biodiversity. Our knowledge about acarofauna of such ecosystems is still far from complete. On the other hand, the gradual shrinking of rain forest observed during the last couple of decades on a global scale should encourage experts to further carry out more extensive research. If it is true that these mites indeed have such a strong natural tendency towards endemism, then every loss of a bit of area of a rain forest in different geographical regions means loss of many species of Uropodina mites.

4. Conclusions

One of the aims of the current study was to show that such problems occur not only in the taxonomy of Uropodina. Our intention was to make other researchers pay more attention to the need for more thorough acarological research that would not be confined only to describing new species, but it would also allow us to ascertain the actual geographical distribution of already classified species, and would provide more information about the biology and ecology of the species. Due to the high level of biodiversity among Uropodina mites, any further research should be more extensive and it should be conducted on all continents. Recording as many new species as possible will also allow us to verify the systematic classification of this group at the level of higher taxonomic classes (i.e., genera and families). Moreover, there is also an urgent need for providing new descriptions of most species that would be based on a large series of specimens, especially those described until the end of the first half of the last century, and they would rest on both morphological data obtained with SEM techniques and on the molecular biology techniques, as DNA barcoding, to identify species. To do that, researchers should organize special expeditions to places which are particularly endangered, the so-called hot-spots, and then the collected material should be deposited in natural history museums, which later could make them available to other acarologists. In Poland, there is such a natural history museum department at the Faculty of Biology at Adam Mickiewicz University in Poznań, and it is called the Natural History Collections.

Author Contributions: Conceptualization, J.B. and A.N.; Methodology, J.B.; Software, J.B.; Validation, J.B., A.N.; Formal Analysis, J.B. and A.N.; Investigation, J.B. and A.N.; Resources, J.B. and A.N.; Data Curation, J.B.; Writing—Original Draft Preparation, J.B. and A.N.; Writing—Review & Editing, A.N.; Visualization, J.B. and A.N.; Supervision, A.N.; Project Administration, J.B.; Funding Acquisition, J.B. and A.N. All authors have read and agreed to the published version of the manuscript.

Funding: This work was partially funded by the Prometeo Project of the Secretariat for Higher Education, Science, Technology and Innovation of the Republic of Ecuador (grant to Łukasz Kaczmarek, permit No 001-15IC-FLO-FAU-DNB/MA from Ministry of Environment in Ecuador) and by Polish Ministry of Science and Higher Education, project number N303 091 32/3082. This work was also partially funded from financial resources of the Faculty of Biology at Adam Mickiewicz University in Poznań.

Acknowledgments: The authors would like to thank CSIRO (Canberra, Australia) for providing facilities for laboratory and field research, and the National Parks authorities of Victoria and New South Wales for permission to collect samples in their respective states. We also thank Łukasz Kaczmarek for collecting and sharing samples in Ecuador, thank Zbigniew Adamski from Electron and Confocal Microscope Laboratory, Faculty of Biology, Adam Mickiewicz University in Poznań, for taking the SEM images.

Conflicts of Interest: The authors declare no conflict of interest.

Appendix A

Sites with Penicillaturopoda pennicillata mentioned in the literature and found in our own materials:

Panama, dead wood collected near Colon, 9°20″ N 79°53″ W. Brazil, Santos, Sao Paulo State, 29. IX 1967, transect on steep hills of a plateau near a coast, at approx. 400–500 m, secondary forest on a steep hill, 23°57″ S 46°21″ W, leg. J. Balogh: from touchwood near a watercourse [BRB44]; damp moss on stones [BRB45]; dense moss with touchwood [BRB46]; Material examined. Ecuador. 16 Female, CIPCA (Centro de Investigación Posgrado y Conservación Amazónica ang. Centre of Research, Graduate and Amazon Conservation), 28 XI 2014, 01.48200° S 78.00205° W, corn and papa china field, quantitative sample, leg. Ł. Kaczmarek [EKW-041]; 1 DN, 1 LV, Urban, Puyo, 19 X 2014, 01.48834° S 78.00355° W, near a watercourse/stream, leg. Ł. Kaczmarek [EKW-001]; 1 female, 1DN, 1 LV, Agriculture, CIPCA, 14 XI 2014, 01.23710° S 77.88805° W, between banana and cocoa trees, leg. Ł. Kaczmarek, [EKW-002]; 2 females, 1 DN, Urban, Puyo, 14 X 2014, 01.49291° S 78.00056° W, greenbelt between road lanes near Megaldaz, leg. Ł. Kaczmarek, [EKW-006]; 1DN, Urban, Puyo, 4 XI 2014, 01.47986° S 78.00068° W, near a sawmill, leg. Ł. Kaczmarek, EKW-007; 1 famale, Agriculture, CIPCA, 14 XI 2014, 01.23710° S 77.88805° W, between banana and cocoa trees, leg. Ł. Kaczmarek, EKW-011; 1 female, 3 DN, Urban Puyo, 14 X 2014, 01.48535° S 77.99717° W, in a park at a palm tree, leg. Ł. Kaczmarek, [EKW-013]; 1 female, 1 DN, Agriculture, CIPCA, 24 XI 2014, 01.23710° S 77.88805° W, among banana trees, leg. Ł. Kaczmarek, [EKW-015]; 3 females, Urban, Puyo, 14 X 2014, 01.49017° S 77.99918° W, city center, near a watercourse, at a tree, leg. Ł. Kaczmarek, [EKW-016]; 1 DN, 1 LV, Urban, Puyo, 19 X 2014, 01.48062° S 78.00282° W, a park along a river, leg. Ł. Kaczmarek, [EKW-019]; 7 females, 1 DN, 1 PN, Urban, Puyo, 4 XI 2014, 01.48200° S 78.00205° W, city of Puyo, a small green area between two streets Manabí and Galapagos, near a tree, leg. Ł. Kaczmarek, [EKW-024]; 2 females, Urban, Puyo, 27 X 2015, 01.48018° S 78.00538° W, at a tree. Leg. Ł. Kaczmarek, [EKW-026]; 2 females, 1 DN, piedmont tropical forest, 8 I 2015, 00.83029° N 78.40355° W, near road E10, approx. 10 km SE from Lita, litter from deciduous forest, leg. Ł. Kaczmarek, [EKW-029]; 1 female, Agriculture, CIPCA, 24 XI 2014, 01.23710° S 77.88805° W, among banana trees, [EKW-042]; 9 females, 2 DN, Agriculture, CIPCA, 28 XI 2014, 01.23710° S 77.88805° W, on a corn and papachina field, leg. Ł. Kaczmarek, [EKW-045]; 11 females, 1 DN, Agriculture, CIPCA, 28 XI 2014, 01.23710° S 77.88805° W, on a corn and papachina field, leg. Ł. Kaczmarek, [EKW-048]; 1 female, Agriculture, CIPCA, 20 XI 2014, 01.23710° S 77.88805° W, among cocoa trees, leg. Ł. Kaczmarek, [EKW-050]; 1 PN, Agriculture, CIPCA, 20 XI 2014, 01.23710° S 77.88805° W, among cocoa trees, leg. Ł. Kaczmarek, [EKW-051].

References

1. Müller, O.F. *Zoologiae Danicae Prodromus, seu Animalism Daniae et Norvegien Indigenarum. Characteres, Nomina et Synonyma Imprimis Populariarum*; Havniae, Typis Hallageriis: Copenhagen, Denmark, 1776; pp. 186–187.
2. Wiśniewski, J.; Hirschmann, W. Katalog der Ganggattungen, Untergattungen, Gruppen und Arten der Uropodiden der Erde (Taxonomie, Literatur, Grősse, Verbreitung, Vorkommen). Gangsystematic der Parasitiformes 548. *Acarologie* **1993**, *40*, 1–220.
3. Błoszyk, J. *Geograficzne i Ekologiczne Zróżnicowanie Zgrupowań Roztoczy z Kohorty Uropodina (Acari: Mesostigmata) w Polsce. 1. Uropodina Lasów Grądowych (Carpinion betuli)*; Kontekst: Poznań, Poland, 1999; p. 245.
4. Mašán, P. Mites of the cohort Uropodina (Acarina, Mesostigmata) in Slovakia. *Annot. Zool. Bot. Bratislava.* **2001**, *223*, 1–320.
5. Błoszyk, J.; Bajaczyk, R.; Markowicz, M.; Gulvik, M. Geographical and ecological variability of mites of the suborder Uropodina (Acari: Mesostigmata) in Europe. *Biol. Lett.* **2003**, *40*, 15–35.
6. Błoszyk, J. East Carpathian geographical elements in the mite fauna of Poland. *Biol. Lett.* **1998**, *35*, 137–147.
7. Błoszyk, J.; Bajaczyk, R.; Błoszyk, G.; Napierała, A. Uropodina (Acari: Mesostigmata) parków narodowych Polski na tle innych obszarów. *Kosmos* **2002**, *51*, 463–470.
8. Napierała, A.; Błoszyk, J.; Kozak, J.; Bruin, J. Spatial distribution of mites of the suborder Uropodina (Acari: Mesostigmata) in small isolated forest area. *Exp. Appl. Acarol.* **2006**, *39*, 289–295. [CrossRef]
9. Athias-Binche, F.; Błoszyk, J. Australian Uropodina (Acari: Anactonotrichida). 1. Australocilliba gen.n. (Cillibidae). *J. Aust. Ent. Soc.* **1988**, *27*, 1–8. [CrossRef]
10. Błoszyk, J.; Halliday, R.B.; Dylewska, M. *Acroseius womersleyi* gen. nov., sp. nov., a new genus and species of Uropodina from Australia (Acari: Trachytidae). *Syst. Appl. Acarol.* **2005**, *10*, 41–60. [CrossRef]
11. Błoszyk, J.; Halliday, R.B.; Napierała, A. *Acroseius weiri* sp. nov. (Acari: Trachytidae), a new species of Uropodina from eastern Australia, with notes on the biogeography of the genus. *Syst. Appl. Acarol.* **2013**, *18*, 273–290.
12. Błoszyk, J.; Halliday, R.B.; Adamski, Z.; Książkiewicz-Parulska, Z. *Capricornella bicornuta*, a new genus and species of mite from eastern Australia (Acari: Uropodina). *Zootaxa* **2017**, *4244*, 321–338. [CrossRef]
13. Błoszyk, J.; Szymkowiak, P. *Trachytes kaliszewskii* n. sp. (Acari: Uropodina) from Great Basin (Utah, USA), with remarks on the habitats and distribution of the members of the genus *Trachytes*. *Great Basin Nat.* **1996**, *56*, 59–72.
14. Hirschmann, W.; Wiśniewski, J. Weltweite Revision der ganggattung *Trichouropoda* Berlese1916. II. Die interstructura-Gruppe (Trichouropodini, Uropodinae). GDP T. 492. *Acarologie* **1986**, *33*, 81–109.
15. Hirschmann, W.; Wiśniewski, J. Weltweite Revision der ganggattung *Trichouropoda* Berlese1916. IV. Die longiseta-Gruppe (Trichouropodini, Uropodinae). GDP T.496. *Acarologie* **1987**, *34*, 1–50.
16. Hirschmann, W.; Wiśniewski, J. Weltweite Revision der ganggattung Trichouropoda Berlese1916 Nachträge zu den von 1986 bis 1988 revidieren Gruppen (Trichouropodini, Uropodinae). GDP T. 503. *Acarologie* **1988**, *35*, 85–115.
17. Pečina, P. Czechoslovak Uropodid mites of the Genus *Trachytes* Michael, 1894 (Acari: Mesostigmata). *Acta Univ. Carol.-Biol. Praha* **1970**, *1*, 39–59.
18. Trägårdh, I. Acaridienaus dem Sarekgebirge. *Naturwiss. Untersuch. Sarek-Geb. Schwed. Lappland* **1910**, *4*, 375–586.
19. Willmann, C. Milben aus Mineralquellen (2. Mitteilund). *Zool. Anz.* **1950**, *145*, 186–195.
20. Chapple, D.G.; Hoskin, C.J.; Chapple, S.N.J.; Thompson, M.B. Phylogeographic divergence in the widespread delicate skink (*Lampropholis delicata*) corresponds to dry habitat barriers in eastern Australia. *BMC Evol. Biol.* **2011**, *11*, 191. [CrossRef]
21. Baker, C.H.; Graham, G.C.; Scott, K.D.; Cameron, S.L.; Yeates, D.K.; Merritt, D.J. Distribution and phylogenetic relationships of Australian glow-worms Arachnocampa (Diptera, Keroplatidae). *Mol. Phylogenet. Evol.* **2008**, *48*, 506–514. [CrossRef]
22. Stansic, J. The distribution and patterns of species diversity of land snails in eastern Australia. *Mem. Queensl. Mus.* **1994**, *36*, 207–214.
23. Floyd, A.G. *Rainforest Trees of Mainland South-eastern Australia. Forestry Commission of New South Wales*; Inkata Press: Melbourne, Australia; Sydney, Australia, 1989; pp. 1–420.

24. Błoszyk, J.; Halliday, R.B.; Dylewska, M.; Napierała, A. An interesting case of vicariance in the endemic mite genus *Acroseius* in eastern Australia (Acari: Uropodina: Trachytidae). In *Integrative Acarology. Proceedings of the 6th European Congress, European Association of Acarologists*; Bertrand, M., Kreiter, S., McCoy, A., Migeon, M., Navajas, M., Tixier, M.S., Vial, L., Eds.; European Association of Acarologists: Montpellier, France, 2008; pp. 44–46.
25. Hirschmann, W. Gangsystematik der Parasitiformes. Teil 536. 41 Rotundabaloghia-Arten aus Sudamerika (Venezuela, Ekuador, Peru, Bolivien, Brasilien) und Mittelamerika (Guatemala) (Dinychini, Uropodinae). Schriftenreihe für Vergleichende Milbenkunde. *Acarologie* **1992**, *39*, 69–95.
26. Hirschmann, W.; Zirngiebl-Nicol, I. Uropodiden Das Gangsystem der Fanilie Uropodide (Berlese, 1882) Hirschmann u. Zirngiebl-Nicol nov. comb. Brstimmungstabellen Kurzdiagnosen Operculumbestimmungstabellen. GDP T.7. *Acarologie* **1964**, *6*, 2–22.
27. Hirschmann, W.; Zirngiebl-Nicol, I. Uropodiden bestimmungstabellen von 300 Uropodiden-Arten (Larven, Protonymphen, Deutonymphen, Weibchen, Männchen). GDP T.9. *Acarologie* **1965**, *8*, 2–33.
28. Hirschmann, W.; Zirngiebl-Nicol, I. Neunzehn neue Uropoda-Arte. DGP T. 39. *Acarologie* **1969**, *12*, 20.
29. Hirschmann, W. Adulten-Gruppen und Bestimmungstabelle von 63 Uropoda-Arten (Uropodini, Uropodinae). GDP T. 120. *Acarologie* **1972**, *18*, 67–74.
30. Hirschmann, W. Teilgange, Stadien von 21 nneuen Uropoda (Phalodinychus)-Arten (Uropodini, Uropodinae). GDP T. 123. *Acarologie* **1972**, *18*, 79–90.
31. Hirschmann, W. Stadienfamilien und Stadiengattungender Atrichopygidiina, erstellt im Vergleich zum Gangsystem Hirschmann 1979. SDP T.1. *Acarologie* **1979**, *26*, 57–70.
32. Hirschmann, W.; Hiramatsu, N. Weibchen einer neuen Uropoda (*Phaulodinychus*) -Art der penicillata-Gruppe aus den Philippinen (Uropodini, Uropodinae). GDP T.517. *Acarologie* **1990**, *37*, 86–88.
33. Niedbała, W. *Brachychthoniidae Polski (Acari, Oribatei): Studium Ekologiczno-Faunistyczne. Monografie Fauny Polski*; Państ. Wydaw Naukowe: Warszawa, Poland; Kraków, Poland, 1976; pp. 1–144.
34. Olszanowski, Z. *A Monograph of the Nothridae and Camisiidae of Poland (Acari: Oribatida: Crotonioidea)*; Genus (Supplement); BS: Wrocław, Poland, 1996; pp. 1–201.
35. Niedbała, W. Ptyctimous mites (Acari, Oribatida) of Poland. Fauna Poloniae. *Nat. Opt. Dux. Found.* **2008**, *3*, 1–242.
36. Błaszak, C. *Zerconidae (Acari, Mesostigmata) Polski*; Państwowe Wydawn. Naukowe: Kraków, Poland, 1974; pp. 1–315.
37. Błoszyk, J.; Książkiewicz-Parulska, Z.; Adamski, Z.; Napierała, A. Influence of Pleistocene glaciation on the distribution of three species of Labidostomma in Europe (Acari: Labidostommatidae). *Syst. Appl. Acarol.* **2017**, *22*, 841–857. [CrossRef]

diversity

MDPI

Article

The Biodiversity of Demodecid Mites (Acariformes: Prostigmata), Specific Parasites of Mammals with a Global Checklist and a New Finding for *Demodex sciurinus*

Joanna N. Izdebska * and Leszek Rolbiecki

Department of Invertebrate Zoology and Parasitology, Faculty of Biology, University of Gdańsk, Wita Stwosza 59, 80-308 Gdańsk, Poland; leszek.rolbiecki@ug.edu.pl

* Correspondence: biojni@ug.edu.pl

Received: 31 May 2020; Accepted: 24 June 2020; Published: 30 June 2020

Abstract: Demodecidae are the most specialized parasitic mites of mammals; they typically inhabit the skin, but they have been found in other tissues and organs. They can cause demodecosis (a disease which is hazardous and difficult to cure) in humans, domestic animals and livestock. They are parasites with high host and topical specificity. They have been found for most orders of mammals, and they are common in the populations of numerous host species. Therefore, they not only constitute an important subject of veterinary and medical study, but also comprise an excellent model for faunistic and parasitological analyses concerning different aspects of functioning and evolution of the host–parasite relationship. The current level or knowledge of demodecid mites is irregular and fragmentary, and numerous questions require elaboration and ordering, from the taxonomic diversity to geographic distribution and relations with hosts. Such data may be of use i.a. for the development of more efficient and reliable diagnostic methods, as well as understanding the etiology and pathogenesis mechanisms of demodecosis, currently a contentious issue. The present paper lists all formally-described valid species of demodecid mites, together with other functioning specific names, verified and with comments on their status. This is significant for correct species identification and demodecosis diagnostics. The list has been drawn up on the basis of data acquired in the period 1842–2020. It contains 122 valid species of parasite, including their hosts and geographic distribution, data on parasitism, as well as only the second record of *Demodex sciurinus* in Eurasian red squirrel *Sciurus vulgaris* in over 100 years since its initial discovery.

Keywords: Acariformes; Demodecidae; Prostigmata; checklist; diversity; parasites

1. Introduction

The members of the Demodecidae are specialized, typically monoxenic mammal parasites, and are likely abundant within host populations. They are stationary parasites, with their whole life cycle spent on the host; however, depending on the species, its topography, seasonal dynamics and transmission mechanism, they may exhibit a variable level of infestation prevalence, which may reach up to 100% [1,2]. Typically, the presence of demodecid mites does not produce disease symptoms, even at high infestation intensity and high density on the skin [3–5]. However, under favorable host circumstances, the high density of these mites may be linked to the development of demodecosis (formerly demodicosis, demodicidosis). Demodecosis often has a complicated course, depending on various factors including the species constituting the etiologic factor. Its symptoms typically include the presence of various skin lesions with different topography, hair loss, eyelid margin inflammation and conjunctivitis, and changes within gum mucous membranes; however, it also can occur in generalized form. Complications may also occur in the form of secondary bacterial or fungal infections. It is suspected that the possibility of developing the parasitosis may be increased by reduced host immunity, poor condition or incorrect diet, such as deficiencies in vitamins and micronutrients [6,7]. Despite the high infestation prevalence in wild mammal populations, demodecosis is observed relatively rarely. However, it constitutes a serious problem for humans, domestic animals and livestock. *Demodecosis canina*, a domestic dog disease, typically caused by *D. canis* is characterized by a particularly hazardous course, with a chronic or even fatal result [2,8]. It has a highly variable course and symptoms, which on the one hand may be linked to the vast diversity of its hosts, with different breeds demonstrating variable susceptibility, as well as etiological factors, as three new species of these mites have been identified in dogs relatively recently (Table 1). Another burdensome and resilient parasitosis is *demodecosis hominis*, caused in humans by the synhospital human demodecid mite *D. brevis*, associated with skin sebaceous glands, and *D. folliculorum*, found in hair follicles. It manifests there typically in the form of skin lesions (e.g., pityriasis folliculorum, rosacea-like demodecidosis, pustular folliculitis, papulo-pustular scalp eruptions, perioral dermatitis and hyperpigmented patches of the face) in the head area (skin and facial hair), hair loss and eyelid margin inflammation, as well as conjunctivitis. The nature of the pathogenesis depends on various factors, such as the etiological agent (i.e., demodecid species), and the symptoms are reminiscent of other dermatoses; therefore, diagnostics, including demodecid species identification, is important for efficient treatment [6]. A number of demodecosis variants, dictated by the host inhabited by the demodecid mite, have been described for cats and cattle, or laboratory animals such as mice and hamsters (Table 1).

Due to their morphological modifications, including miniaturization, or various adaptations to parasitism within the different microhabitats, offered by the skin, tissues and organs of the host, the Demodecidae can claim to be the most specialized parasitic mites of mammals. Moreover, they demonstrate high host and topical specificity, representing the majority of orders of mammals [9]. Their long evolutionary relationship with hosts makes them the perfect model for faunistic or parasitological analyses concerning different aspects of the functioning of a host–parasite relationship and the co-evolution of these mites and their hosts. Although the family includes species of high medical and veterinary importance, their level of recognition is insufficient, and the available information regarding them is fragmentary and dispersed. A series of issues need to be studied and ordered, from taxonomic diversity, to geographic distribution and relations with hosts. In particular, data on their distribution, including their occurrence in different areas of the host range, remains incomplete. The majority of records describe cases of these parasites found in relation to parasitosis (*demodecosis*) symptoms, and these are relatively rare, being restricted mostly to domestic mammals and humans, and are only sporadically observed in wild animals [2,4,10]. Detecting an asymptomatic infestation is complicated by the miniature size of the mite, with the smallest species reaching only 70–80 μm in length, and its secretive life history; in addition, the parasites inhabit a range of skin structures, including normal and sensory hair follicles and different types of glands, as well as a variety of organ tissue types, such as the tongue, gums, anterior segments of the digestive tract or auditory canals [11]. Therefore, certain species, despite inhabiting widely distributed and common hosts, are known only from singular case studies or in sparse records [4,12].

Furthermore, the literature, particularly parasitological and veterinarian studies, encompasses a series of unverified data, including species with an actual *nomen nudum* status, invalid-unauthorized synonyms or information based on doubtful identification [2,10]: The numbers of species classified into this family varies from several dozen to over one hundred depending on the source [10,13,14]. Originally, one comprehensive study existed, covering the 16 then-known species [15], including full data on their distribution and documented records. Only one global checklist, covering 100 species and their hosts, has been published in modern times [13], this being a part of a more extensive study of the Eleutherengona as obligatory mammal parasites. However, it does not include data on records or any validation of the functioning of the names in the literature, and Demodecidae checklists for the selected host group, i.e., rodents and soricomorphs, that have been published [16–18].

The present paper serves as a comprehensive study of all known species, and includes a verification of the unauthorized names, including the *nomen nudum*. Furthermore, it also includes a new record: The second finding of *Demodex sciurinus* globally, confirming the existence of this species close to one hundred years from its original discovery. Another significant objective of the revision is to organize currently-available Demodecidae records, not only for faunistic purposes, but also for parasitological, veterinary and medical research. A key value of such a summary of the current state of research is that it also highlights the absence of information from numerous countries where demodecid mites, and its relationship with demodecosis, are a significant area of study: In some of these areas, no information on the distribution of demodecid mite species has been published. Our global data also constitute a significant starting point for future, more comprehensive regional analyses, as well as the development of diversity models in the context of host–parasite relationships. These more specific findings would be of great value in the development of more efficient and reliable diagnostic methods, and in improving our understanding of the etiological mechanisms and pathogenesis of demodecosis, which is currently a contentious issue.

2. Materials and Methods

2.1. Detection of Demodecidae in Sciurus vulgaris

One squirrel, *Sciurus vulgaris,* originating from northern Poland (Gdynia 54°30′ N 18°32′ E), collected in 2017, was examined for demodecid mites.

Demodecid mites were isolated using skin digestion methods [19]. Skin fragments of 1 cm^2 were collected from several body regions, including the head (around eyes, ear pinnae, nose, lips, chin, cheeks and vertex), neck, abdomen, back, limbs, tail and genital-anal area. Skin samples were preserved in 70% ethanol and digested in 10% potassium hydroxine solution. The obtained samples were decanted and analyzed using phase-contrast microscopy; an examination of 1 cm^2 of skin was equal to the analysis of approximately 100 wet preparations. The mites were mounted in polyvinyl-lactophenol solution and photographed. The following measurements (μm) were taken as follows: Total body length equals length of gnathosoma, podosoma and opisthosoma; gnathosomal width equals width at base; and podosomal and opisthosomal width equals maximum width. The specimens were deposited in scientific collections within the framework of the Collection of Extant Invertebrates in Department of Invertebrate Zoology and Parasitology, University of Gdańsk, Gdańsk, Poland (UGDIZP).

2.2. Literature Review—The Checklist Structure, Biogeographic and Parasitological Data Analysis

The checklist has been drawn up based on manuscripts published during the period 1842–2020 (278 items). It also contains a new record, marked in the Table 1 as the present study. Demodecidae species have been listed in systematic order, and in alphabetical order within the genera. The list includes all formally described species and other functioning specific names; all of which are verified and provided with comments on their status. Information on dates of host species, as well as the occurrence have been also included. Wherein, for cosmopolitan demodecid mite species, selected records from various range regions were given. Host records related to unidentified *Demodex* spp. have not been included.

The scientific names, common names, and systematics of the hosts follow Wilson and Reeder [20] and the Taxonomic Information System [21].

3. Results

3.1. A New Record of Demodex sciurinus

The examined squirrel specimens were found to have *D. sciurinus* (Table 2, Figure 1). A total number of 13 females and 8 males were identified, as well as several specimens at nymphal stages; male and immature stages were demonstrated for the first time. All mites were found in the skin of the penis. The presence of demodecid mites was not associated with demodecosis symptoms.

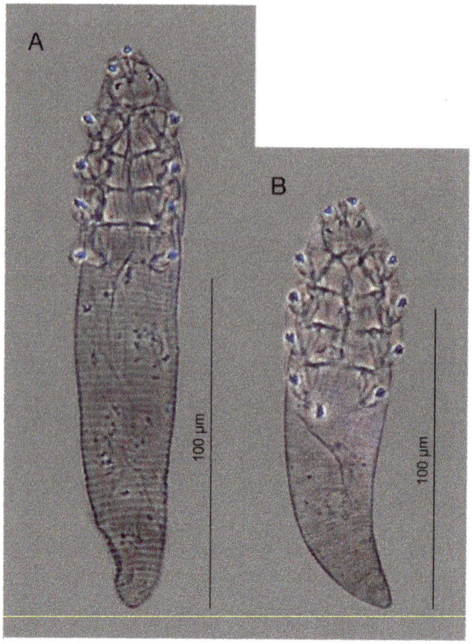

Figure 1. *Demodex sciurinus*: (**A**) female. (**B**) male.

3.2. Biodiversity and Geographic Distribution of Demodecidae Mites

A total of 122 demodecid mite species with verified systematic status are presently known, of which one represents *Apodemodex*, 106 *Demodex*, one *Glossicodex*, seven *Ophthalmodex*, one *Pterodex*, one *Rhinodex*, one *Soricidex* and four *Stomatodex* (Table 1). Representatives of the Demodecidae have been recorded on all continents outside of the polar regions (Figure 2), and their presence is typically dictated by the presence of a typical host; however, no studies of this group have been conducted in numerous areas of its range, even for common mammal species. Many species of hosts have wide distribution ranges and are also considered cosmopolitan.

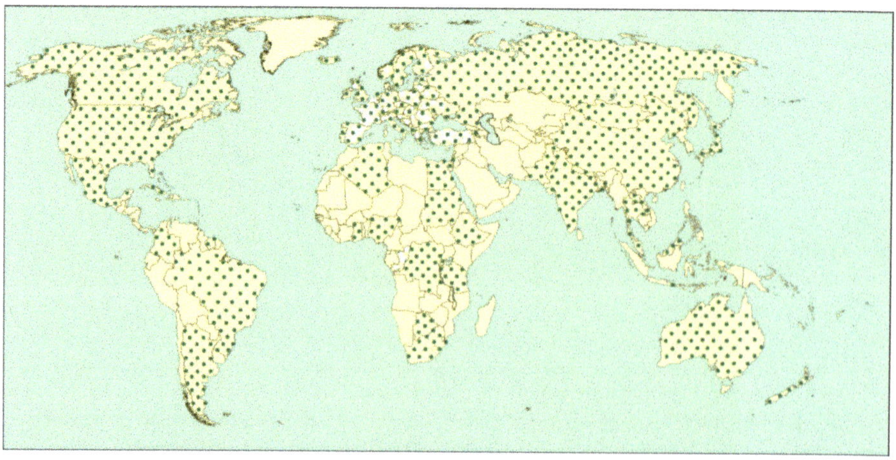

Figure 2. Geographis distribution (green dots) of Demodecidae in the world.

3.3. *Demodecidae Parasitism and Relationships with Hosts*

The greatest diversity of Demodecidae has, thus far, been described for bats (five genera), followed by rodents (three genera) and soricomorphs (two genera) as hosts; only *Demodex* representatives have been recorded in the remaining mammal orders. In turn, the species diversity in the individual host groups typically corresponds to the species diversity of the host (Figures 3 and 4). Hence the highest number of Demodecidae species have been described from the most abundant groups of mammals, e.g., 43 from rodents, 27 from bats, 17 from ungulates and 15 from carnivorans. In contrast, the Demodecidae from primates (six species) or marsupials (three species) have been especially poorly studied, which may be associated with limited access to the material. More than one species of Demodecidae has been recorded in 27 species of mammals, and the greatest number of synhospital species have been described among rodents and bats (Table 3).

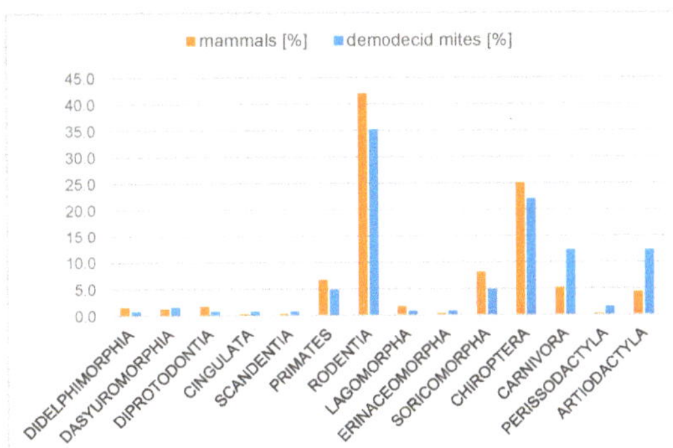

Figure 3. Share (%) of species from particular orders within mammals in comparison with the participation (%) of related demodecid mites.

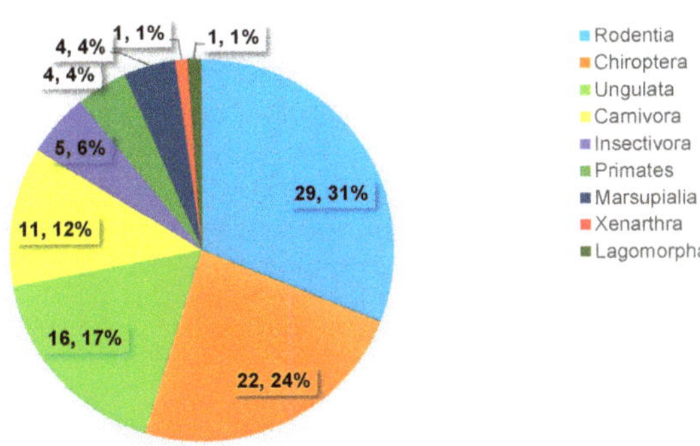

Figure 4. Mammals—hosts of Demodecidae (species number and % of mammals from different orders).

4. Discussion

Although the Demodecidae are associated with almost all of the modern mammalian orders (Table 1), their distribution and occurrence in host populations have been poorly and unevenly studied. This paucity of information has been attributed to the difficulty in detecting asymptomatic infestation, the low number of faunistic studies and issues associated with correct species identification. This is confirmed by the high number of records for demodecid mites without species identification, listed as *Demodex* sp. e.g., [22–29], including valuable new records of hosts.

4.1. Demodecidae Biodiversity Analysis in the Light of Taxonomic Identification Problems

The number of demodecid mite species described thus far (122) does not seem to be very high when considering that over 5000 mammal species could act as potential hosts. This primarily stems from technical issues linked to their detection (e.g., miniature size, secretive life history with rare manifesting of their presence in the form of demodecosis), and the strict species description criteria used for the group: Representative series of specimens of both sexes are used, often with juvenile stages [9,11].

Demodecidae species associated with human, domestic and livestock mammals are an important issue in the comprehensive elaboration of this group. Although demodecid mites have been listed in handbooks and other overview papers, in lists with parasites with data on their pathogenic importance for hosts in different countries and continents, comparatively few published records exist. Therefore, verification of whether these purported Demodecidae indeed occur in these sources, or can be potentially detected in them, has proven difficult. A series of demodecosis descriptions, case studies or clinical studies exist indicating occurrence of individual demodecid mites in a given area; however, these descriptions typically lack the information needed to determine the geographic locality of the parasite or its frequency of occurrence, or whether earlier records exist. Often the identification of the demodecid mite is limited only to the genus e.g., [30–35].

In addition, a number of species with unverified status or *nomen nudum* exist in the parasitological or veterinary literature. As the Demodecidae are monoxenic parasites, hosts are often assigned species solely on the basis of an alleged host specificity. Recording the presence of a *Demodex* species in a new host species suggests a high probability that this species is new to Science; however, this requires an appropriate taxonomic analysis to be conducted and a description in accordance with the International Commission on Zoological Nomenclature (ICZN) requirements to be published [2,10]. Unfortunately, such newly-discovered taxa have been assigned unsupported names with suitable descriptions; these have also been copied in other publications, thus becoming established in the specialist literature. For instance, the lists of the parasitofauna of the gerbil, a commonly-used laboratory animal, frequently contain a reference to *Demodex merioni* [36,37]: An alleged species not supported with a description and not assigned to any concrete host species, since the term gerbil is used to refer to a multitude of taxa. A similar situation concerned *D. cornei* from the domestic dog. The name had functioned for many years to describe an alleged species referred to as "short form" from the dog epidermis. Although morphological and morphometric research [8,38] has confirmed the existence of such species, the lack of a formal, unambiguous description meant that it was impossible to verify records from different parts of the world, as these were mostly based on the criterion of length, possibly confirmed with topical distinctiveness. Although, eventually, the specific status has been explained and confirmed with an appropriate species description [2], this does not provide any possibility to verify earlier records not supported with morphological characteristics. The groundlessness of using size as a criterion to identify species within the same host has been further confirmed by the discovery of another "short" canine demodecid species, *D. cyonis*. Furthermore, literature data including the *nomina nuda D. araneae* and *D. bonaparti* has been published without an appropriate description [39], despite being correctly distinguished by the author: A specialist on Demodecidae research. Likewise, the description of *D. myotidis, D. sciurei, D. sylvilagi* or *D. transitionalis* were included in an unpublished dissertation, a procedure that does not meet the ICZN criterion on the publication of species descriptions (Table 1).

At present, numerous directions in Demodecidae research have employed molecular methods; however, their outcomes are not reliable given the lack of correlation between morphological taxonomy and molecular divergence. An example here would be *D. "felis"*, which was recorded for the domestic cat in a study solely based on molecular analyses [40]. The authors, who assigned it a temporary name, giving the impression of a species name, stipulate that it is only a working name proposal, intended for the purpose of distinguishing the alleged species, which, according to those authors, differs from other feline demodecid mites e.g., [41–44]. Such a study where cladistics is based solely on molecular data without confirming that the inferred genetic distance is a reliable evidence at the infraspecific level, cannot provide a sufficient basis to assign a species name for an identified demodicid mite; this is especially true as the results, in this case, have not been supported with any other evidence, including morphology, and it cannot be said which taxa they concern. However, despite this irregularity, the species already functions in the veterinary literature at the specific name level.

Another issue in the study of Demodecidae distribution concerns the existence of uncertainties with regards to correct identification. Individual species can differ with regard to sets of characters and small morphological elements which may be only several micrometers in size, sometimes less than 1 μm; such minute variation requires the use of suitable preparation methods, phase contrast techniques and immersion microscopy, as well as experience in such taxonomic analyses. At the same time, some studies use alleged host specificity or, sometimes, size as the basis for identification; however, the study methods described in the works do not leave any doubt that a correct identification had not been possible. As a host may be associated with different specific Demodecidae species with similar sizes and proportions, any application of the host specificity criterion in species identification is not only insufficient, but also groundless. An accidental transfer onto atypical hosts cannot be excluded, which may happen under favorable conditions, even in the case of highly-specialized parasites.

4.2. State and Perspectives for the Study on Geographic Distribution

In view of presented data, the highest number of species (confirmed records) have been recorded in Poland (51 species), the USA (23), Czech Republic (18) and Great Britain (18). Naturally, this does not stem from any special preferences of the Demodecidae for the hosts occurring in those countries, but it is consequence of sampling bias. A clear contrast can be seen between Demodecidae records obtained from wild, domestic and livestock mammals. Detections in wild animals are rare, because demodecid mites rarely manifest their presence in the form of *demodecosis*.

It is also possible that the occurrence of the known Demodecidae species is considerably wider than that indicated by the published data, and likely coincides with the ranges of their hosts. This has been confirmed by the latest records of *D. chiropteralis*, *D. melesinus* and, the present record of, *D. sciurinus*: Species formerly known only from individual records from England, and a single observation recorded a hundred years previously from distant Poland. The currently identified individuals of *D. sciurinus* exhibited traits complying with the description and figures published by Hirst [45], despite the description deviating from the modern standards assumed for the Demodecidae taxonomy. Therefore, a redescription will definitely be necessary in the future; this would include an initial description of the juvenile stages, which will be possible after collecting a wider range of material from a greater number of hosts. Current intensive research conducted within the area of Poland has further confirmed the presence of almost all species formerly described from the Czech Republic, the Netherlands or the USA, provided that typical hosts are to be found. It should be added that many species exhibit very high infestation prevalence in the host populations, reaching up to 100% e.g., [1,4,12,18,19].

It is currently important to organize the diverse body of data concerning the occurrence of species of medical and veterinary significance, i.e., to verify and correlate data on the occurrence of demodecosis in various host species with information on the occurrence of the agent species, as confirmed by taxonomic identification.

4.3. Host-Parasite Relationships

The diversity of Demodecidae in individual host groups is typically convergent with the species richness of the host group (Figures 2 and 3). Moreover, the study of this group of species is also related to the incidence rate of demodecosis and the practical importance of hosts, thus the number of species described from carnivores is relatively higher (with regards to the biodiversity of this group) than for other mammals. Among the 17 demodecid mite species described for carnivorous mammals, six originate from domestic dogs and cats. In addition, the availability of material for study, and the technical issues related to detection, are other significant factors. The detection of asymptomatic infestation is labor-intensive, with an efficiency that is inversely proportionate to the host size. It is therefore unsurprising that the highest number of species have been described from small mammals. In contrast, the Demodecidae from primates (six species), or marsupials (three species) have been especially poorly studied, which may be associated with the limited access to the material, with only individual specimens of selected species obtained from zoological gardens being tested, and the fact that certain areas of the world, such as Asia and Australia, are absent from the body of data (Tables 1 and 2). Therefore, it is possible that the number of existing species is considerably higher than presented herein.

The Demodecidae exhibit high specificity towards hosts (monoxeny). Only five species are considered oligoxenic, i.e., recorded from more than one host; however, all host species were closely related. Of this group, *D. apodemi* and *S. corneti* have been described according to insufficient criteria, based on the current state of knowledge, and hence require redescription. In turn, *D. sabani* and *D. kutzeri* also require taxonomic revision, as they may constitute aggregate species, with their taxa being difficult to distinguish according to morphological criteria; such revision should include additional criteria such as ontogeny, molecular characteristics and parasitological testing. Furthermore, the host status is not always clear, e.g., *D. kutzeri* has been recorded from various deer species, with ambiguous species status (*Cervus canadensis* or *C. elaphus canadensis*).

Demodecid mites are generally characterized by monoxeny and a strict co-evolutionary relationship with their hosts, which is linked with the development of advanced adaptations to parasitism [9]. The mites also tend to inhabit different microhabitats within the host species, and synhospital (co-occurring) species are known to occur [46]. Although such species have been described for rodents, soricomorphs, bats, carnivores, ruminants and primates, the amount of knowledge regarding these species corresponds more to the amount of research performed on their hosts, rather than their actual distribution in the environment (Table 3). Until recently, the highest number of Demodecidae species were known from *C. perspicillata*, a species of bat that has been thoroughly studied for the presence of these mites (Table 3); in contrast, until the end of the 20th Century, only two demodecid species were known for the house mouse: A cosmopolitan, synanthropic animal used as a laboratory subject, pet or as food for other animals and one of the best-studied mammals. Intensive research conducted in recent years has revealed five further species of parasite, including one from a new genus: *Glossicodex* (Figure 5) [11,47].

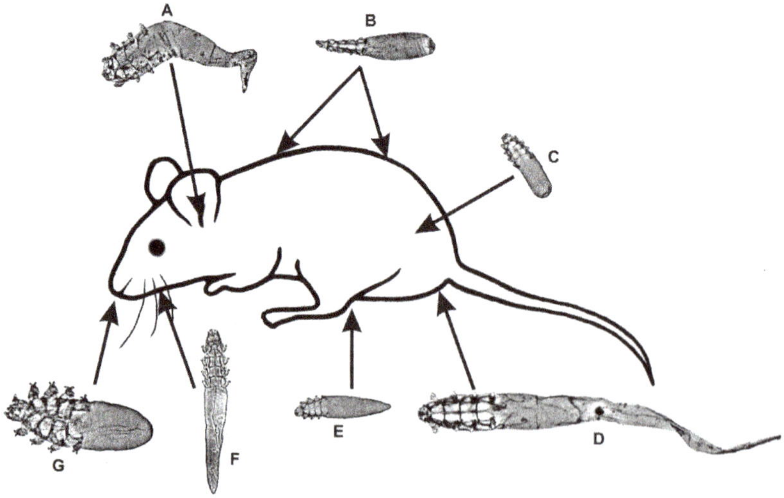

Figure 5. Topographic preferences of demodecid mites in house mouse: (**A**) *Demodex conicus*. (**B**) *Demodex musculi*. (**C**) *Demodex marculus*. (**D**). *Demodex flagellurus*. (**E**) *Demodex fusiformis*. (**F**) *Demodex vibrissae*. (**G**) *Glossicodex musculi*.

Table 1. Checklist of demodecid mites.

Demodecid Mites	Host Species (Ordo, Family)	Occurrence	Comments to the Status of Demodecid Mite Species
Apodemodex cornutus Bukva, 1996	*Neomys anomalus* Cabrera, 1907 (Soricomorpha, Soricidae)	Czech Republic loc. class. [48]	Valid
Demodex acutipes Bukva and Preisler, 1988	*Cervus elaphus* Linnaeus, 1758 (Artiodactyla, Cervidae)	Czech Republic loc. class. [49], Poland [50]	Valid
Demodex aelleni Fain, 1960	*Myotis daubentonii* (Kuhl, 1817) (Chiroptera, Vespertilionidae)	Switzerland loc. class. [51]	Valid
Demodex agrarii Bukva, 1994	*Apodemus agrarius* (Pallas, 1771) (Rodentia, Muridae)	Poland [46,52], Slovak Republic loc. class. [53]	Valid
Demodex ailuropodae Xu, Xie, Liu, Zhou and Shi, 1986	*Ailuropoda melanoleuca* (David, 1869) (Carnivora, Ursidae)	China, zoological garden ex situ [54]	Valid
Demodex antechini Nutting and Sweatman, 1970	*Antechinus stuartii* Macleay, 1841 (Dasyuromorphia, Dasyuridae)	Australia loc. class. [55]	Valid
Demodex apodemi Hirst, 1918	*Apodemus agrarius* (Rodentia, Muridae)	Poland [52], Russia [56]	Valid; described by Hirst [57], next considered as a subspecies *D. arvicolae* var. *apodemi* [15], and verified by Izdebska [16] as *D. apodemi*; specimens from *A. agrarius*, probably belongs to the separate species
	Apodemus sylvaticus (Linnaeus, 1758) (Linnaeus, 1758)	Great Britain loc. class [15,57], Poland [58], Russia [56]	
Demodex araneae Nutting, 1950	*Ateles* sp. (Primates, Atelidae)	Nutting [39] after Nutting [59]	Nom. nud.; description not published within the meaning of the ICZN
Demodex aries Desch, 1986	*Ovis aries* Linnaeus, 1758 (Artiodactyla, Bovidae)	New Zealand loc. class [60], Czech Republic [61]	Valid
Demodex artibei Vargas, Bassols, Desch, Quintero and Polaco, 1995	*Artibeus aztecus* K. Andersen, 1906 (Chiroptera, Phyllostomidae)	Mexico loc. class. [62]	Valid
Demodex arvicolae Zschokke, 1888	*Microtus agrestis* (Linnaeus, 1761) (Rodentia, Cricetidae)	Astrahan/Europe on the border with Asia [56], Europe loc. class. [15]	Valid; the record from the *M. arvalis* is questionable; maybe *D. microti* was found
	Microtus arvalis (Pallas, 1778) (Rodentia, Cricetidae)	Astrahan/Europe on the border with Asia [56]	
Demodex aurati Nutting, 1961	*Mesocricetus auratus* (Waterhouse, 1839) (Rodentia, Cricetidae)	Described and finding in laboratory animals, e.g., ex situ [37,63–69]	Valid
Demodex auricularis Izdebska, Rolbiecki and Fryderyk, 2014	*Apodemus sylvaticus* (Rodentia, Muridae)	Poland loc. class. [58]	Valid
Demodex bandicotae Izdebska, Rolbiecki, Morand and Ribas, 2017	*Bandicota indica* (Beschstein, 1800) (Rodentia, Muridae)	Laos loc. class. [17]	Valid
Demodex bantengi Firda, Nutting and Sweatman, 1987	*Bos javanicus* d'Alton, 1823 (Artiodactyla, Bovidae)	Bali loc. class. [70]	Valid
Demodex bicaudatus Kniest and Lukoschus, 1981	*Macroglossus minimus* (E. Geoffroy, 1810) (Chiroptera, Pteropodidae)	Australia loc. class. [71]	Valid
Demodex bisonianus Kadulski and Izdebska, 1996	*Bison bonasus* (Linnaeus, 1758) (Artiodactyla, Bovidae)	Poland loc. class. [72–78]	Valid
Demodex bonapartei Nutting, 1950	*Mustela erminea cicognanii* Bonaparte, 1838 (Carnivora, Mustelidae)	Nutting [39] after Nutting [59]	Nom. nud.; description not published within the meaning of the ICZN
Demodex bovis Stiles, 1892 (redescription Desch and Nutting 1971)	*Bos taurus* Linnaeus, 1758 (Artiodactyla, Bovidae)	Probably cosmopolitan, e.g., s. loc. [79], Egypt [80], Ethiopia [81], Nigeria [82], Sudan [83–85], Canada [86], USA [87], Argentina [88], Colombia [89], India [90], Mongolia [91], New Zealand [92,93], Czech Republic [94], Germany [95], Hungary [96], Italy [97], Poland [78,98]	Valid; European bison - an accidental finding in a closed farm
	Bos taurus indicus Linnaeus, 1758 (Artiodactyla, Bovidae) [*B. indicus* given by the Authors [99] is actually *B. t. indicus*]	Brazil [99]	
	Bison bonasus (Artiodactyla, Bovidae)	Poland, in the breeding condition ex situ [74,77]	

Table 1. *Cont.*

Demodecid Mites	Host Species (Ordo, Family)	Occurrence	Comments to the Status of Demodecid Mite Species
Demodex brevis Akbulatova, 1963 (redescription, Desch and Nutting 1972)	*Homo sapiens* Linnaeus, 1758 (Primates, Hominidae)	Cosmopolitan, e.g., USA [100–102], China [103,104], Australia [105], New Zealand [93], Poland [106], Russia loc. class. [107], Turkey [108–110]	Valid
Demodex buccalis Bukva, Vítovec and Vlček 1985	*Myodes glareolus* (Schreber, 1780) (Rodentia, Cricetidae)	Czech Republic loc. class [111], Poland [112]	Valid
Demodex caballi (Railliet, 1895) (redescription, Desch and Nutting 1978)	*Equus caballus* Linnaeus, 1758 (Perissodactyla, Equidae)	Probably cosmopolitan, e.g., s. loc. [113], USA [114,115], New Zealand [116]	Valid
Demodex cafferi Nutting and Guilfoy, 1979	*Syncerus caffer* (Sparrman, 1779) (Artiodactyla, Bovidae)	Botswana loc. class. [117], Republic of South Africa [118,119]	Valid
Demodex canis (Leydig, 1859) (redescription, Nutting and Desch, 1978)	*Canis lupus familiaris* Linnaeus, 1758 (Carnivora, Canidae)	Probably cosmopolitan, e.g., s. loc. [120], USA [121], Colombia [122], Cuba [123], India [124–126], Nepal [127], Pakistan [128], Thailand [129], New Zealand [93,130], Bangladesh [131], Poland [2,8,38,132,133], Russia [134], Turkey [135]	Valid
Demodex caprae Railliet, 1895	*Capra hircus* Linnaeus, 1758 (Artiodactyla, Bovidae)	Probably cosmopolitan, e.g., Ethiopia [136], China [137], New Zealand [93], Czech Republic [138], France loc. class. [113], Poland [139], Switzerland [113,140]	Valid
Demodex carolliae Desch, Lebel, Nutting and Lukoschus, 1971	*Carollia perspicillata* (Linnaeus, 1758) (Chiroptera, Phyllostomidae)	Republic of Suriname loc. class. [141]	Valid
Demodex castoris Izdebska, Fryderyk and Rolbiecki, 2016	*Castor fiber* Linnaeus, 1758 (Rodentia, Castoridae)	Poland loc. class. [142]	Valid
Demodex cati Megnin, 1877 (redescription, Desch and Nutting, 1979)	*Felis catus* Linnaeus, 1758 (Carnivora, Felidae)	Probably cosmopolitan, e.g., s. loc. [143], USA [144,145], Brazil [146], New Zealand [116], Bulgaria [147], Germany [148], Great Britain [15], Italy [149], Poland [132], Spain [150]	Valid
Demodex caviae Bacigalupo and Roveda, 1954	*Cavia porcellus* (Linnaeus, 1758) (Rodentia, Caviidae)	Described and finding in laboratory animals, e.g., ex situ [151–153]	Valid
Demodex cervi Prietsch, 1886	*Rusa unicolor* (Kerr, 1792) (Artiodactyla, Cervidae)	Germany ex situ [154]	Valid; hom. for *D. cervi* sensu Kutzer and Grünberg, 1972 (see *D. kutzeri*)
Demodex chiropteralis Hirst, 1921	*Plecotus auritus* (Linnaeus, 1758) (Chiroptera, Vespertilionidae)	Great Britain loc. class. [155], Poland [12]	Valid
Demodex conicus Izdebska and Rolbiecki, 2015	*Mus musculus* Linnaeus, 1758 (Rodentia, Muridae)	Poland loc. class. [156]	Valid
Demodex cornei: Izdebska and Rolbiecki, 2018	*Canis lupus familiaris* Linnaeus, 1758 (Carnivora, Canidae)	Probably cosmopolitan, Poland loc. class. [2]	Valid; records before 2018 have uncertain status, except Izdebska [8] and Izdebska and Fryderyk [38]
Demodex corniculatus Izdebska, 2012	*Apodemus flavicollis* (Melchior, 1834) (Rodentia, Muridae)	Poland loc. class. [1,16]	Valid
Demodex criceti Nutting and Rauch, 1958	*Mesocricetus auratus* (Waterhouse, 1839) (Rodentia, Cricetidae)	Described and finding in laboratory animals, e.g., ex situ [37,65,66,68,69,157]	Valid
Demodex cricetuli Hurley and Desch, 1994	*Cricetulus migratorius* (Rodentia, Cricetidae)	Described in laboratory animals ex situ [158]	Valid
Demodex cuniculi Pfeiffer, 1903	*Oryctolagus cuniculus* (Linnaeus, 1758) (Lagomorpha, Leporidae)	In the breeding condition, e.g., China [15], Great Britain loc. class. [15,159]	Valid
Demodex cyonis Morita, Ohmi, Kiwaki, Ike and Nagata, 2018	*Canis lupus familiaris* Linnaeus, 1758 (Carnivora, Canidae)	Japan loc. class. [160]	Valid
Demodex dasypodi Desch and Stewart, 2002	*Dasypus novemcinctus* Linnaeus, 1758 (Cingulata, Dasypodidae)	USA loc. class. [161]	Valid
Demodex desmodi Desch, 1994	*Desmodus rotundus* (E. Geoffroy, 1810) (Chiroptera, Phyllostomidae)	Republic of Suriname loc. class. [162]	Valid

Table 1. *Cont.*

Demodecid Mites	Host Species (Ordo, Family)	Occurrence	Comments to the Status of Demodecid Mite Species
Demodex erinacei Hirst, 1917	*Erinaceus europaeus* Linnaeus, 1758 (Erinaceomorpha, Erinaceidae)	Great Britain loc. class. [163]	Valid
Demodex erminae Hirst, 1919	*Mustela erminea* Linnaeus, 1758 (Carnivora, Mustelidae)	New Zealand [93], Great Britain loc. class. [15]	Valid
Demodex equi Railliet, 1895	*Equus caballus* Linnaeus, 1758 (Perissodactyla, Equidae)	Probably cosmopolitan, e.g., s. loc. [113], USA [164], Great Britain [15], Poland [165]	Valid
Demodex felis	*Felis catus* Linnaeus, 1758 (Carnivora, Felidae)	[40]	Nom. nud.
Demodex flagellurus Bukva, 1985	*Mus musculus* (Rodentia, Muridae)	Czech Republic loc. class. [166], Poland [167–169]	Valid
Demodex folliculorum (Simon, 1842) (redescription Desch and Nutting, 1972)	*Homo sapiens* (Primates, Hominidae)	Cosmopolitan, e.g., Algeria [170], Egypt [171], USA [100,101], China [103], India [172,173], Australia [105], New Zealand [93,130], Belgium [174], Croatia [175], Germany loc. class. [176], Great Britain [177], Greece [178], Iceland [179], Ireland [180], Poland [106], Turkey [108–110,181,182]	Valid
Demodex folliculorum sinensis Xie, Liu, Hsu and Hsu, 1982	*Homo sapiens* (Primates, Hominidae)	China [183]	
Demodex foveolator Bukva, 1984	*Crocidura suaveolens* (Pallas, 1811) (Soricomorpha, Soricidae)	Czech Republic loc. class. [184], Poland [18]	Valid
Demodex fusiformis Izdebska and Rolbiecki, 2015	*Mus musculus* (Rodentia, Muridae)	Poland loc. class. [169]	Valid
Demodex gapperi Nutting, Emejuaiwe and Tisolel, 1971	*Myodes gapperi* (Rodentia, Cricetidae)	USA loc. class. [185]	Valid
Demodex gatoi Desch and Stewart, 1999	*Felis catus* Linnaeus, 1758 (Carnivora, Felidae)	USA loc. class. [186,187], Austria [188], Finland [189], Poland [44,132], Spain [150])	Valid
Demodex ghanensis Oppong, Lee and Yasin, 1975	*Bos taurus* (Artiodactyla, Bovidae)	Ghana loc. class. [190], Sudan [83]	Valid
Demodex glareoli Hirst, 1919	*Myodes glareolus* (Rodentia, Cricetidae)	Great Britain loc. class. [15], Poland [112]	Valid; described by Hirst [15] as a subspecies *D. arvicolae* var. *glareoli*, than verified by Izdebska [112] as *D. glareoli*
Demodex gliricolens Hirst, 1921	*Arvicola amphibius* (Linnaeus, 1758) (Rodentia, Cricetidae)	Great Britain loc. class. [155]	Valid
Demodex gracilentus Izdebska and Rolbiecki, 2013	*Apodemus agrarius* (Rodentia, Muridae)	Poland loc. class. [46]	Valid
Demodex huttereri Mertens, Lukoschus and Nutting, 1983	*Apodemus agrarius* (Rodentia, Muridae)	Germany loc. class. [191], Poland [192]	Valid
Demodex injai Desch and Hillier, 2003	*Canis lupus familiaris* Linnaeus, 1758 (Carnivora, Canidae)	Probably cosmopolitan, USA loc. class. [193], Brazil [194], Spain [195], Poland [8,38]	Valid
Demodex intermedius Lukoschus, Mertesn, Nutting and Nadchatram, 1984	*Tupaia glis* (Diard, 1820) (Scandentia, Tupaiidae)	Malaysia loc. class. [196]	Valid
Demodex kutzeri Bukva, 1987	*Alces alces* (Linnaeus, 1758) (Atriodactyla, Cervidae)	Poland [197]	Valid; (=*D. cervi* sensu Kutzer and Grünberg, 1972; hom. for *D. cervi* Prietsch, 1886)
	Capreolus capreolus (Linnaeus, 1758) (Artiodactyla, Cervidae)	Poland [198,199]	
	Cervus elaphus Linnaeus, 1758 (Artiodactyla, Cervidae)	Austria [200], Czech Republic loc. class. [154], Poland [198,201]	
	Cervus elaphus nelsoni Nelson, 1902 (Artiodactyla, Cervidae)	USA [202,203]	
	Cervus nippon pseudaxis Gervais, 1841 (Artiodactyla, Cervidae)	Berlin, zoological garden ex situ [154]	
	Dama dama (Linnaeus, 1758)	Poland [204]	
	Odocoileus hemionus hemionus (Rafinesque, 1817) (Artiodactyla, Cervidae)	USA [202,203]	

Table 1. *Cont.*

Demodecid Mites	Host Species (Ordo, Family)	Occurrence	Comments to the Status of Demodecid Mite Species
	Odocoileus virginianus (Zimmermann, 1780) (Artiodactyla, Cervidae)	USA [202]	
Demodex lacrimalis Lukoschus and Jongman, 1974	*Apodemus sylvaticus* (Rodentia, Muridae)	Italy [205], Netherlands loc. class. [205], Poland [206]	Valid
Demodex leucogasteri Hughes and Nutting, 1981	*Onychomys leucogaster* (Wied-Neuwied, 1841) (Rodentia, Cricetidae)	USA loc. class. [207]	Valid
Demodex longior Hirst, 1918	*Apodemus sylvaticus* (Rodentia, Muridae)	Great Britain loc. class. [15,57], Poland [46], Russia [56]	Valid
Demodex longissimus Desch, Nutting and Lukoschus, 1972	*Carollia perspicillata* (Chiroptera, Phyllostomidae)	Republic of Suriname loc. class. [208]	Valid
Demodex lutrae Izdebska and Rolbiecki, 2014	*Lutra lutra* (Linnaeus, 1758) (Carnivora, Mustelidae)	Poland loc. class. [3]	Valid
Demodex macaci Karjala, Desch and Starost, 2005	*Macaca mulatta* (Zimmermann, 1780) (Primates, Cercopithecidae)	USA, laboratory colony ex situ [209]	Valid
Demodex macroglossi Desch, 1981	*Macroglossus minimus* (Chiroptera, Pteropodidae)	Australia loc. class. [210]	Valid
Demodex marculus Izdebska and Rolbiecki, 2015	*Mus musculus* (Rodentia, Muridae)	Poland loc. class. [169]	Valid
Demodex marsupiali Nutting, Lukoschus and Desch, 1980	*Didelphis marsupialis* Linnaeus, 1758 (Didelphimorphia, Didelphidae)	Republic of Surinam loc. class. [211]	Valid
Demodex melanopteri Lukoschus, Jongman and Nutting, 1972	*Eptesicus brasiliensis melanopterus* (Jentink, 1904) (Chiroptera, Vespertilionidae) [*E. melanopterus* given by the Authors [212] is actually *E. b. melanopterus*]	Republic of Suriname loc. class. [212]	Valid
Demodex melesinus Hirst, 1921	*Meles meles* (Linnaeus, 1758) (Carnivora, Mustelidae)	Great Britain loc. class. [213], Poland [4]	Valid
Demodex merioni (=*meriones*)	*Meriones* spp. (Rodentia, Muridae)	Finding in laboratory animals, e.g., [36,37]	Nom. nud.
Demodex mexicanus Vargas, Bassols, Desch, Quintero and Polaco, 1995	*Corynorhinus mexicanus* G. M. Allen, 1916 (Chiroptera, Vespertilionidae)	Mexico loc. class. [62]	Valid
Demodex microti Izdebska and Rolbiecki, 2013	*Microtus arvalis* (Rodentia, Cricetidae)	Poland loc. class. [214]	Valid
Demodex mollis Izdebska, Rolbiecki, Fryderyk and Mierzyński, 2017	*Apodemus flavicollis* (Rodentia, Muridae)	Poland loc. class. [1]	Valid
Demodex molossi Desch, Nutting and Lukoschus, 1972	*Molossus molossus* (Pallas, 1766) (Chiroptera, Molossidae)	Republic of Suriname loc. class. [208]	Valid
Demodex muscardini Hirst, 1917	*Muscardinus avellanarius* (Linnaeus, 1758) (Rodentia, Gliridae)	Armenia [56], Great Britain loc. class. [15,163]	Valid
Demodex musculi Oudemans, 1897 (redescription, Izdebska and Rolbiecki, 2015)	*Mus musculus* (Rodentia, Muridae)	Europe [15], ds loc. class. [215], Poland [167,169], Russia [56], Spain [216]; laboratory animals, e.g., ex situ [7,217,218]	Valid
Demodex myotidis	*Myotis lucifugus lucifugus* (Le Conte, 1831) (Chiroptera, Vespertilionidae)	Nutting [39] after Nutting [59]	Nom. nud.; description not published within the meaning of the ICZN
	Myotis septentrionalis Trouessart, 1897 (Chiroptera, Vespertilionidae)	Nutting [39] after Di Benedetto [219]	
	Eptesicus fuscus (Beauvois, 1796) (Chiroptera, Vespertilionidae)	Nutting [39] after Di Benedetto [219]	
Demodex mystacina Desch, 1989	*Mystacina tuberculata* Gray, 1843 (Chiroptera, Mystacinidae)	New Zealand loc. class. [220]	Valid
Demodex nanus Hirst, 1918 (redescription Desch, 1987)	*Rattus norvegicus* (Berkenhout, 1769) (Rodentia, Muridae)	Great Britain [57], Poland [19,221,222], Russia [56]; laboratory animals ex situ [223]	Valid
	Rattus rattus (Linnaeus, 1758) (Rodentia, Muridae)	New Zealand [223], Great Britain loc. class. [57], Russia [56]	

Table 1. *Cont.*

Demodecid Mites	Host Species (Ordo, Family)	Occurrence	Comments to the Status of Demodecid Mite Species
Demodex neomydis Bukva, 1995	*Neomys anomalus* (Soricomorpha, Soricidae)	Czech Republic loc. class. [224]	Valid
Demodex neoopisthosomae Desch, Lukoschus and Nadchatram, 1986	*Eonycteris spelaea* (Dobson, 1871) (Chiroptera, Pteropodidae)	Malaysia loc. class. [225]	Valid
Demodex norvegicus Bukva, 1995	*Rattus norvegicus* (Rodentia, Muridae)	Czech Republic loc. class. [226], Poland [19,221,222]	Valid
Demodex novazelandica Desch, 1989	*Mystacina tuberculata* (Chiroptera, Mystacinidae)	New Zealand loc. class. [220]	Valid
Demodex nycticeii Desch, 1996	*Nycticeius humeralis* (Rafinesque, 1818) (Chiroptera, Vespertilionidae)	USA loc. class. [227]	Valid
Demodex odocoilei Desch and Nutting, 1974	*Odocoileus virginianus* (Artiodactyla, Cervidae)	USA loc. class. [228]	Valid
	Odocoileus hemionus columbianus (Richardson, 1829) (Artiodactyla, Cervidae)	USA [229]	
Demodex ovis Railliet, 1895 (redescription, Desch 1986)	*Ovis aries* (Artiodactyla, Bovidae)	s. loc. [113], Australia [60], New Zeland [60], Czech Republic [61], Israel [230], Poland [231]	Valid
Demodex peromysci Lambert, Lukoschus and Whitaker, 1983	*Peromyscus leucopus* (Rafinesque, 1818) (Rodentia, Cricetidae)	USA loc. class. [232]	Valid
Demodex phocidi Desch, Dailey and Tuomi, 2003	*Phoca vitulina* Linnaeus, 1758 (Carnivora, Phocidae)	USA, sealife center ex situ [233], Poland in situ [5]	Valid
Demodex phodopi Desch, Davis and Klompen, 2006	*Phodopus sungorus* (Pallas, 1773) (Rodentia, Cricetidae)	Described in laboratory animals ex situ [234]	Valid
Demodex phylloides Csokor, 1879	*Sus scrofa scrofa* Linnaeus, 1758 (Artiodactyla, Suidae)	Poland [235–238]	Valid
	Sus scrofa domesticus Erxleben, 1777 (Artiodactyla, Suidae)	Probably cosmopolitan, e.g., Tanzania [239], Canada [240], USA [87], Brasil [241,242], New Zealand [93,130], historical Galicia loc. class. [243], Italy [244]	
Demodex phyllostomatis Leydig, 1859	*Phyllostomus hastatus* (Pallas, 1767) (Chiroptera, Phyllostomidae)	Republic of Suriname loc. class. [51,120]	Valid
Demodex plecoti Izdebska Rolbiecki, Mierzyński and Bidziński, 2019	*Plecotus auritus* (Chiroptera, Vespertilionidae)	Poland loc. class. [10]	Valid
Demodex ponderosus Izdebska and Rolbiecki, 2014	*Rattus norvegicus* (Rodentia, Muridae)	Poland loc. class. [222]	Valid
Demodex pseudaxis Schpringol'ts-Schmidt, 1937	*Cervus nippon hortulorum* Swinhoe, 1864 (Artiodactyla, Cervidae)	Russia/Far east [245]	Valid; need verification
Demodex ratti Hirst, 1917 (redescription Bukva, 1995)	*Rattus norvegicus* (Rodentia, Muridae)	Czech Republic [226], Europe, s. loc. [15,246], Poland [19,221,222,246–248], Russia [56]	Valid
Demodex ratticola Bukva, 1995	*Rattus norvegicus* (Rodentia, Muridae)	Czech Republic loc. class. [226], Poland [222,247,248]	Valid
Demodex rosus Bukva, Vítovec and Vlček, 1985	*Apodemus flavicollis* (Rodentia, Muridae)	Czech Republic loc. class. [111], Poland [249]	Valid
Demodex sabani Desch, Lukoschus and Nadchatram, 1984	*Leopoldamys edwardsi* (Thomas, 1882) (Rodentia, Muridae)	Malaysia [250]	Valid
	Leopoldamys sabanus (Thomas, 1887) (Rodentia, Muridae)	Malaysia loc. class. [250]	
	Niviventer cremoriventer (Miller, 1900) (Rodentia, Muridae)	Malaysia [250]	
	Niviventer rapit (Bonhote, 1903) (Rodentia, Muridae)	Malaysia [250]	
	Rattus annandalei (Bonhote, 1903) (Rodentia, Muridae)	Malaysia [250]	
	Rattus tiomanicus (Miller, 1900) (Rodentia, Muridae)	Malaysia [250]	
	Sundamys muelleri (Jentink, 1879) (Rodentia, Muridae)	Malaysia [250]	
Demodex saimiri Lebel and Nutting 1973	*Saimiri sciureus* (Linnaeus, 1758) (Primates, Cebidae)	Biological supply houses ex situ [251]	Valid
Demodex sciurei Lebel, 1970	*Saimiri sciureus* (Primates, Cebidae)	[252]	Nom. nud.; description not published within the meaning of the ICZN

Table 1. *Cont.*

Demodecid Mites	Host Species (Ordo, Family)	Occurrence	Comments to the Status of Demodecid Mite Species
Demodex sciurinus Hirst, 1923	*Sciurus vulgaris* Linnaeus, 1758 (Rodentia, Sciuridae)	Great Britain loc. class. [45], Poland [present study]	Valid
Demodex sinocricetuli Desch and Hurley, 1997	*Cricetulus barabensis* (Pallas, 1773) (Rodentia, Cricetidae)	laboratory animals ex situ [253]	Valid
Demodex soricinus Hirst, 1918 (redescription Bukva, 1993	*Plecotus auritus* (Chiroptera, Vespertilionidae)	Great Britain [57,155]	Valid; *P. auritus* -probably wrong host record
	Sorex araneus Linnaeus, 1758 (Soricomorpha, Soricidae) [*S. vulgaris* given by the Author [57] is actually *S. araneus*]	Czech Republic [254], Great Britain loc. class. [57], Poland [255]	
Demodex spelaea Desch, Lukoschus and Nadchatram, 1986	*Eonycteris spelaea* (Chiroptera, Pteropodidae)	Malaysia loc. class. [225]	Valid
Demodex suis (Kadlec, 1975)	*Sus scrofa domesticus* (Artiodactyla, Suidae)	s. loc. [256], s. loc. [257], Czech Republic [258]	Nom abort.; syn. *D. phylloides*
Demodex sungori Desch, Davis and Klompen, 2006	*Phodopus sungorus* (Rodentia, Cricetidae)	Described in laboratory animals ex situ [234]	Valid
Demodex sylvilagi Maravelas, 1962	*Sylvilagus transitionalis* (Bangs, 1895) (Lagomorpha, Leporidae)	Nutting [39] after Maravelas [259]	Nom. nud.; description not published within the meaning of the ICZN
Demodex talpae Hirst, 1921	*Talpa europaea* Linnaeus, 1758 (Soricomorpha, Talpidae)	Great Britain loc. class. [155], Poland [260]	Valid
Demodex tauri Bukva, 1986	*Bos taurus* (Artiodactyla, Bovidae)	Czech Republic loc. class. [261]	Valid
Demodex tigris Shi, Xie and Hsu, 1985	*Panthera tigris amoyensis* (Hilzheimer, 1905) (Carnivora, Felidae)	China, zoological garden ex situ [262]	Valid
Demodex tortellinioides Desch and Holz, 2006	*Antechinus agilis* Dickman, Parnaby, Crowther and King, 1998 (Dasyuromorphia, Dasyuridae)	Australia loc. class. [263]	Valid
Demodex transitionalis Moravelas, 1962	*Sylvilagus transitionalis* (Bangs, 1895) (Lagomorpha, Leporidae)	Nutting [39] after Maravelas [259]	Nom. nud.; description not published within the meaning of the ICZN
Demodex uncii Desch, 1993	*Uncia uncia* (Schreber, 1775) (Carnivora, Felidae)	USA, zoological garden ex situ Desch [264]	Valid
Demodex ursi Desch, 1995	*Ursus americanus* Pallas, 1780 (Carnivora, Ursidae)	USA loc. class. [265,266]	Valid
Demodex vibrissae Izdebska, Rolbiecki and Fryderyk, 2016	*Mus musculus* (Rodentia, Muridae)	Poland loc. class. [47]	Valid
Demodex zalophi Dailey and Nutting, 1979	*Zalophus californianus* (Lesson, 1828) (Carnivora, Otariidae)	USA, Australia loc. class. [267]	Valid
Glossicodex musculi Izdebska and Rolbiecki, 2016	*Mus musculus* (Rodentia, Muridae)	Poland loc. class. [11]	Valid
Ophthalmodex apodemi Bukva, Nutting and Desch, 1992	*Apodemus sylvaticus* (Rodentia, Muridae)	Czech Republic loc. class. [268]	Valid
Ophthalmodex artibei Lukoschus and Nutting, 1979	*Artibeus lituratus* (Chiroptera, Phyllostomidae)	Republic of Surinam loc. class. [269]	Valid
Ophthalmodex australiensis Woeltjes and Lukoschus, 1981	*Rhinonicteris aurantia* (Gray, 1845) (Chiroptera, Hipposideridae)	Australia loc. class. [270]	Valid
Ophthalmodex carolliae Lukoschus, Woeltjes, Desch and Nutting, 1980	*Carollia perspicillata* (Chiroptera, Phyllostomidae)	Republic of Surinam loc. class. [271]	Valid
Ophthalmodex juniatae Veal, Giesen and Whitaker, 1984	*Myotis lucifugus* (Le Conte, 1831) (Chiroptera, Vespertilionidae)	USA loc. class. [272]	Valid
Ophthalmodex molossi Lukoschus, Woeltjes, Desch and Nutting, 1980	*Molossus molossus* (Chiroptera, Molossidae)	Republic of Surinam loc. class. [271]	Valid
Ophthalmodex wilsoni Woeltjes and Lukoschus, 1981	*Vespadelus pumilus* (Gray, 1841 (Chiroptera, Vespertilionidae)	Australia loc. class. [270]	Valid
Pterodex carolliae Lukoschus, Woeltjes, Desch and Nutting, 1980	*Carollia perspicillata* (Chiroptera, Phyllostomidae)	Republic of Suriname loc. class. [273]	Valid
Rhinodex baeri Fain, 1959	*Galago moholi* A. Smith, 1836 (Primates, Galagidae)	Rwanda loc. class. [274]	Valid

Table 1. *Cont.*

Demodecid Mites	Host Species (Ordo, Family)	Occurrence	Comments to the Status of Demodecid Mite Species
Soricidex dimorphus Bukva, 1982	*Sorex araneus* Linnaeus, 1758 (Soricomorpha, Soricidae)	Czech Republic loc. class. [275,276], Poland [255]	Valid
Stomatodex cercarteti Desch, 1991	*Cercartetus nanus* (Desmarest, 1818) (Diprotodontia, Burramyidae)	Australia loc. class. [277]	Valid
Stomatodex corneti Fain, 1960			Valid
Stomatodex corneti corneti Fain, 1960	*Barbastella barbastellus* (Schreber, 1774) (Chiroptera, Vespertilionidae)	Belgium loc. class. [51], Great Britain [278]	
	Nycteris sp. (Chiroptera, Nycteridae)	Rwanda [51]	
Stomatodex corneti myotis Fain, 1960	*Myotis dasycneme* (Boie, 1825) (Chiroptera, Vespertilionidae)	Belgium [51]	
	Myotis myotis (Borkhausen, 1797) (Chiroptera, Vespertilionidae)	Belgium [51]	
Stomatodex galagoensis Fain, 1959	*Galago moholi* A. Smith, 1836 (Primates, Galagidae)	Rwanda loc. class. [274]	Valid
Stomatodex rousetti Fain, 1960	*Rousettus aegyptiacus* (Geoffroy, 1810) (Chiroptera, Pteropodidae)	Democratic Republic of the Congo loc. class. [51]	Valid

Hom.: homonym, loc. class.: locus classicus, nom. abort.: nomen abortium, nom. nud.: nomen nudum, s. loco: sine loco, syn.: synonymum.

Table 2. Body size (micrometers) for adults of *Demodex sciurinus*.

Character	♂($n = 8$)	♀($n = 13$)
Length of gnathosoma	15 (13–18), SD 2	17 (15–18), SD 1
Width of gnathosoma (at base)	18 (15–23), SD 2	18 (14–20), SD 2
Length of podosoma	49 (43–55), SD 5	54 (50–60), SD 3
Width of podosoma	31 (28–35), SD 3	33 (30–38), SD 2
Length of opisthosoma	72 (55–90), SD 11	97 (75–125), SD 16
Width of opisthosoma	31 (25–36), SD 4	33 (30–38), SD 2
Aedeagus	20 (16–23), SD 3	–
Vulva	–	6 (4–8), SD 1
Total length of body	135 (120–158), SD 14	168 (143–193), SD 17

Table 3. Synhospital (co-occurring) demodecid mites in the same host species.

Mammals Ordo	Mammals Species	Demodecid Mites
PRIMATES	*Galago moholi*	*Rhinodex baeri* *Stomatodex galagoensis*
	Homo sapiens	*Demodex brevis* *Demodex folliculorum*
RODENTIA	*Apodemus agrarius*	*Demodex agrarii* *Demodex apodemi* *Demodex gracilentus* *Demodex huttereri*
	Apodemus flavicollis	*Demodex corniculatus* *Demodex mollis* *Demodex rosus*
	Apodemus sylvaticus	*Demodex apodemi* *Demodex auricularis* *Demodex lacrimalis* *Demodex longior* *Ophthalmodex apodemi*
	Mus musculus	*Demodex conicus* *Demodex flagellurus* *Demodex fusiformis* *Demodex marculus* *Demodex musculi* *Demodex vibrissae* *Glossicodex musculi*
	Rattus norvegicus	*Demodex nanus* *Demodex norvegicus* *Demodex ponderosus* *Demodex ratti* *Demodex ratticola*
	Mesocricetus auratus	*Demodex aurati* *Demodex criceti*
	Myodes glareolus	*Demodex buccalis* *Demodex glareoli*
	Phodopus sungorus	*Demodex phodopi* *Demodex sungori*
SORICOMORPHA	*Neomys anomalus*	*Apodemodex cornutus* *Demodex neomydis*
	Sorex araneus	*Demodex soricinus* *Soricidex dimorphus*
CHIROPTERA	*Carollia perspicillata*	*Demodex carolliae* *Demodex longissimus* *Ophthalmodex carolliae* *Pterodex carolliae*
	Eonycteris spelaea	*Demodex neoopisthosomae* *Demodex spelaea*
	Macroglossus minimus	*Demodex bicaudatus* *Demodex macroglossi*
	Molossus molossus	*Demodex molossi* *Ophthalmodex molossi*
	Mystacina tuberculata	*Demodex mystacina* *Demodex novazelandica*

Table 3. *Cont.*

Mammals Ordo	Mammals Species	Demodecid Mites
	Plecotus auritus	*Demodex chiropteralis* *Demodex plecoti*
CARNIVORA	*Canis lupus familiaris*	*Demodex canis* *Demodex cornei* *Demodex cyonis* *Demodex injai*
	Felis catus	*Demodex cati* *Demodex gatoi*
PERISSODACTYLA	*Equus caballus*	*Demodex caballi* *Demodex equi*
ARTIODACTYLA	*Cervus elaphus*	*Demodex acutipes* *Demodex kutzeri*
	Cervus nippon	*Demodex kutzeri* *Demodex pseudaxis*
	Odocoileus hemionus	*Demodex kutzeri* *Demodex odocoilei*
	Odocoileus virginianus	*Demodex kutzeri* *Demodex odocoilei*
	Bos taurus	*Demodex bovis* *Demodex ghanensis* *Demodex tauri*
	Ovis aries	*Demodex aries* *Demodex ovis*

5. Conclusions

The Demodecidae have high veterinary and medical importance, and these aspects have directed the majority of research into the family. Despite the fact that such research stretches back to the 19th Century, appropriate zoological studies (taxonomy, fauna) are scarce and limited to species descriptions, typically based on singular records from one locality and a single host. More detailed biodiversity studies, based on the analysis of the occurrence of the parasite in different populations of individual host species, or the co-occurrence of different Demodecidae in one host species, have been published only in Poland. Even those studies have been limited by the availability of the host or the need to conduct comprehensive studies of its entire parasitofauna.

Unfortunately, no comparable data on Demodecidae are available from other regions; in addition, the group has never been included in holistic parasitofauna studies, or even studies of parasitic arthropods. The greatest obstacles to such studies are associated with parasite detection, particularly since most infestations are asymptomatic, and problems with species identification. This stands in contrast with the multitude of global medical and veterinary reports on demodecosis, or routine testing for this disease conducted at diagnostic laboratories.

There is a clear need to integrate zoological and parasitological research with medical and veterinary studies, or to perform further interdisciplinary studies. Only, such, broader approaches will provide a greater understanding of the key issues concerning Demodecidae parasitism, allow the development of efficient diagnostic methods and deepen our understanding of the causes and mechanisms of demodecosis.

Author Contributions: Conceptualization, J.N.I. and L.R.; Sampling, J.N.I. and L.R.; Microscopy Analysis, J.N.I. and L.R.; Data Analysis, J.N.I. and L.R.; Writing-Original Draft Preparation, J.N.I. and L.R.; Writing Review and Editing J.N.I. and L.R. All authors have read and agreed to the published version of the manuscript.

Funding: This research received no external funding.

Conflicts of Interest: The authors declare no conflict of interest.

References

1. Izdebska, J.N.; Rolbiecki, L.; Fryderyk, S.; Mierzyński, Ł. Adult and immature stages of the new species of the genus *Demodex* (Acariformes: Demodecidae) with data on parasitism, topography, and topical specificity of demodecid mites in the yellow-necked mouse, *Apodemus flavicollis* (Rodentia: Muridae). *J. Parasitol.* **2017**, *103*, 320–329. [CrossRef]
2. Izdebska, J.N.; Rolbiecki, L. The status of *Demodex cornei*: Description of the species and developmental stages, and data on demodecid mites in the domestic dog *Canis lupus familiaris*. *Med. Vet. Entomol.* **2018**, *32*, 346–357. [CrossRef] [PubMed]
3. Izdebska, J.N.; Rolbiecki, L. *Demodex lutrae* n. sp. (Acari) in European otter *Lutra lutra* (Carnivora: Mustelidae) with data from other demodecid mites in carnivores. *J. Parasitol.* **2014**, *100*, 784–789. [CrossRef] [PubMed]
4. Izdebska, J.N.; Cierocka, K.; Rolbiecki, L.; Kozina, P.; Kołodziej-Sobocińska, M. *Demodex melesinus* (Acariformes: Demodecidae)—The forgotten European badger parasite, rediscovered after 100 years. *Acta Parasitol.* **2018**, *63*, 665–668. [CrossRef] [PubMed]
5. Izdebska, J.N.; Rolbiecki, L.; Cierocka, K.; Pawliczka, I. *Demodex phocidi* (Acariformes: Demodecidae) from *Phoca vitulina* (Carnivora: Phocidae)—The second observation in the world and a supplement to the species description. *Oceanol. Hydrobiol. Stud.* **2020**, *49*, 49–55. [CrossRef]
6. Izdebska, J.N. New data of demodecosis hominis—Ethiology, pathogenesis and diagnosis. In *Arthropods. Ecological and Pathological Aspects of Parasite-Host Relationships*; Buczek, A., Błaszak, C., Eds.; Akapit: Lublin, Poland, 2010; pp. 137–145.
7. Nashat, M.A.; Luchins, K.R.; Riedel, E.R.; Izdebska, J.N.; Lepherd, M.L.; Lipman, N.S. Characterization of *Demodex musculi* infestation, associated co-morbidities, and its topographical distribution in a mouse strain with defective adaptive immunity. *Comp. Med.* **2017**, *67*, 315–329.
8. Izdebska, J.N. *Demodex* spp. (Acari, Demodecidae) and demodecosis in dogs: Characteristics, symptoms, occurrence. *Bull. Vet. Inst. Puławy* **2010**, *54*, 335–338.
9. Izdebska, J.N. Adaptation to Parasitism in Skin Mites From The Demodecidae Family (Acari, Prostigmata). In *Arthropods. Epidemiological Importance*; Buczek, A., Błaszak, C., Eds.; Koliber: Lublin, Poland, 2006; pp. 31–36.
10. Izdebska, J.N.; Rolbiecki, L.; Mierzyński, Ł.; Bidziński, K. Morphological and ontogenetic characteristics of *Demodex plecoti* sp. nov. (Acariformes: Demodecidae) from the brown long-eared bat *Plecotus auritus* (Chiroptera: Vespertilionidae), with comments on parasitism. *Syst. Appl. Acarol.* **2019**, *24*, 377–388. [CrossRef]
11. Izdebska, J.N.; Rolbiecki, L. A new genus and species of demodecid mites from the tongue of a house mouse *Mus musculus*: Description of adult and immature stages with data on parasitism. *Med. Vet. Entomol.* **2016**, *30*, 135–143. [CrossRef]
12. Izdebska, J.N.; Rolbiecki, L.; Mierzyński, Ł.; Bidziński, K. Demodecid mites (Acariformes, Demodecidae) in brown long-eared bat *Plecotus auritus* (Chiroptera, Vespertilionidae)—Second record in the world and systematic status of *Demodex chiropteralis* Hirst, 1921. *Ann. Parasitol.* **2018**, *64*, 109–113. [PubMed]
13. Bochkov, A.V. A review of mites of the Parvorder Eleutherengona (Acariformes: Prostigmata)—Permanent parasites of mammals. *Acarina* **2009**, *1*, 1–149.
14. Rather, P.A.; Hassan, I. Human *Demodex* mite: The versatile mite of dermatological importance. *Indian J. Dermatol.* **2014**, *59*, 60–66. [CrossRef]
15. Hirst, S. *Studies on Acari. No. 1. the Genus Demodex, Owen*; British Museum (Natural History): London, UK, 1919; pp. 1–44.
16. Izdebska, J.N. A new Demodecidae species (Acari) from the yellow-necked mouse *Apodemus flavicollis* (Rodentia, Muridae)—Description with data on parasitism. *J. Parasitol.* **2012**, *98*, 1101–1104. [CrossRef]
17. Izdebska, J.N.; Rolbiecki, L.; Morand, S.; Ribas, A. A new species and new host record of Demodecidae (Acariformes: Prostigmata) associated with the bandicoot rat (Rodentia: Muridae) from Lao PDR with data on parasitism and a checklist of the demodecid mites of rodents. *Syst. Appl. Acarol.* **2017**, *22*, 1910–1923. [CrossRef]

18. Cierocka, K.; Izdebska, J.N.; Rolbiecki, L. *Demodex foveolator* (Acariformes: Demodecidae) from *Crocidura suaveolens* (Soricomorpha: Soricidae)—The second observation worldwide, and a checklist of the demodecid mites of soricomorphs. *Ann. Parasitol.* **2019**, *65*, 329–332.
19. Izdebska, J.N. *Demodex* spp. (Acari: Demodecidae) in brown rat (Rodentia: Muridae) in Poland. *Wiad. Parazytol.* **2004**, *50*, 333–335.
20. Wilson, D.E.; Reeder, D.M. *Mammals Species of the World: A Taxonomic and Geographic Reference*, 3rd ed.; The Johns Hopkins University Press: Baltimore, MD, USA, 2005; pp. 1–2142.
21. Taxonomic Information System (ITIS). Available online: http://www.itis.gov (accessed on 25 April 2020).
22. Carpenter, J.W.; Freeny, J.C.; Patton, C.S. Occurrence of *Demodex* Owen 1843 on a white-tailed deer from Oklahoma. *J. Wildl. Dis.* **1972**, *8*, 112–114. [CrossRef]
23. Hamir, A.N.; Snyder, D.E.; Hanlon, C.A.; Rupprecht, C.E. First Report of a *Demodex* sp. in raccoons (*Procyon lotor*). *J. Wildl. Dis.* **1993**, *29*, 139–141. [CrossRef]
24. Waggie, K.S.; Marion, P.L. *Demodex* sp. in California ground squirrels. *J. Wildl. Dis.* **1997**, *33*, 368–370. [CrossRef] [PubMed]
25. Vestweber, J.G.; Ridley, R.K.; Nietfeld, J.C.; Wilkerson, M.J. Demodicosis in an American bison. *Can. Vet. J.* **1999**, *40*, 417–418. [PubMed]
26. Vogelnest, L.J.; Vogelnest, L.; Mueller, R.S. An undescribed *Demodex* sp. and demodicosis in a captive koala (*Phascolarctos cinereus*). *J. Zoo Wildl. Med.* **2000**, *31*, 100–106.
27. Gentes, M.L.; Proctor, H.; Wobeser, G. Demodicosis in a mule deer (*Odocoileus hemionus hemionus*) from Saskatchewan, Canada. *J. Wildl. Dis.* **2007**, *43*, 758–761. [CrossRef] [PubMed]
28. Bornstein, S. Important ectoparasites of alpaca (*Vicugna pacos*). *Acta Vet. Scand.* **2010**, *52* (Suppl. 1), S17. [CrossRef]
29. Jimenez, I.A.; Odom, M.R.; Childs-Sanford, S.E.; Lucio-Forster, A.; Bowman, D.D. *Lynxacarus* and *Demodex* infestation in a captive jaguar (*Panthera onca*) in La Democracia, Belize. *Vet. Rec. Case Rep.* **2020**, *8*, e001037. [CrossRef]
30. Zhao, Y.E.; Peng, Y.; Wang, X.I.; Wu, L.P.; Wang, M.; Yan, H.I.; Xiao, S.X. Facial dermatosis associated with *Demodex*: A case-control study. *J. Zhejiang Univ. Sci. B* **2011**, *12*, 1008–1015. [CrossRef] [PubMed]
31. Fourie, J.J.; Liebenberg, J.E.; Horak, I.G.; Taenzler, J.; Heckeroth, A.R.; Frénais, R. Efficacy of orally administered fluralaner (BravectoTM) or topically applied imidacloprid/moxidectin (Advocate®) against generalized demodicosis in dogs. *Parasit. Vectors* **2015**, *8*, 1–7. [CrossRef] [PubMed]
32. Yasine, A.; Kumsa, B.; Hailu, Y.; Ayana, D. Mites of sheep and goats in Oromia Zone of Amhara Region, North Eastern Ethiopia: Species, prevalence and farmers awareness. *BMC Vet. Res.* **2015**, *11*, 1–6. [CrossRef]
33. Six, R.H.; Becskei, C.; Mazaleski, M.M.; Fourie, J.J.; Mahabir, S.P.; Myers, M.R.; Slootmans, N. Efficacy of sarolaner, a novel oral isoxazoline, against two common mite infestations in dogs: *Demodex* spp. and *Otodectes cynotis*. *Vet. Parasitol.* **2016**, *222*, 62–66. [CrossRef]
34. Yun, C.H.; Yun, J.H.; Baek, J.O.; Roh, J.Y.; Lee, J.R. *Demodex* mite density determinations by standardized skin surface biopsy and direct microscopic examination and their relations with clinical types and distribution patterns. *Ann. Dermatol.* **2017**, *29*, 137–142. [CrossRef]
35. Salem, N.Y.; Abdel-Saeed, H.; Farag, H.S.; Ghandour, R.A. Canine demodicosis: Hematological and biochemical alterations. *Vet. World EISSN* **2020**, *13*, 68–72. [CrossRef]
36. Wilkerson, J.D.; Berger, D.M.P. Parasites of Gerbils. In *Flynn's Parasites of Laboratory Animals*; Baker, D.G., Ed.; Blackwell Publishing: Ames, IA, USA, 2007; pp. 413–420.
37. Fehr, M.; Koestlinger, S. Ectoparasites in small exotic mammals. *Vet. Clin. Exot. Anim.* **2013**, *16*, 611–657. [CrossRef]
38. Izdebska, J.N.; Fryderyk, S. Diversity of three species of the genus *Demodex* (Acari, Demodecidae) parasitizing dogs in Poland. *Pol. J. Environ. Stud.* **2011**, *20*, 565–569.
39. Nutting, W.B. Demodicidae—Status and prognostics. *Acarologia* **1964**, *6*, 441–454.
40. Silbermayr, K.; Horvath-Ungerboeck, C.; Eigner, B.; Joachim, A.; Ferrer, L. Phylogenetic relationships and new genetic tools for the detection and discrimination of the three feline *Demodex* mites. *Parasitol. Res.* **2015**, *114*, 747–752. [CrossRef] [PubMed]
41. Ferreira, D.; Sastre, N.; Ravera, I.; Altet, L.; Francino, O.; Bardagí, M.; Ferrer., L. Identification of a third feline *Demodex* species through partial sequencing of the 16S rDNA and frequency of *Demodex* species in 74 cats using a PCR assay. *Vet. Dermatol.* **2015**, *26*, 239–246. [CrossRef]

42. Taffin, E.R.; Casaert, S.; Claerebout, E.; Vandekerkhof, T.J.; Vandenabeele, S. Morphological variability of *Demodex cati* in a feline immunodeficiency virus-positive cat. *J. Am. Vet. Med. Assoc.* **2016**, *249*, 1308–1312. [CrossRef]
43. Jańczak, D.; Ruszczak, A.; Kaszak, I.; Gołąb, E.; Barszcz, K. Clinical aspects of demodicosis in veterinary and human medicine. *Med. Weter.* **2017**, *73*, 265–271.
44. Jańczak, D.; Gołąb, E.; Borkowska-Bąkała, D.; Barszcz, K. Inwazja *Demodex gatoi* u kota rasy brytyjskiej krótkowłosej leczonego przewlekle lekami immunosupresyjnymi. *Med. Weter.* **2017**, *73*, 248–251.
45. Hirst, S. On some new or little-known species of Acari. *Proc. Zool. Soc. Lond.* **1923**, *2*, 971–1000. [CrossRef]
46. Izdebska, J.N.; Rolbiecki, L. A new species of *Demodex* (Acari, Demodecidae) with data on topical specificity and topography of demodectic mites in the striped field mouse *Apodemus agrarius* (Rodentia, Muridae). *J. Med. Entomol.* **2013**, *50*, 1202–1207. [CrossRef]
47. Izdebska, J.N.; Rolbiecki, L.; Fryderyk, S. A new species of *Demodex* (Acari: Demodecidae) from the skin of the vibrissal area of the house mouse *Mus musculus* (Rodentia: Muridae), with data on parasitism. *Syst. Appl. Acarol.* **2016**, *21*, 1031–1039. [CrossRef]
48. Bukva, V. *Apodemodex cornutus* gen. n. et sp. n. (Acari: Demodecidae): New genus and new species of the hair follicle mite from the Mediterranean water shrew, *Neomys anomalus* (Insectivora: Soricidae). *Folia Parasitol.* **1996**, *43*, 312–316.
49. Bukva, V.; Preisler, J. Observations on the morphology of the hair follicle mites (Acari: Demodicidae) from *Cervus elaphus* L. 1758 including description of *Demodex acutipes* sp. n. *Folia Parasitol.* **1988**, *35*, 67–75.
50. Izdebska, J.N.; Fryderyk, S. *Demodex acutipes* Bukva et Preisler, 1988 (Acari, Demodecidae)—A rare parasite of red deer (*Cervus elaphus* L.). *Ann. Parasitol.* **2012**, *58*, 161–166.
51. Fain, A. Les acarines psoriques parasites des chauves-souris. XIII. La famille Demodecidae Nicolet. *Acarologia* **1960**, *2*, 80–87.
52. Izdebska, J.N.; Cydzik, K. Occurrence of *Demodex* spp. (Acari, Demodecidae) in the striped field mouse *Apodemus agrarius* (Rodentia, Muridae) in Poland. *Wiad. Parazytol.* **2010**, *56*, 59–61.
53. Bukva, V. *Demodex agrarii* sp. n. (Acari: Demodecidae) from cerumen and the sebaceous glands in the ears of the striped field mouse, Apodemus agrarius (Rodentia). *Folia Parasitol.* **1994**, *41*, 305–311.
54. Xu, Y.; Xie, H.; Liu, S.; Zhou, Z.; Shi, X. A new species of the genus *Demodex* (Acariformes: Demodicidae). *Acta Zool. Sin.* **1986**, *32*, 163–167.
55. Nutting, W.B.; Sweatman, G.K. *Demodex antechini* sp.nov. (Acari, Demodicidae) parasitic on *Antechinus stuartii* (Marsupialia). *Parasitology* **1970**, *60*, 425–429. [CrossRef]
56. Bregetova, N.G.; Bulanova-Zahvatkina, E.M.; Volgin, V.I.; Dubinin, V.B.; Zahvatkin, A.A.; Zemskaâ, A.A.; Lange, A.B.; Pavlovskij, E.N.; Serdûkova, G.V.; Šluger, E.G. *Kleŝi Gryzunov Fauny SSSR*; Izdatel'stvo Akademii Nauk SSSR: Moskva/Leningrad, Russia, 1955; pp. 1–459.
57. Hirst, S. XI.—On four new species of the genus *Demodex*, Owen. *Ann. Mag. Nat. Hist. Ser. 9* **1918**, *2*, 145–146. [CrossRef]
58. Izdebska, J.N.; Rolbiecki, L.; Fryderyk, S. *Demodex auricularis* sp. nov. (Acari: Demodecidae) from the ear canal of the European wood mouse *Apodemus sylvaticus* (Rodentia: Muridae). *Int. J. Acarol.* **2014**, *40*, 214–219. [CrossRef]
59. Nutting, W.B. Studies on the Genus *Demodex* Owen (Acari, Demodicoidea, Demodicidae). Ph.D. Thesis, Cornell University, Ithaca, NY, USA, 1950.
60. Desch, C.E. *Demodex aries* sp. nov., a sebaceous gland inhabitant of the sheep, *Ovis aries*, and a redescription of *Demodex ovis* Hirst, 1919. *N. Z. J. Zool.* **1986**, *13*, 367–375. [CrossRef]
61. Bukva, V. Three species of the hair follicle mites (Acai: Demodecidae) parasitizing the sheep, *Ovis aries* L. *Folia Parasitol.* **1990**, *37*, 81–91.
62. Vargas, M.; Bassols, I.B.; Desch, C.E.; Quintero, M.T.; Polaco, O.J. Description of two new species of the genus *Demodex* Owen, 1843 (Acari: Demodecidae) associated with Mexican bats. *Int. J. Acarol.* **1995**, *21*, 75–82. [CrossRef]
63. Nutting, W.B. *Demodex aurati* sp. nov. and *D. criceti*, ectoparasites of the golden hamster (*Mesocricetus auratus*). *Parasitology* **1961**, *51*, 515–522. [CrossRef] [PubMed]
64. Nutting, W.B.; Rauch, H. Distribution of *Demodex aurati* in the host (*Mesocricetus auratus*) skin complex. *J. Parasitol.* **1963**, *49*, 323–329. [CrossRef]

65. Owen, D.; Young, C. The occurrence of *Demodex aurati* and *Demodex criceti* in the Syrian hamster (*Mesocricetus auratus*) in the United Kingdom. *Vet. Rec.* **1973**, *17*, 282–284. [CrossRef]
66. Retnasabapathy, A.; Lourdusamy, D. *Demodex aurati* and *Demodex criceti* in the golden hamster (*Mesocricetus auratus*). *Southeast. Asian J. Trop. Med. Public Health.* **1974**, *5*, 460. [CrossRef]
67. Cardoso, M.J.L.; Franco, S.R.V.S. Demodicosis in golden hamster (*Mesocricetus auratus*)—Case in Brasil. *Ars Vet.* **2003**, *19*, 126–128.
68. Karaer, Z.; Kurtdede, A.; Ural, K.; Sari, B.; Cingi, C.C.; Karakurum, M.C.; Haydardedeoğlu, A.E. Demodicosis in a golden (Syrian) hamster (*Mesocricetus auratus*). *Ankara Univ. Vet. Fak. Derg.* **2009**, *56*, 227–229.
69. Brosseau, G. Oral fluralaner as a treatment for *Demodex aurati* and *Demodex criceti* in a golden (Syrian) hamster (*Mesocricetus auratus*). *Can. Vet. J.* **2020**, *61*, 135–137. [PubMed]
70. Firda, K.E.; Nutting, W.B.; Sweatman, G.K. *Demodex bantengi* n. sp. from *Bos javanicus* (D'alton) with notes on gross pathology (Acari: Demodicidae). *Int. J. Acarol.* **1987**, *13*, 227–231. [CrossRef]
71. Kniest, F.M.; Lukoschus, F.S. Parasites of Western Australia. XIII. A new species of demodicid mite from the meibomian glands of the bat *Macroglossus minimus*. *Rec. West. Aust. Mus.* **1981**, *9*, 111–118.
72. Kadulski, S.; Izdebska, J.N. *Demodex bisonianus* sp. nov. (Acari, Demodicidae) a new parasite of the bison (*Bison bonasus* L.). *Wiad. Parazytol.* **1996**, *42*, 103–110. [PubMed]
73. Izdebska, J.N. Zmienność Adulti i Form Juwenilnych *Demodex bisonianus* (Acari, Demodecidae). In *Stawonogi Pasożytnicze i Alergogenne*; Buczek, A., Błaszak, C., Eds.; KGM: Lublin, Poland, 2000; pp. 47–56.
74. Izdebska, J.N. The occurrence of parasitic arthropods in two groups of European bison in the Białowieża Primeval Forest. *Wiad. Parazytol.* **2001**, *47*, 801–804. [PubMed]
75. Izdebska, J.N. Stawonogi pasożytnicze żubrów z Bieszczad. *Sci. Messenger Lviv State Acad. Vet. Med. S.Z. Gzhytskyj* **2001**, *3*, 208–211.
76. Izdebska, J.N. European bison arthropod parasites from closed Polish breeding facilities. *Acta Parasitol.* **2001**, *46*, 135–137.
77. Izdebska, J.N. Skin mites (Acari: Demodecidae, Psoroptidae and Sarcoptidae) of the European bison, *Bison bonasus*. *Biol. Lett.* **2006**, *43*, 169–174.
78. Izdebska, J.N. Roztocze skórne (Acari: Demodecidae, Psoroptidae, Sarcoptidae) żubra na tle akarofauny ssaków kopytnych—Problemy specyficzności żywicielskiej i topicznej. In *Rola Hodowli Ex Situ w Procesie Restytucji Zubra*; Olech, W., Ed.; O.K.L.: Gołuchów, Poland, 2007; pp. 22–27.
79. Stiles, C. On *Demodex folliculorum* var. *bovis* in American cattle. *Can. Entomol.* **1892**, *24*, 286–290. [CrossRef]
80. Salib, F.A. First report of bovine demodicosis in native Egyptian cow. *Biomed. J. Sci. Tech. Res.* **2018**, *11*, 8424–8426. [CrossRef]
81. Chanie, M.; Tadesse, S.; Bogale, B. Prevalence of bovine Demodicosisin Gondar Zuria District, Amhara Region, Northwest Ethiopia. *Glob. Vet.* **2013**, *11*, 30–35.
82. Slingenbergh, J.; Mohammed, A.N.; Bida, S.A. Studies on bovine demodecosis in northern Nigeria. Specification and host parasite relationships. *Vet. Q.* **1980**, *2*, 90–94. [CrossRef]
83. Abu-Samra, M.T.; Shuaib, Y.A. Pathology and pathogenesis of bovine skin and meibomian gland demodicosis. *Rev. Elev. Med. Vet. Pays Trop.* **2014**, *67*, 77–85. [CrossRef]
84. Abu-Samra, M.T.; Shuaib, Y.A. Bovine demodicosis: Leather from the raw material to the finished product. *J. Soc. Leather Techol. Chem.* **2015**, *99*, 80–90.
85. Abu-Samra, M.T.; Shuaib, Y.A. Morphometric and morphologic characteristics of *Demodex bovis* and *Demodex ghanensis* (Acarina: Trombidiformes: Demodicidae) isolated from cattle in the Sudan. *J. Anim. Vet. Sci.* **2017**, *3*, 25–33.
86. Kennedy, M.J. Mange in cattle: Demodectic mange. *Agri Facts* **2001**, *4*, 1–2.
87. Desch, D.E.; Nutting, W.B. Demodicids (Trombidiformes: Demodicidae) of medical and veterinary importance. In Proceedings of the 3d International Congress of Acarology, Prague, Czech Republic, 31 August–6 September 1971; pp. 499–505.
88. Trucco, T.; Ahibe, H.; Arribillaga, M. Sarna demodécica (*Demodex bovis*) en terneros en la provincia de Entre Ríos. Descripción de un cuadro clínico. *Vet. Arg.* **2019**, *36*, 1–8.
89. Cardona, J.; Vargas, M.; Perdomo, S. Descripción clínica de la demodicosis bovina (*Demodex bovis*) en Córdoba, Colombia. *Rev. Investig. Vet. Perú* **2013**, *24*, 125–129. [CrossRef]
90. Reddy, B.S.; Sivajothi, S. Clinical management of demodicosis in Ongole cattle. *J. Parasit. Dis.* **2016**, *40*, 1311–1312. [CrossRef]

91. Matthes, H.F.; Bukva, V. Features of bovine demodecosis (*Demodex bovis* Stiles, 1892) in Mongolia: Preliminary observations. *Folia Parasitol.* **1993**, *40*, 154–155.
92. Helson, G.A.H. Some arthropods affecting man and livestock in New Zealand. *N. Z. Vet. J.* **1956**, *4*, 11–18. [CrossRef]
93. Nutting, W.B.; Kettle, P.R.; Tenquist, J.D.; Whitten, L.K. Hair follicle mites (*Demodex* spp.) in New Zealand. *N. Z. J. Zool.* **1975**, *2*, 219–222. [CrossRef]
94. Bukva, V.; Vítovec, J.; Schandl, V. První nálezy demodikózy skotu v Československu. *Vet. Med. Praha* **1985**, *30*, 515–520. [PubMed]
95. Hoffmann, G. Grössenunterschiede bei *Demodex bovis* (Acari). *Angew. Parasitol.* **1989**, *30*, 141–143.
96. Nemeséri, L.; Széky, A. Die Rinderdemodikose. *Berl. Munch. Tierarztl. Wochenschr.* **1962**, *16*, 304–307.
97. Villa, L.; Gazzonis, A.L.; Perlotti, C.; Zanzani, S.A.; Sironi, G.; Manfredi, M.T. First report of *Demodex bovis* infestation in bovine besnoitiosis co-infected dairy cattle in Italy. *Parasitol. Int.* **2020**, *79*. in press. [CrossRef]
98. Grzywiński, L.; Kliszewski, E.; Piotrowski, R. Demodekoza bydła i jej zwalczanie. *Wiad. Parazytol.* **1986**, *32*, 585–590.
99. de Araújo Wanderley, J.N.; Athayde, A.C.R.; de Moura, J.F.P.; Bezerra, L.R.; de Melo Vaz, A.F.; de Lima, E.Q.; de Oliveira, J.P.F.; Silva, W.W. Factors affecting occurrence of demodecosis by *Demodex bovis* in Sindhi cattle (*Bos indicus*). *Trop. Anim. Health Prod.* **2020**, in press.
100. Desch, C.; Nutting, W.B. *Demodex folliculorum* (Simon) and *D. brevis* Akbulatova of man: Redescription and reevaluation. *J. Parasitol.* **1972**, *58*, 169–177. [CrossRef]
101. Sengbusch, H.G.; Hauswirth, J.W. Prevalence of hair follicle mites, *Demodex folliculorum* and *D. brevis* (Acari: Demodicidae), in a selected human population in Western New York, USA. *J. Med. Entomol.* **1986**, *23*, 384–388. [CrossRef]
102. Gao, Y.Y.; Di Pascuale, M.A.; Li, W.; Liu, D.T.; Baradaran-Rafii, A.; Elizondo, A.; Kawakita, T.; Raju, V.K.; Tseng, S.C. High prevalence of *Demodex* in eyelashes with cylindrical dandruff. *Investig. Ophthalmol. Vis. Sci.* **2005**, *46*, 3089–3094. [CrossRef]
103. Zhao, Y.E.; Guo, N.; Wu, L.P. The effect of temperature on the viability of *Demodex folliculorum* and *Demodex brevis*. *Parasitol. Res.* **2009**, *105*, 1623–1628. [CrossRef] [PubMed]
104. Liang, L.; Ding, X.; Tseng, S.C.G. High prevalence of *Demodex brevis* infestation in Chalazia. *Am. J. Ophthalmol.* **2014**, *157*, 342–348. [CrossRef] [PubMed]
105. Nutting, W.B.; Green, A.C. Pathogenesis associated with hair follicle mites (*Demodex* spp.) in Australian Aborigines. *Br. J. Dermatol.* **1976**, *94*, 307–312. [CrossRef] [PubMed]
106. Izdebska, J.N.; Jankowski, Z. Demodex brevis and *D. folliculorum* (Actinedida, Demodecidae): Specific human parasites. A comparative study of the effectiveness of diagnostic methods involving autopsy. In *Advances in Polish Acarology*; Gabryś, G., Ignatowicz, S., Eds.; SGGW: Warszawa, Poland, 2006; pp. 128–136.
107. Akbulatova, L.H. Demodekoza čeloveka. *Vestnik Dermat. Venerol.* **1963**, *38*, 34–42.
108. Kaya, S.; Selimoglu, M.A.; Kaya, O.A.; Ozgen, U. Prevalence of *Demodex folliculorum* and *Demodex brevis* in childhood malnutrition and malignancy. *Pediatr. Int.* **2013**, *55*, 85–89. [CrossRef]
109. Sari, Y.; Zeytun, E.; Doğan, S.; Karakurt, Y. Oküler dandrufflu hastalarda *Demodex folliculorum* ve *D. brevis* (Acari: Demodicidae) yaygınlığı ve yoğunluğu. *Acarol. Stud.* **2019**, *1*, 23–43.
110. Zeytun, E.; Karakurt, Y. Prevalence and load of *Demodex folliculorum* and *Demodex brevis* (Acari: Demodicidae) in patients with chronic blepharitis in the province of Erzincan, Turkey. *J. Med. Entomol.* **2019**, *8*, 2–9. [CrossRef]
111. Bukva, V.; Vítovec, J.; Vlček, M. *Demodex rosus* sp. n. and *D. buccalis* sp. n. (Acari: Demodicidae) parasitizing the upper digestive tract of rodents. *Folia Parasitol.* **1985**, *32*, 151–162.
112. Izdebska, J.N.; Kozina, P.; Gólcz, A. The occurrence of *Demodex* spp. (Acari, Demodecidae) in bank vole *Myodes glareolus* (Rodentia, Cricetidae) with data on its topographical preferences. *Ann. Parasitol.* **2013**, *59*, 129–133.
113. Railliet, A. *Traité de Zoologie Médicale et Agricole*; Asselin et Houzeau: Paris, France, 1895; pp. 1–1303.
114. Desch, C.E.; Nutting, W.B. Redescription of *Demodex caballi* (= *D. folliculorum* var. *equi* Railliet, 1895) from the horse, *Equus caballus*. *Acarologia* **1978**, *20*, 235–240.
115. Mathes, R.L.; Paige Carmichael, K.; Peroni, J.; Anthony Moore, P. Primary lacrimal gland adenocarcinoma of the third eyelid in a horse. *Vet. Ophthalmol.* **2011**, *14*, 48–54. [CrossRef] [PubMed]
116. Tenquist, J.D.; Charleston, W.A.G. An annotated checklist of ectoparasites of terrestrial mammals in New Zealand. *J. R. Soc. N. Z.* **2001**, *31*, 481–542. [CrossRef]

117. Dräger, N.; Paine, G.D. Demodicosis in African buffalo (*Syncerus caffer caffer*) in Botswana. *J. Wildl. Dis.* **1980**, *16*, 521–524. [CrossRef]
118. Nutting, W.B.; Guilfoy, F.M. *Demodex cafferi* n. sp. from the African buffalo, *Syncerus cafer*. *Int. J. Acarol.* **1979**, *5*, 9–14. [CrossRef]
119. Wolhuter, J.; Bengis, R.G.; Reilly, B.K.; Cross, P.C. Clinical demodicosis in African buffalo (*Syncerus caffer*) in the Kruger National Park. *J. Wildl. Dis.* **2009**, *45*, 502–504. [CrossRef]
120. Leydig, F. Ueber Haarsackmilben und Krätzmilben. *Arch. Naturgesch.* **1859**, *25*, 338–354.
121. Nutting, W.B.; Desch, C.E. *Demodex canis*: Redescription and reevaluation. *Cornell Vet.* **1978**, *68*, 139–149. [PubMed]
122. Arroyo-Munive, Y.J.; Hincapié-Gutiérrez, L.C. Demodicosis generalizada canina tratada con Fluralaner: Reporte de un caso. *Vet. Zoot.* **2018**, *12*, 62–71.
123. Guerra, Y.; Mencho, J.D.; Rodríguez Diego, J.G.; Marín, E.; Olivares, J.L. *Demodex* spp. en perros con demodicosis, en una región de Cuba. *Rev. Salud Anim.* **2010**, *32*, 37–41.
124. Sivajothi, S.; Sudhakara Reddy, B.; Kumari, K.N.; Rayulu, V.C. Morphometry of *Demodex canis* and *Demodex cornei* in dogs with demodicosis in India. *Int. J. Vet. Health Sci. Res.* **2013**, *1*, 1–4.
125. Sivajothi, S.; Sudhakara Reddy, B.; Rayulu, V.C. Demodicosis caused by *Demodex canis* and *Demodex cornei* in dogs. *J. Parasit. Dis.* **2015**, *39*, 673–676. [CrossRef]
126. Veena, M.; Dhanalakshmi, H.; Kavitha, K.; Souza, P.E.D.; Puttalaksmamma, G.C. Morphological characterization of *Demodex* mites and its therapeutic management with neem leaves in canine demodicosis. *J. Entomol. Zool. Stud.* **2017**, *5*, 661–664.
127. Sharma, S.; Pokharel, S. Diagnosis and therapeutic management of mixed *Demodex* and *Sarcoptes* mite infestation in dog. *Acta Sci. Agric.* **2019**, *3*, 163–166.
128. Ashfaq, K.; Aqib, A.I.; Fakhar-e-Alam Kulyar, M.; Naeem, R.F.; Shoaib, M.; Bhutta, Z.A.; Tanveer, Q.; Asif, M. Alternative therapeutic approach to treat canine demodicosis. *EC Vet. Sci.* **2019**, *4*, 251–256.
129. Sakulploy, R.; Sangvaranond, A. Canine demodicosis caused by *Demodex canis* and short opisthosomal *Demodex cornei* in Shi Tzu dog from Bangkok Metropolitan Thailand. *Kasetsart Vet.* **2010**, *20*, 27–35.
130. Thomson, G.M. *The Naturalisation of Animals and Plants in New Zealand*; Cambridge University Press: Cambridge, UK, 1922; pp. 1–607.
131. Ali, M.H.; Begum, N.; Azam, M.G.; Roy, B.C. Prevalence and pathology of mite infestation in street dogs at Dinajpur municipality area. *J. Bangladesh Agril. Univ.* **2011**, *9*, 111–119. [CrossRef]
132. Izdebska, J.N. Demodecid mites (Acari, Actinedida) in carnivorous mammals (Mammalia, Carnivora) in Poland. In *Arthropods. A Variety of Forms and Interactions*; Buczek, A., Błaszak, C., Eds.; Koliber: Lublin, Poland, 2005; pp. 121–125.
133. Pawełczyk, O.; Pająk, C.; Solarz, K. The risk of exposure to parasitic mites and insects occurring on pets in Southern Poland. *Ann. Parasitol.* **2016**, *62*, 337–344.
134. Moskvina, T.V. Two morphologically distinct forms of *Demodex* mites found in dogs with canine demodicosis from Vladivostok, Russia. *Acta Vet. Beogr.* **2017**, *67*, 82–91. [CrossRef]
135. Pekmezci, G.Z.; Pekmezci, D.; Bolukbas, C.S. Molecular characterization of *Demodex canis* (Acarina: Demodicidae) in domestic dogs (*Canis familiaris*). *Kocatepe Vet. J.* **2018**, *11*, 430–433. [CrossRef]
136. Mulugeta, Y.; Yacob, H.T.; Ashenafi, H. Ectoparasites of small ruminants in three selected agro-ecological sites of Tigray Region, Ethiopia. *Trop. Anim. Health Prod.* **2010**, *42*, 1219–1224. [CrossRef] [PubMed]
137. Zhao, Y.E.; Cheng, J.; Hu, L.; Ma, J.X. Molecular identification and phylogenetic study of *Demodex caprae*. *Parasitol. Res.* **2014**, *113*, 3601–3608. [CrossRef] [PubMed]
138. Fleischer, P.; Lukešová, D.; Skřivánek, M.; Hofírek, B.; Štursa, I. První nálezy demodikózy koz v České Republice. *Vet. Med. (Praha)* **1996**, *41*, 289–293.
139. Kaba, J.; Fagasiński, A.; Krawiec, M. Nużyca u kóz. *Życie Wet.* **1999**, *74*, 24–28.
140. Strabel, D.; Schweizer, G.; Gansohr, B.; Braun, U. Der einsatz von avermectinen bei zwei ziegen mit demodikose. *Schweiz. Arch. Tierheilkd.* **2003**, *145*, 585–587. [CrossRef]
141. Desch, C.; Lebel, R.R.; Nutting, W.B.; Lukoschus, F. Parasitic mites of Surinam: I. *Demodex carolliae* sp. nov. (Acari: Demodicidae) from the fruit bat *Carollia perspicillata*. *Parasitology* **1971**, *62*, 303–308. [CrossRef]
142. Izdebska, J.N.; Fryderyk, S.; Rolbiecki, L. *Demodex castoris* sp. nov. (Acari: Demodecidae) parasitizing *Castor fiber* (Rodentia), and other parasitic arthropods associated with *Castor* spp. *Dis. Aquat. Org.* **2016**, *118*, 1–10. [CrossRef] [PubMed]

143. Megnin, P. Memoire sur le *Demodex folliculorum* Owen. *J. Aanat. Physiol.* **1877**, *13*, 97–122.
144. Desch, C.; Nutting, W.B. *Demodex cati* Hirst 1919: A redescription. *Cornell Vet.* **1979**, *69*, 280–285.
145. Bizikova, P. Localized demodicosis due to *Demodex cati* on the muzzle of two cats treated with inhalant glucocorticoids. *Vet. Dermatol.* **2014**, *25*, 222–225. [CrossRef]
146. Valandro, M.A.; da Exaltação Pascon, J.P.; de Arruda Mistieri, M.L.; Gallina, T. Demodiciose felina por *Demodex cati*. *Acta Sci. Vet.* **2016**, *44* (Suppl. 1), 1–4.
147. Iliev, P.T.; Zhelev, G.; Ivanov, A.; Prelezov, P. *Demodex cati* and feline immunodeficiency virus co-infection in a cat. *Bulg. J. Vet. Med.* **2019**, *22*, 237–242. [CrossRef]
148. Löwenstein, C.; Beck, W.; Bessmann, K.; Mueller, R.S. Feline demodicosis caused by concurrent infestation with *Demodex cati* and an unnamed species of mite. *Vet. Rec.* **2005**, *157*, 290–292. [CrossRef] [PubMed]
149. Matricoti, I.; Maina, E. The use of oral fluralaner for the treatment of feline generalized demodicosis: A case report. *J. Small Anim. Pract.* **2017**, *58*, 476–479. [CrossRef] [PubMed]
150. Ortúñez, A.; Verde, M.T.; Navarro, L.; Real, L.; Vilela, C. Demodicosis felina: A propósito de tres casos clínicos. *Clin. Vet. Peq. Anim.* **2009**, *29*, 165–171.
151. Bacigalupo, J.; Roveda, R.J. *Demodex caviae* n. sp. *Rev. Med. Vet.* **1954**, *36*, 149–153.
152. Ballweber, L.R.; Harkness, J.E. Parasites of guinea pigs. In *Flynn's Parasites of Laboratory Animals*; Baker, D.G., Ed.; Blackwell Publishing: Ames, IA, USA, 2007; pp. 421–449.
153. Schönfelder, J.; Henneveld, K.; Schönfelder, A.; Hein, J.; Müller, R. Concurrent infestation of *Demodex caviae* and *Chirodiscoides caviae* in a guinea pig. A case report. *Tierarztl. Prax. Kleintiere* **2010**, *38*, 28–30.
154. Bukva, V. *Demodex kutzeri* sp. n. (Acari: Demodicidae), an identical parasite of two species of deer, *Cervus elaphus* and *C. nippon pseudaxis*. *Folia Parasitol.* **1987**, *34*, 173–181.
155. Hirst, S. On some new or little-known species of Acari, mostly parasitic in habit. *Proc. Zool. Soc. Lond.* **1921**, *1*, 357–378.
156. Izdebska, J.N.; Rolbiecki, L. A new species of the genus *Demodex* Owen, 1843 (Acari: Demodecidae) from the ear canals of the house mouse *Mus musculus* L. (Rodentia: Muridae). *Syst. Parasitol.* **2015**, *91*, 167–173. [CrossRef]
157. Nutting, W.B.; Rauch, H. *Demodex criceti* n. sp. (Acarina: Demodicidae) with notes on its biology. *J. Parasitol.* **1958**, *44*, 328–333. [CrossRef] [PubMed]
158. Hurley, R.J.; Desch, C.E. *Demodex cricetuli*: New species of hair follicle mite (Acari: Demodecidae) from the Armenian hamster, *Cricetulus migratorius* (Rodentia: Cricetidae). *J. Med. Entomol.* **1994**, *31*, 529–533. [CrossRef]
159. Harvey, R.G. *Demodex cuniculi* in dwarf rabbits (*Oryctolagus cuniculus*). *J. Small Anim. Pract.* **1990**, *31*, 204–207. [CrossRef]
160. Morita, T.; Ohmi, A.; Kiwaki, A.; Ike, K.; Nagata, K. A new stubby species of Demodectic mite (Acari: Demodicidae) from the domestic dog (Canidae). *J. Med. Entomol.* **2018**, *55*, 323–328. [CrossRef] [PubMed]
161. Desch, C.E.; Stewart, B. First description of a hair follicle mite from the host order Xenarthra: *Demodex dasypodi* n. sp. (Acari: Demodecidae) from the nine-banded armadillo, *Dasypus novemcinctus* Linnaeus, 1758 (Dasypodidae). *Int. J. Acarol.* **2002**, *28*, 169–174. [CrossRef]
162. Desch, C.E. A new species of *Demodex* Owen, 1843 (Acari: Demodecidae) from the meibomian glands of the vampire bat *Desmodus rotundus* (E. Geoffroy, 1810) (Chiroptera: Phyllostomidae: Desmodontinae) from Surinam. *Int. J. Acarol.* **1994**, *20*, 39–43. [CrossRef]
163. Hirst, S. LII.—On three new parasitic Acari. *Ann. Mag. Nat. Hist. Ser. 8* **1917**, *20*, 431–434. [CrossRef]
164. Besch, E.D.; Griffiths, H.J. Demonstration of *Demodex equi* (Railliet, 1895) from a horse from Minnesota. *J. Am. Vet. Med. Assoc.* **1956**, *128*, 82–83.
165. Izdebska, J.N. Symptomless Skin Mite Infestation of Horses in Northern Poland. In *Advances in Polish Acarology*; Gabryś, G., Ignatowicz, S., Eds.; SGGW: Warszawa, Poland, 2006; pp. 123–127.
166. Bukva, V. *Demodex flagellurus* sp. n. (Acari: Demodicidae) from the preputial and clitoral glands of the house mouse, *Mus musculus* L. *Folia Parasitol.* **1985**, *32*, 73–81.
167. Izdebska, J.N. Nowe gatunki *Demodex* spp. (Acari, Demodecidae) u *Mus musculus* w Polsce. *Wiad. Parazytol.* **2000**, *46*, 277–280.
168. Izdebska, J.N.; Rolbiecki, L. Correlation between the occurrence of mites (*Demodex* spp.) and nematodes in the house mice (*Mus musculus* Linnaeus, 1758) in the Gdańsk urban agglomeration. *Biol. Lett.* **2006**, *43*, 175–178.

169. Izdebska, J.N.; Rolbiecki, L. Two new species of *Demodex* (Acari: Demodecidae) with a redescription of *Demodex musculi* and data on parasitism in *Mus musculus* (Rodentia: Muridae). *J. Med. Entomol.* **2015**, *52*, 604–613. [CrossRef] [PubMed]
170. Yassine, M.; Thorraya, L.; Haiet, A.H. *Demodex folliculIorum* mimicking fungal infection: A case report. *J. Dermat.* **2018**, *3*, 1–2. [CrossRef]
171. Salem, D.A.; El-Shazly, A.; Nabih, N.; El-Bayoumy, Y.; Saleh, S. Evaluation of the efficacy of oral ivermectin in comparison with ivermectin–metronidazole combined therapy in the treatment of ocular and skin lesions of *Demodex folliculorum*. *Int. J. Infect. Dis.* **2013**, *17*, e343–e347. [CrossRef] [PubMed]
172. Tandon, N.; Hati, A.K.; Thammayya, A. *Demodex folliculorum* in man. *Rec. Zool. Surv. India* **1980**, *76*, 139–142.
173. Hallur, V.; Singh, G.; Rudramurthy, S.M.; Kapoor, R.; Chakrabarti, A. *Demodex* mite infestation of unknown significance in a patient with rhinocerebral mucormycosis due to *Apophysomyces elegans* species complex. *J. Med. Microbiol.* **2013**, *62*, 926–928. [CrossRef]
174. Forton, F.; Seys, B. Density of *Demodex folliculorum* in rosacea: A case-control study using standardized skin-surface biopsy. *Br. J. Dermatol.* **1993**, *128*, 650–659. [CrossRef]
175. Basta-Juzbašić, A.; Šubić, J.Š.; Ljubojević, S. *Demodex folliculorum* in development of dermatitis rosaceiformis steroidica and rosacea-related diseases. *Clin. Dermatol.* **2002**, *20*, 135–140. [CrossRef]
176. Simon, G. Ueber eine in den kranken und normalen Haarsäcken des Menschen lebende Milbe. *Arch. Anat. Physiol. Wissensch. Med.* **1842**, *11*, 218–237.
177. Spickett, S.G. Studies on *Demodex folliculorum* Simon (1842). I. Life history. *Parasitology* **1961**, *51*, 181–192. [CrossRef]
178. Larios, G.; Alevizos, A.; Perimeni, D.; Rigopoulos, D.; Katsambas, A. Rosacea-like demodicidosis. *Lancet Infect. Dis.* **2008**, *8*, 804. [CrossRef]
179. Gunnarsdottir, S.; Kristmundsson, A.; Freeman, M.A.; Bjornsson, O.M.; Zoega, G.M. *Demodex folliculorum* (hair follicle mite), a hidden cause of blepharitis. *Laeknabladid* **2016**, *102*, 231–235. [PubMed]
180. Murphy, O.; O'Dwyer, V.; Lloyd-McKernan, A. The efficacy of tea tree face wash, 1, 2-Octanediol and microblepharoexfoliation in treating *Demodex folliculorum* blepharitis. *Contact Lens Anterior Eye* **2018**, *41*, 77–82. [CrossRef] [PubMed]
181. Eser, A.; Erpolat, S.; Kaygusuz, I.; Balci, H.; Kosus, A. Investigation of *Demodex folliculorum* frequency in patients with polycystic ovary syndrome. *An. Bras. Dermatol.* **2017**, *92*, 807–810. [CrossRef] [PubMed]
182. Doğan, S.; Doğan, S. An accidental finding of a demodectic mite on a slide: *Demodex folliculorum* (Simon) (Acari: Demodecidae). *Syst. Appl. Acarol.* **2019**, *24*, 962–964. [CrossRef]
183. Xie, H.X.; Liu, S.L.; Hsu, Y.H.; Hsu, Y.C. Taxonomy of the family Demodicidae and a new subspecies (Acarina: Demodicidae). *Acta Zootaxon. Sin.* **1982**, *7*, 265–269.
184. Bukva, V. *Demodex foveolator* sp. n. (Acari: Demodicidae), a new epidermis-dwelling parasite of *Crocidura suaveolens* (Pallas, 1821). *Folia Parasitol.* **1984**, *31*, 42–52.
185. Nutting, W.B.; Emejuaiwe, S.O.; Tisdel, M.O. *Demodex gapperi* sp. n. (Acari: Demodicidae) from the red-backed vole, *Clethrionomys gapperi*. *J. Parsitol.* **1971**, *57*, 660–665. [CrossRef]
186. Desch, C.E.; Stewart, T.B. *Demodex gatoi*: New species of hair follicle mite (Acari: Demodecidae) from the domestic cat (Carnivora: Felidae). *J. Med. Entomol.* **1999**, *36*, 167–170. [CrossRef]
187. Short, J.; Gram, D. Successful treatment of *Demodex gatoi* with 10% Imidacloprid/1% Moxidectin. *J. Am. Anim. Hosp. Assoc.* **2016**, *52*, 68–72. [CrossRef]
188. Silbermayr, K.; Joachim, A.; Litschauer, B.; Panakova, L.; Sastre, N.; Ferrer, L.; Horvath-Ungerboeck, C. The first case of *Demodex gatoi* in Austria, detected with fecal flotation. *Parasitol. Res.* **2013**, *112*, 2805–2810. [CrossRef] [PubMed]
189. Saari, S.A.M.; Juuti, K.H.; Palojärvi, J.H.; Väisänen, K.M.; Rajaniemi, R.L.; Saijonmaa-Koulumies, L.E. *Demodex gatoi*–associated contagious pruritic dermatosis in cats—A report from six households in Finland. *Acta Vet. Scand.* **2009**, *51*, 1–8. [CrossRef]
190. Oppong, E.N.W.; Lee, R.P.; Yasin, S.A. *Demodex ghanensis* sp. nov. (Acari, Demodicidae) parasitic on West African cattle. *Ghana J. Sci.* **1975**, *15*, 39–43.
191. Mertens, L.A.J.M.; Lukoschus, F.S.; Nutting, W.B. *Demodex huttereri* spec. nov. (Acarina: Prostigmata: Demodicidae) from the meibomian glands of *Apodemus agrarius* (Rodentia: Muridae). *Bonn. Zool. Beitr.* **1983**, *34*, 489–498.

192. Izdebska, J.N.; Rolbiecki, L.; Fryderyk, S. New data on distribution of *Demodex huttereri* Mertens, Lukoschus et Nutting, 1983 and topical specificity and topography of demodectic mites in striped field mouse *Apodemus agrarius*. *Wiad. Parazytol.* **2011**, *57*, 261–264.
193. Desch, C.E.; Hillier, A. *Demodex injai*: A new species of hair follicle mite (Acari: Demodecidae) from the domestic dog (Canidae). *J. Med. Entomol.* **2003**, *40*, 146–149. [CrossRef] [PubMed]
194. Amaral Silva Sgarbossa, R.S.; Vieira Sechi, G.; Duarte Pacheco, B.; Buba Lucina, S.; Roberto Paulo, M.; dos Santos Monti, F.; Rodriges de Farias, M. The epidemiological and clinical aspects of *Demodex injai* demodicosis in dogs: A report of eight cases. *Semin Cienc. Agrar.* **2017**, *38*, 3387–3394. [CrossRef]
195. Ordeix, L.; Bardagi, M.; Scarampella, F.; Ferrer, L.; Fondati, A. *Demodex injai* infestation and dorsal greasy skin and hair in eight wirehaired fox terrier dogs. *Vet. Dermatol.* **2009**, *20*, 267–272. [CrossRef]
196. Lukoschus, F.S.; Mertens, L.J.A.M.; Nutting, W.B.; Nadchatram, M. *Demodex intermedius* sp. nov. (Acarina: Prostigmata: Demodicidae) from the meibomian glands of the tree-shrew *Tupaia glis* (Mammalia: Scandentia). *Malay. Nat. J.* **1984**, *36*, 233–245.
197. Kadulski, S. Dalsze badania nad stawonogami pasożytniczymi łosia *Alces alces* w Polsce. *Wiad. Parazytol.* **1996**, *42*, 349–355.
198. Kadulski, S. Ectoparasites of Cervidae in north-east Poland. *Acta Parasitol.* **1996**, *41*, 204–210.
199. Izdebska, J.N.; Fryderyk, S. *Demodex* spp. (Acari, Demodecidae) in wild Artiodactyla in Poland. *Acta Parasitol.* **2000**, *45*, 163–164.
200. Kutzer, E.; Grünberg, W. Demodikose beim Rothirsch (*Cervus elaphus*). *Z. Parasitenkd.* **1972**, *40*, 9–17. [CrossRef] [PubMed]
201. Izdebska, J.N.; Kozina, P.; Fryderyk, S. The occurrence of *Demodex kutzeri* Bukva, 1987 (Acari, Demodecidae) in red deer (*Cervus elaphus* L.) in Poland. *Ann. Parasitol.* **2013**, *59*, 85–88. [PubMed]
202. Desch, C.E.; Andrews, J.J.; Baeten, L.A.; Holder, Z.; Powers, J.G.; Weber, D.; Ballweber, L.R. New records of hair follicle mites (Demodecidae) from North American Cervidae. *J. Wildl. Dis.* **2010**, *46*, 585–590. [CrossRef] [PubMed]
203. Palopoli, M.F.; Tra, V.; Matoin, K.; Mac, P.D. Evolution of host range in the follicle mite *Demodex kutzeri*. *Parasitology* **2017**, *144*, 594–600. [CrossRef] [PubMed]
204. Kadulski, S.; Szczurek, B. Ectoparasites on fallow deer, *Dama dama* (L.) in Pomerania, Poland. *Acta Parasitol.* **2004**, *49*, 80–86.
205. Lukoschus, F.S.; Jongman, R.G.H. *Demodex lacrimalis* spec. nov. (Demodicidae: Trombidiformes) from the Meibomian glands of the European wood mouse *Apodemus sylvaticus*. *Acarologia* **1974**, *16*, 274–281.
206. Izdebska, J.N.; Fryderyk, S. New data on the occurrence of *Demodex lacrimalis* (Acari, Demodecidae) of the wood mouse *Apodemus sylvaticus* (Rodentia, Muridae). *Ann. UMCS Biol.* **2012**, *67*, 7–11. [CrossRef]
207. Hughes, S.E.; Nutting, W.B. *Demodex leucogasteri* n. sp. from *Onychomys leucogaster*—With notes on its biology and host pathogenesis. *Acarologia* **1981**, *22*, 181–186.
208. Desch, C.; Nutting, W.B.; Lukoschus, F.S. Parasitic mites of Surinam VII: *Demodex longissimus* n. sp. from *Carollia perspicillata* and *D. molossi* n. sp. from *Molossus molossus* (Demodicidae: Trombidiformes); meibomian complex inhabitants of Neotropical bats (Chiroptera). *Acarologia* **1972**, *14*, 35–53.
209. Karjala, Z.; Desch, C.E.; Starost, M.F. First description of a new species of *Demodex* (Acari: Demodecidae) from rhesus monkey. *J. Med. Entomol.* **2005**, *42*, 948–952. [CrossRef]
210. Desch, C.E. A new species of demodicid mite (Acari: Prostigmata) from Western Australia parasitic on *Macroglossus minimus* (Chiroptera: Pteropodidae). *Rec. West. Aust. Mus.* **1981**, *9*, 41–47.
211. Nutting, W.B.; Lukoschus, F.S.; Desch, C.E. Parasitic mites of Surinam. XXXVII. *Demodex marsupiali* sp. nov. from *Didelphis marsupialis*: Adaptation to glandular habitat. *Zool. Meded. Leiden* **1980**, *56*, 83–92.
212. Lukoschus, F.S.; Jongman, R.H.G.; Nutting, W.B. Parasitic mites of Surinam. XII. *Demodex melanopteri* sp. n. (Demodicidae: Trombidiformes) from the meibomian glands of the Neotropical bat *Eptesicus melanopterus*. *Acarologia* **1972**, *14*, 54–58.
213. Hirst, S. On three new parasitic mites (*Leptus*, *Schongastia*, and *Demodex*). *Ann. Mag. Nat. Hist.* **1921**, *7*, 37–39. [CrossRef]
214. Izdebska, J.N.; Rolbiecki, L. *Demodex microti* n. sp. (Acari: Demodecidae) in *Microtus arvalis* (Pallas) (Rodentia, Cricetidae) with a checklist of the demodecid mites of cricetids. *Syst. Parasitol.* **2013**, *86*, 187–196. [CrossRef]
215. Oudemans, A.C. List of Dutch Acari. 7th part: Acarididae Latr. 1806, and Phytoptidae Pagenst. 1861, with synonymical remarks end description of new species etc. *Tijdschr. Entomol.* **1897**, *40*, 250–269.

216. Ventura, J.; Feliu, C.; Foronda, P.; Francino, O.; Sastre, N. First record of the presence of skin mites *(Demodex musculi)* in wild house mice from the Canary Islands (Spain). *Int. J. Acarol.* **2020**, *46*. in press.
217. Smith, P.C.; Zeiss, C.J.; Beck, A.P.; Scholz, J.A. *Demodex musculi* infestation in genetically immunomodulated mice. *Comp. Med.* **2016**, *66*, 278–285.
218. Nashat, M.A.; Ricart Arbona, R.J.; Lepherd, M.L.; Santagostino, S.F.; Livingston, R.S.; Riedel, E.R.; Lipman, N.S. Ivermectin-compounded feed compared with topical moxidectin–imidacloprid for eradication of *Demodex musculi* in laboratory mice. *J. Am. Assoc. Lab. Anim. Sci.* **2018**, *57*, 483–497. [CrossRef] [PubMed]
219. Di Benedetto, S.A. The Biology of *Demodex myotidis* sp.nov. (Acarina: Demodicidae) from Three Species of Bats. Master's Thesis, University of Massachusetts Amherst, Aherst, MA, USA, 1961.
220. Desch, C.E. Two new species of Demodex (Acari: Demodicidae) from the New Zealand shorttailed bat, *Mystacina tuberculata* Gray, 1843 (Chiroptera: Mystacinidae). *N. Z. J. Zool.* **1989**, *16*, 221–230. [CrossRef]
221. Izdebska, J.N.; Rolbiecki, L. Występowanie *Demodex* spp. w Korelacji z Poziomem Zarażenia Helmintami Szczura Wędrownego *Rattus norvegicus* (Berk.) z Aglomeracji Miejskiej Trójmiasta. In *Fauna Miast Europy Środkowej 21. Wieku*; Indykiewiecz, P., Barczak, T., Eds.; LOGO: Bydgoszcz, Poland, 2004; pp. 581–584.
222. Izdebska, J.N.; Rolbiecki, L. New species of *Demodex* (Acari: Demodecidae) with data on parasitism and occurrence of other demodecids of *Rattus norvegicus* (Rodentia, Muridae). *Ann. Entomol. Soc. Am.* **2014**, *107*, 740–747. [CrossRef]
223. Desch, C.E. Redescription of *Demodex nanus* (Acari: Demodicidae) from *Rattus norvegicus* and *R. rattus* (Rodentia). *J. Med. Entomol.* **1987**, *24*, 19–23. [CrossRef] [PubMed]
224. Bukva, V. *Demodex neomydis* sp. n. (Acari: Demodecidae) from the hair follicles of the Mediterranean water shrew, *Neomys anomalus* (Insectivora: Soricidae). *Folia Parasitol.* **1995**, *42*, 299–306. [PubMed]
225. Desch, C.E.; Lukoschus, F.S.; Nadchatram, M. Two new species of *Demodex* (Acari: Demodicidae) from the meibomian glands of the tropical Old Word bat, *Eonycteris spelaea* (Chiroptera). *Int. J. Acarol.* **1986**, *12*, 13–25. [CrossRef]
226. Bukva, V. *Demodex* species (Acari: Demodecidae) parasitizing the brown rat, *Rattus norvegicus* (Rodentia): Redescription of *Demodex ratti* and description of *D. norvegicus* sp. n. and *D. ratticola* sp. n. *Folia Parasitol.* **1995**, *42*, 149–160.
227. Desch, C.E. *Demodex nycticeii*: A new species of hair follicle mite (Acari: Demodecidae) from the evening bat, *Nycticeius humeralis* (Chiroptera: Vespertilionidae). *Int. J. Acarol.* **1996**, *22*, 187–191. [CrossRef]
228. Desch, C.E.; Nutting, W.B. *Demodex odocoilei* sp. nov. from the white-tailed deer, *Odocoileus virginianus*. *Can. J. Zool.* **1974**, *52*, 785–789. [CrossRef]
229. Bildfell, R.J.; Mertins, J.W.; Mortenson, J.A.; Cottam, D.F. Hair-loss syndrome in black-tailed deer of the Pacific Northwest. *J. Wildl. Dis.* **2004**, *40*, 670–681. [CrossRef]
230. Yeruham, I.; Rosen, S.; Hadani, A. Sheep demodecosis (*Demodex ovis*, Railliet, 1895) in Israel. *Rev. Elev. Méd. Vét. Pays Trop.* **1986**, *39*, 363–365.
231. Kamyszek, F.; Wertejuk, M. Badania porównawcze ekstensywności zarażenia ektopasożytami owiec w regionach: Wielkopolskie i szczecińskie w latach 1971–1980. *Wiad. Parazytol.* **1983**, *29*, 335–341. [PubMed]
232. Lombert, H.A.P.M.; Lukoschus, F.S.; Whitaker, J.O. *Demodex peromysci*, n. sp. (Acari: Prostigmata: Demodicidae), from the meibomian glands of *Peromyscus leucopus* (Rodentia: Cricetidae). *J. Med. Entomol.* **1983**, *20*, 377–382. [CrossRef]
233. Desch, C.E.; Dailey, M.D.; Tuomi, P. Description of a hair follicle mite (Acari: Demodecidae) parasitic in the earless seal family Phocidae (Mammalia: Carnivora) from the harbor seal *Phoca vitulina* Linnaeus, 1758. *Int. J. Acarol.* **2003**, *29*, 231–235. [CrossRef]
234. Desch, C.E.; Davis, S.L.; Klompen, H. Two new species of *Demodex* Owen, 1843, the hair follicle mites (Demodecidae), from the dzungarian hamster, *Phodopus sungorus* (Pallas, 1773) (Rodentia: Muridae). *Int. J. Acarol.* **2006**, *32*, 75–80. [CrossRef]
235. Kadulski, S. Ectoparasites of the boar *Sus scrofa* L. in Poland. In Proceedings of the Fourth International Congress of Parasitology, A Conference of the World Federation of Parasitologists, Warszawa, Poland, 19–26 August 1978; pp. 211–212.
236. Fryderyk, S. Pasożytnicze Acari dzika (*Sus scrofa* L.) z Pojezierza Pomorskiego. *Wiad. Parazytol.* **2000**, *46*, 163–168.
237. Fryderyk, S.; Izdebska, J.N. *Demodex phylloides* (Acari, Demodecidae) as a specific parasite of *Sus scrofa* (Mammalia, Artiodactyla). *Wiad. Parazytol.* **2001**, *47*, 797–800.

238. Izdebska, J.N.; Fryderyk, S. Zmienność morfologiczna *Demodex phylloides* (Acari, Demodecidae) u dzika (*Sus scrofa*). In *Stawonogi w Medycynie*; Buczek, A., Błaszak, C., Eds.; Liber: Lublin, Poland, 2002; pp. 25–34.
239. Kabululu, M.L.; Ngowi, H.A.; Kimera, S.I.; Lekule, F.P.; Kimbi, E.C.; Johansen, M.V. Risk factors for prevalence of pig parasitoses in Mbeya Region, Tanzania. *Vet. Parasitol.* **2015**, *15*, 460–464. [CrossRef]
240. Ramsay, W.R. On *Demodex phylloides*, (Csokor), in the skin of Canadian swine. *Proc. Can. Inst.* **1883**, *1*, 275–281.
241. Silveira, R.L.; Ribeiro, R.B.; Cruz, A.C.M.; Almeida, L.G.; Carvalho, E.C.Q. Demodicose suínano norte do estado do Riode Janeiro: Relato de caso. *Arq. Bras. Med. Vet. Zootec.* **2012**, *64*, 555–558. [CrossRef]
242. Bersano, J.G.; Mendes, M.C.; Duarte, F.C.; Del Fava, C.; de Oliveira, S.M.; Filha, E.S.; Pinheiro, E.S.; de Castro Nassar, A.F.; de Vasconcellos Bilynskyj, M.C.; Ogata, R.A.; et al. *Demodex phylloides* infection in swine reared in a periurban family farm located on the outskirts of the Metropolitan Region of São Paulo, Brazil. *Vet. Parasitol.* **2016**, *230*, 67–73. [CrossRef]
243. Csokor, J. Ueber Haarsaekmilben und eine neue Varietät derselben bei Sehweinen, *Demodex phylloides*. *Verh. Zool. Bot. Ges. Österr.* **1879**, *29*, 419–450.
244. Puccini, V.; Colella, G. Prima segnalazione per l'Italia sud-orientale di *Demodex phylloides* Csokor 1879 nel suino. *Acta Med. Vet.* **1975**, *21*, 65–72.
245. Špringol'c Šmidt, A.I. Éktoparazity nekotoryh vidov dal'novostočnyh olenej. *Vestn. Dal'nevostočnyj Fil. Akad. Nauk SSSR* **1937**, *26*, 133–140.
246. Hirst, S. Remarks on certain species of the genus *Demodex*, Owen (the *Demodex* of man, the horse, dog, rat, and mouse). *Ann. Mag. Nat. Hist. Ser. 8* **1917**, *20*, 232–235. [CrossRef]
247. Izdebska, J.N.; Rolbiecki, L. Demodectic mites of the brown rat *Rattus norvegicus* (Berkenhout, 1769) (Rodentia, Muridae) with a new finding of *Demodex ratticola* Bukva, 1995 (Acari, Demodecidae). *Ann. Parasitol.* **2012**, *58*, 71–74. [PubMed]
248. Izdebska, J.N.; Rolbiecki, L. Topical structure and topography of *Demodex* spp. (Acari Demodecidae), in brown rat *Rattus norvegicus* (Rodentia, Muridae). In *Arthropods. The Medical and Economic Importance*; Buczek, A., Błaszak, C., Eds.; Akapit: Lublin, Poland, 2012; pp. 133–141.
249. Izdebska, J.N.; Fryderyk, S. Recent data on *Demodex rosus* Bukva, Vitovec et Vlcek, 1985 (Acari, Demodecidae) from oral cavity of yellow-necked field mouse, *Apodemus flavicollis* (Rodentia, Muridae). *Wiad. Parazytol.* **2011**, *57*, 257–260.
250. Desch, C.E.; Lukoschus, F.S.; Nadchatram, M. A new demodicid (Acari: Demodicidae) from the meibomian glands of Southeast Asian rats (Rodentia: Muridae). *Trop. Biomed.* **1984**, *1*, 55–62.
251. Lebel, R.R.; Nutting, W.B. Demodectic mites of subhuman Primates I: *Demodex saimiri* sp. n. (Acari: Demodicidae) from the squirrel monkey, *Saimiri sciureus*. *J. Parasitol.* **1973**, *59*, 719–722. [CrossRef]
252. Lebel, R.R. *Demodex saimiri* sp. nov. and *Demodex sciurei* sp. nov. ectoparasites of the squirrel monkey, *Saimiri sciureus*, with notes on demodicids from other subhuman primates. Master's Thesis, University of Massachusetts, Amherst, MA, USA, 1970.
253. Desch, C.E.; Hurley, R.J. *Demodex sinocricetuli*: New species of hair follicle mite (Acari: Demodecidae) from the Chinese form of the striped hamster, *Cricetulus barabensis* (Rodentia: Muridae). *J. Med. Entomol.* **1997**, *34*, 317–320. [CrossRef]
254. Bukva, V. Redescription of *Demodex soricinus* (Acari: Demodecidae) from the sebaceous glands of *Sorex araneus* (Insectivora). *Folia Parasitol.* **1993**, *40*, 243–248.
255. Izdebska, J.N. Species of Demodecidae (Acari, Actinedida), new for the fauna of Poland, in common shrew (*Sorex araneus* L.). *Zool. Pol.* **2004**, *49*, 47–51.
256. Kaufmann, J. *Parasitic Infections of Domestic Animals: A Diagnostic Manual*; Birkhäuser Verlag: Basel, Switzerland, 1996; pp. 1–423.
257. Mehlhorn, M. (Ed.) *Encyclopedic Reference of Parasitology*, 4th ed.; Springer Verlag: Berlin/Heidelberg, Germany, 2016; pp. 1–3084.
258. Kadlec, V. *Demodex suis*. *Vet. Med.* **1975**, *20*, 441–443.
259. Maravelas, G.K. Studies of *Demodex sylvilagi* n. sp. (Acarina: Demodicidae) from the New England Cottontail *Sylvilagus transitionalis* Bangs with a Description of *Demodex transitionalis* n. sp. Master's Thesis, University of Massachusetts Amherst, Amherst, MA, USA, 1962.
260. Izdebska, J.N.; Rolbiecki, L. New data on parasites of mole *Talpa europaea* (Mammalia, Insectivora) in northern Poland. *Wiad. Parazytol.* **2003**, *49*, 97–98.

261. Bukva, V. *Demodex tauri* sp. n. (Acari: Demodicidae), a new parasite of cattle. *Folia Parasitol.* **1986**, *33*, 363–369.
262. Shi, X.; Xie, H.; Hsu, Y. A new species of the genus *Demodex* (Acariformes: Demodicidae). *Acta Zool. Sin.* **1985**, *10*, 385–387.
263. Desch, C.E.; Holz, P.H. *Demodex tortellinioides* n. sp. (Acari: Demodecidae) from the agile antechinus, *Antechinus agilis* Dickman, Parnaby, Crowther and King, 1998 (Marsupialia: Dasyuridae) in Australia. *Int. J. Acarol.* **2006**, *32*, 81–84. [CrossRef]
264. Desch, C.E. A new species of hair follicle mite (Acari: Demodecidae) from the snow leopard, *Panthera uncia* (Schreber, 1775) (Felidae). *Int. J. Acarol.* **1993**, *19*, 63–67. [CrossRef]
265. Desch, C.E. A new species of Demodex (Acari: Demodecidae) from the black bear of North America, *Ursus americanus* Pallas, 1780 (Ursidae). *Int. J. Acarol.* **1995**, *21*, 23–26. [CrossRef]
266. Foster, G.W.; Cames, T.A.; Forrester, D.J. Geographical distribution of *Demodex ursi* in black bears from Florida. *J. Wildl. Dis.* **1998**, *34*, 161–164. [CrossRef]
267. Dailey, M.D.; Nutting, W.B. *Demodex zalophi* sp. nov. (Acari: Demodicidae) from *Zalophus californianus*, the California sea lion. *Acarologia* **1979**, *21*, 423–428.
268. Bukva, V.; Nutting, W.B.; Desch, C.E. Description of *Ophthalmodex apodemi* sp.n. (Acari: Demodecidae) from the ocular area of *Apodemus sylvaticus* (Rodentia: Muridae) with notes on pathogenicity. *Int. J. Acarol.* **1992**, *18*, 269–276. [CrossRef]
269. Lukoschus, F.S.; Nutting, W.B. Parasitic mites of Surinam. XIII. *Ophthalmodex artibei* gen. nov., spec. nov. (Prostigmata: Demodicidae) from *Artibeus lituratus* with notes on pathogenesis. *Int. J. Acarol.* **1979**, *5*, 299–304. [CrossRef]
270. Woeltjes, A.G.W.; Lukoschus, F.S. Parasites of Western Australia. XIV. Two new species of *Ophthalmodex* Lukoschus and Nutting (Acarina: Prostigmata: Demodicidae) from the eyes of bats. *Rec. West. Aust. Mus.* **1981**, *9*, 307–313.
271. Lukoschus, F.S.; Woeltjes, A.G.; Desch, C.E.; Nutting, W.B. Parasitic mites of Surinam XXXV: Two new *Ophthalmodex* spp. (*O. carolliae*, *O. molossi*: Demodicidae) from the bats *Carollia perspicillata* and *Molossus molossus*. *Int. J. Acarol.* **1980**, *6*, 45–50. [CrossRef]
272. Veal, R.A.; Giesen, K.M.T.; Whitaker, J.O. A new species of the genus *Ophthalmodex* Lukoschus & Nutting 1979 (Prostigmata: Demodicidae) from *Myotis lucifugus* (Chiroptera: Vespertilionidae). *Acarologia* **1984**, *25*, 347–350.
273. Lukoschus, F.S.; Woeltjes, A.G.W.; Desch, C.E.; Nutting, W.B. Parasitic mites of Surinam XX: *Pterodex carolliae* gen. nov., spec. nov. (Demodicidae) from the fruit bat, *Carollia perspicillata*. *Int. J. Acarol.* **1980**, *6*, 9–14. [CrossRef]
274. Fain, A. Deux nouveaux genres d'Acariens vivant dans l'épaisseur des muqueuses nasale et buccale chez un Lémurien. *Bull. Ann. Soc. R. Belg. Entomol.* **1959**, *95*, 263–273.
275. Bukva, V. *Soricidex dimorphus* g. n., sp. n. (Acari, Demodicidae) from the common shrew, *Sorex araneus*. *Folia Parasitol.* **1982**, *40*, 343–349.
276. Bukva, V. Sexual dimorphism in the hair follicle mites (Acari: Demodecidae): Scanning electron microscopy of *Soricidex dimorphus*. *Folia Parasitol.* **1993**, *40*, 71–79.
277. Desch, C.E. A new Australian species of *Stomatodex* Fain (Acari: Demodecidae) from the oesophagus of *Cercartetus nanus* (Desmarest) (Marsupialia). *J. Aust. Entomol. Soc.* **1991**, *30*, 239–241. [CrossRef]
278. Baker, A.S.; Craven, J.C. Checklist of the mites (Arachnida: Acari) associated with bats (Mammalia: Chiroptera) in the British Isles. *Syst. Appl. Acarol.* **2003**, *14*, 1–20. [CrossRef]

MDPI
St. Alban-Anlage 66
4052 Basel
Switzerland
Tel. +41 61 683 77 34
Fax +41 61 302 89 18
www.mdpi.com

Diversity Editorial Office
E-mail: diversity@mdpi.com
www.mdpi.com/journal/diversity

www.ingramcontent.com/pod-product-compliance
Lightning Source LLC
Chambersburg PA
CBHW040813250926
51850CB00007BA/1153